bloodhoney* SUNCHASER PICTURES

January 17th, 1994

4:30 a.m.

Underneath brightly-lit streets of Reseda, California, the ground suddenly trembled and came apart. Shockwaves spread in all directions, across the San Fernando Valley, all the way to Las Vegas 220 miles away. Some concrete structures merely shook and fractured, others collapsed. Streetlights buckled and crashed to the ground, the power grid began to shut down sector-by-sector, casting half of Southern California into total darkness.

North America had just experienced its strongest-ever instrumentally recorded earthquake in an urban area: magnitude of 6.7. Millions of residents ran onto the streets, away from crumbling structures, then came to a grinding halt, staring at the skies above.

Strange phone calls flooded the 911 Emergency centers, overwhelming the operators. They were told of shimmering lights above, a velvety "cloud" stretching across the sky. Coming of the aliens, or perhaps the long-awaited apocalypse?

The explanation turned out to be much simpler. The electromagnetic curtain of endless photons had been lifted, and what had been there for billions of years was revealed. For some, it was the first time seeing this in their lives, and for others, in decades. With all the artificial lights extinguished in a 200-mile diameter, the Milky Way shone brightly from horizon to horizon. Constellations and star clusters covered every inch of the sky. Meteors streaked as they burned up in the atmosphere.

The Northridge Earthquake, an event that only lasted about 20 seconds, had exposed one of the world's most light-polluted cities to the true night sky.

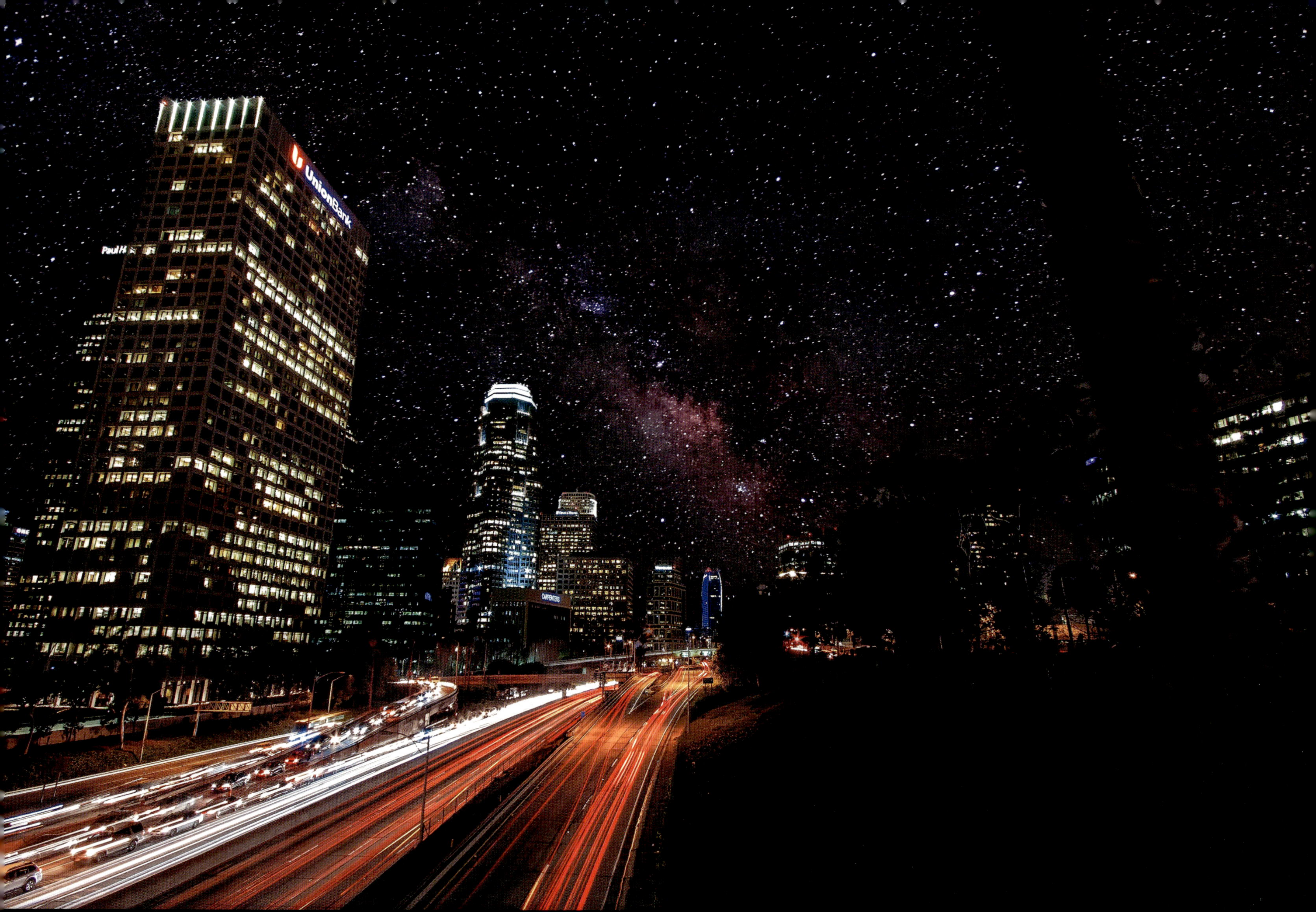

There are no known images of stars above Southern California in the hours following the Northridge Earthquake.

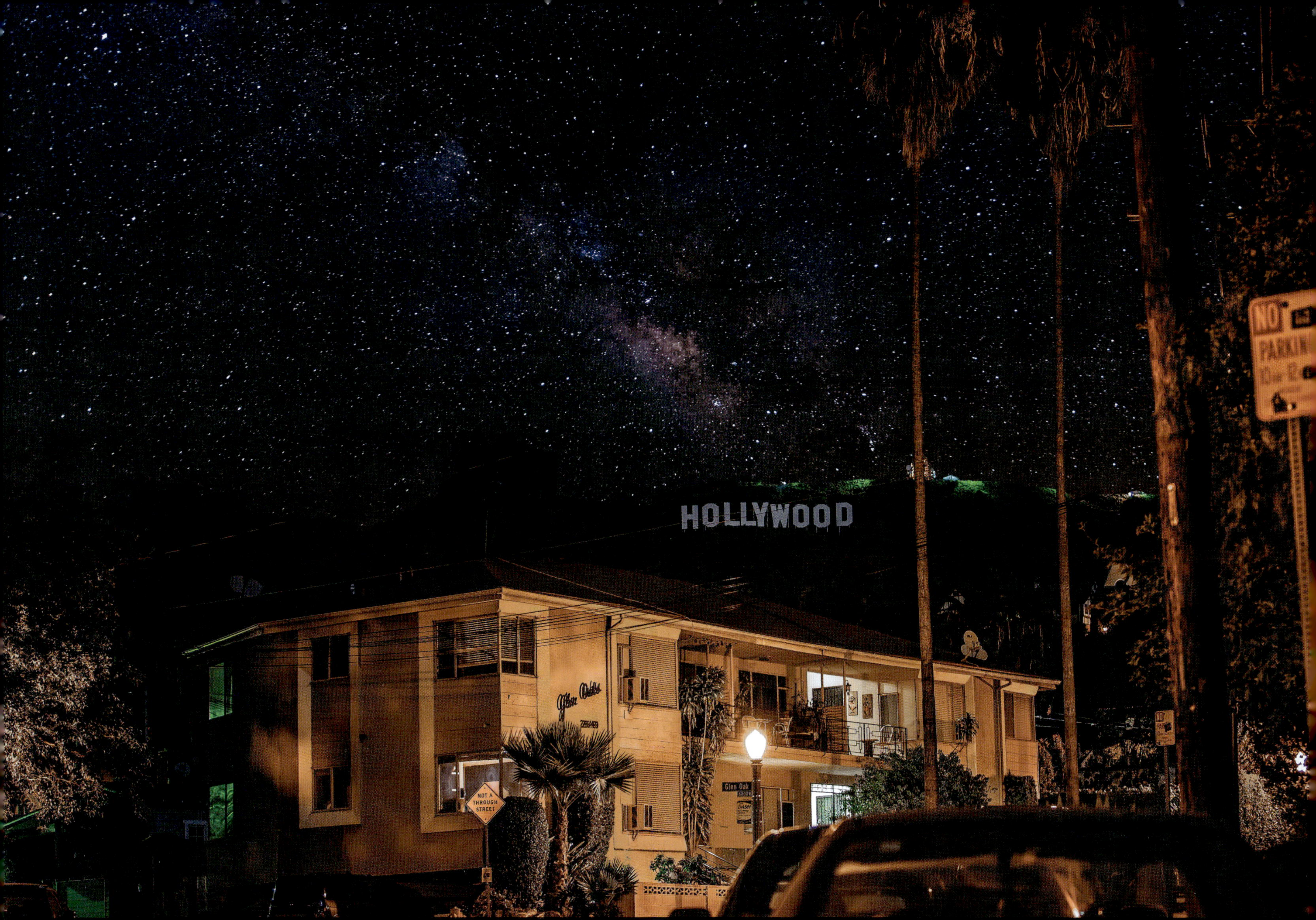

Photographs above are composite simulations of what people of Los Angeles may have been able to see that morning.

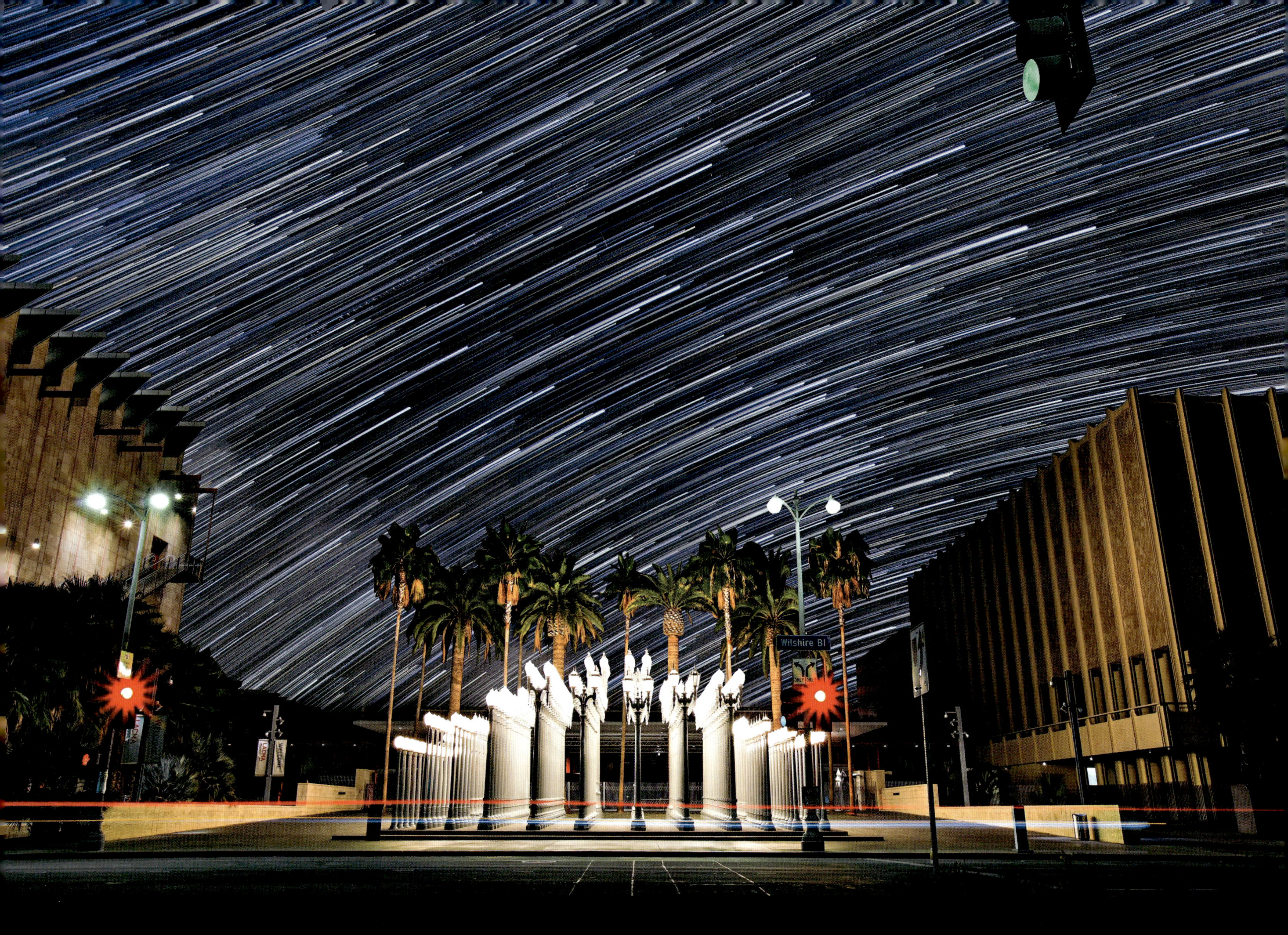

These images juxtapose heavily light-polluted Los Angeles with the pristine night skies of the Grand Canyon National Park

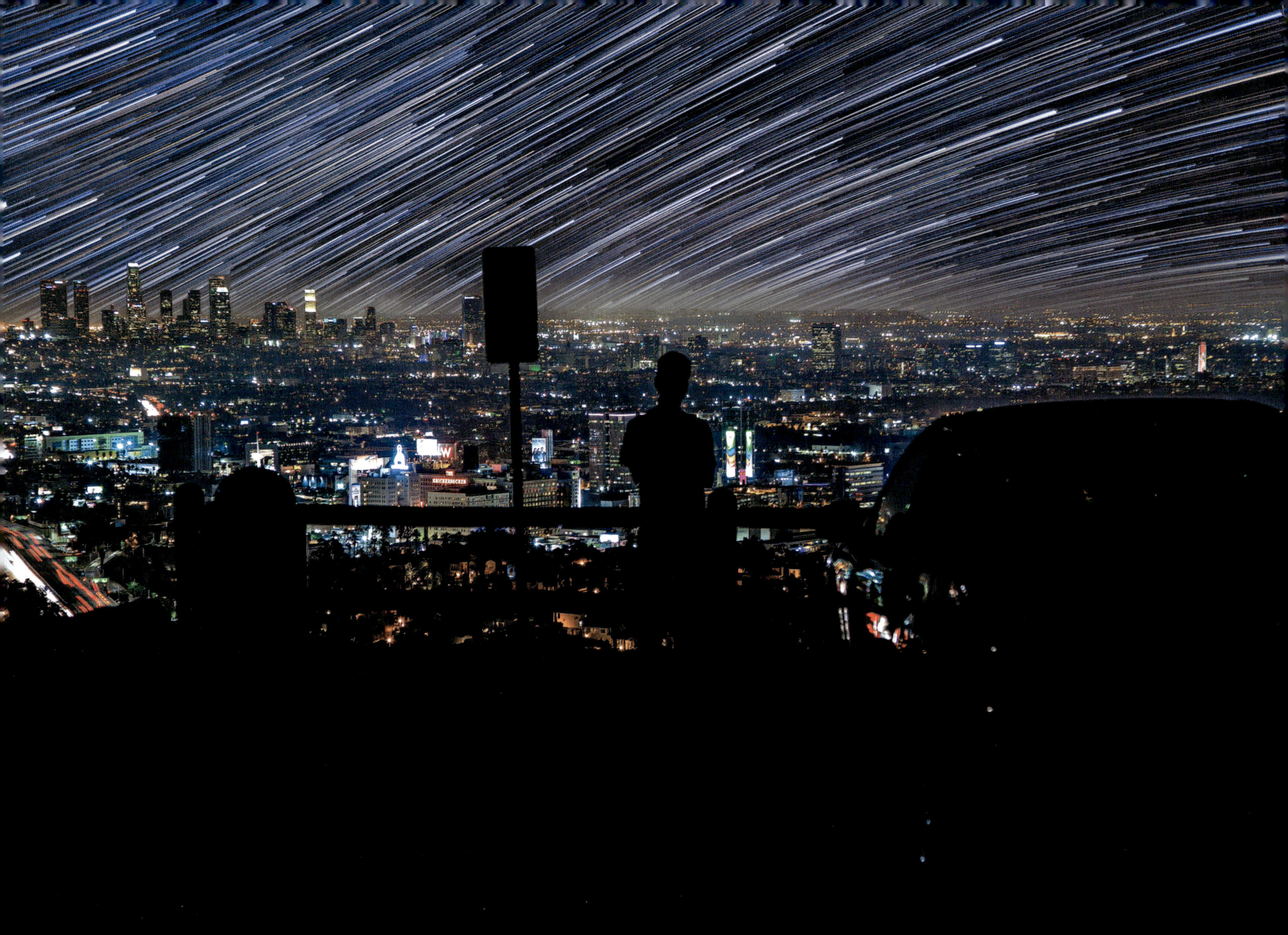

Star trails seen here are portrayed through the use of Earth's rotation axis and multiple image exposures blended together

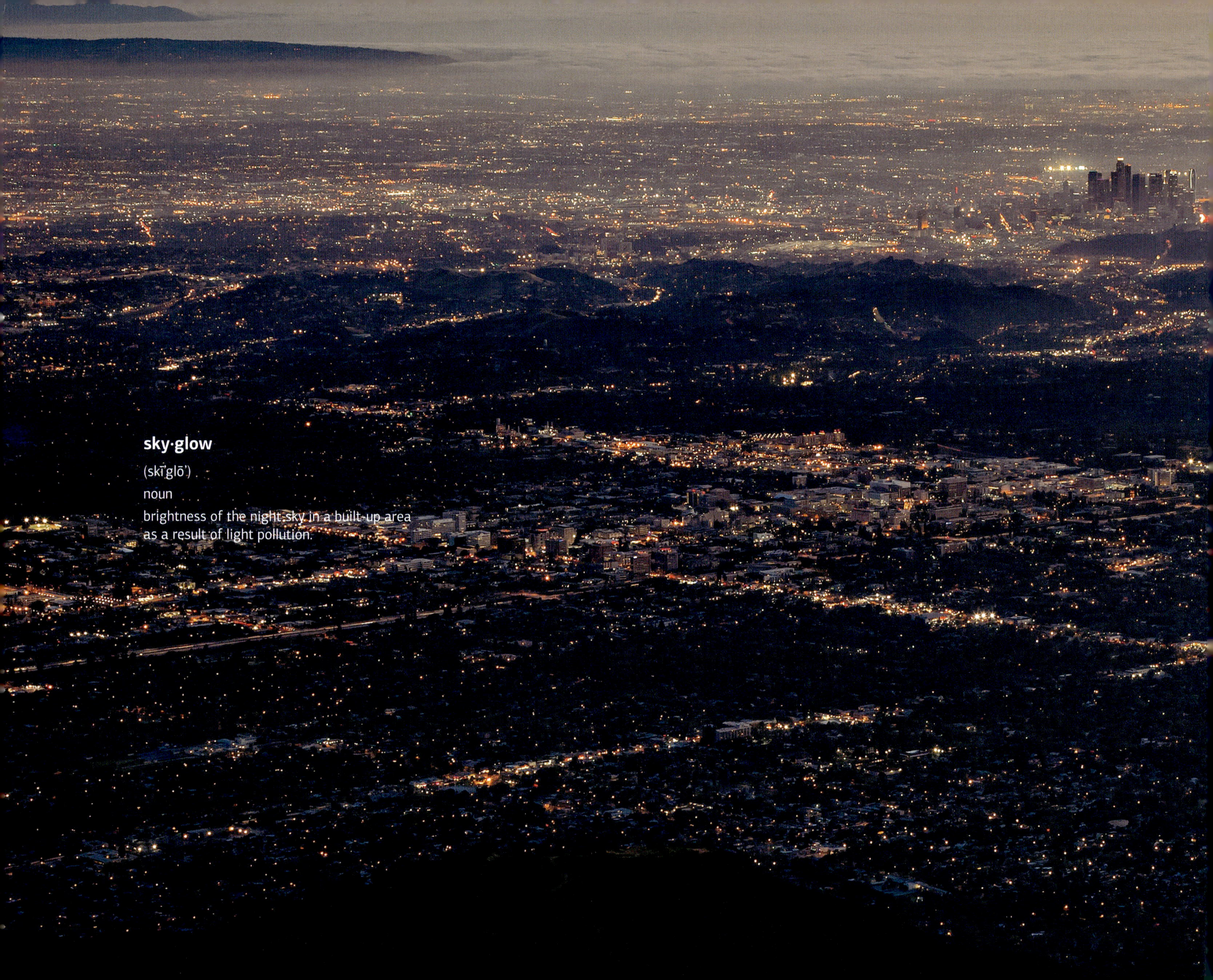

sky·glow

(skī′glō′)

noun

brightness of the night sky in a built-up area as a result of light pollution.

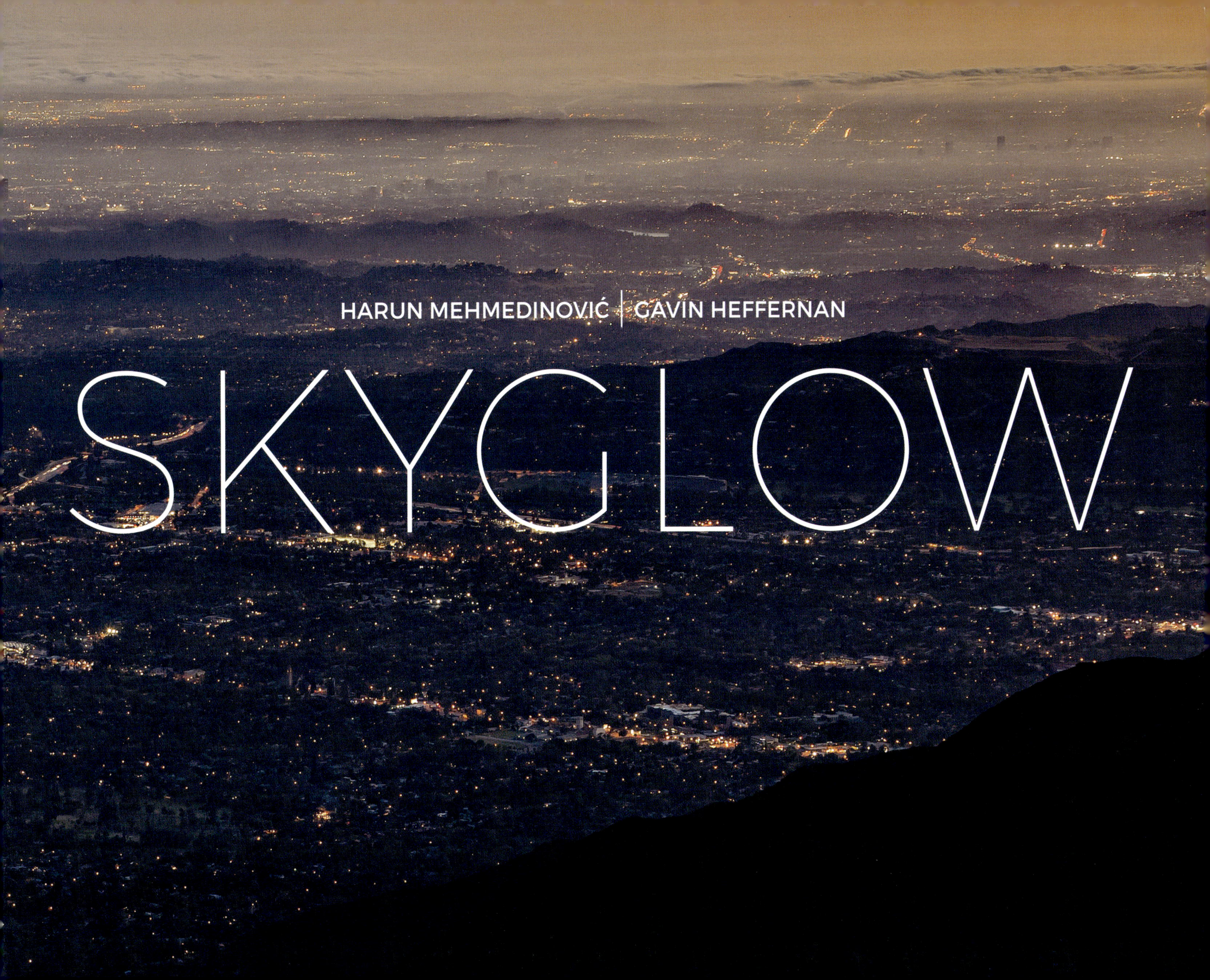

HARUN MEHMEDINOVIĆ | GAVIN HEFFERNAN

SKYGLOW

Bortle Scale

The Bortle Scale is a nine-level numeric measure of the night sky's brightness in a particular location. It quantifies the astronomical visibility of celestial objects and the interference caused by light pollution. John E. Bortle created the scale and published it in the February 2001 edition of *Sky & Telescope* magazine to help amateur astronomers evaluate and compare the darkness of observing sites. The scale ranges from Class 1, the darkest skies available on Earth, to Class 9, daylight-like inner city skies. Throughout this book, Bortle Scale class numbers will appear next to images.

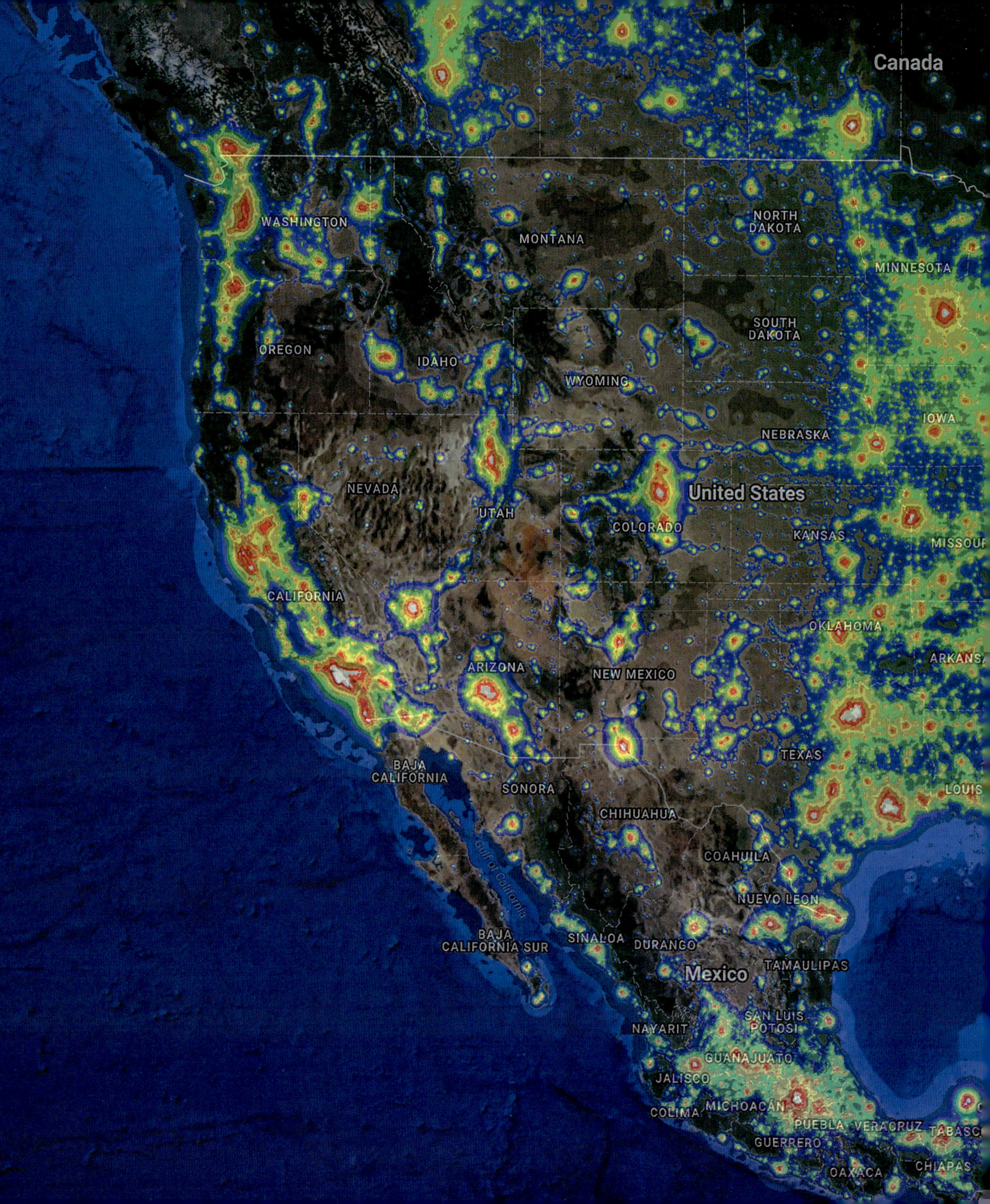

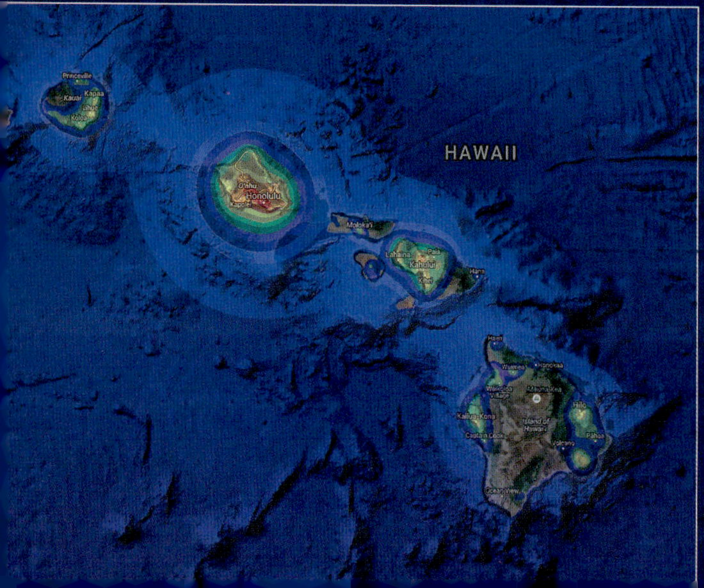

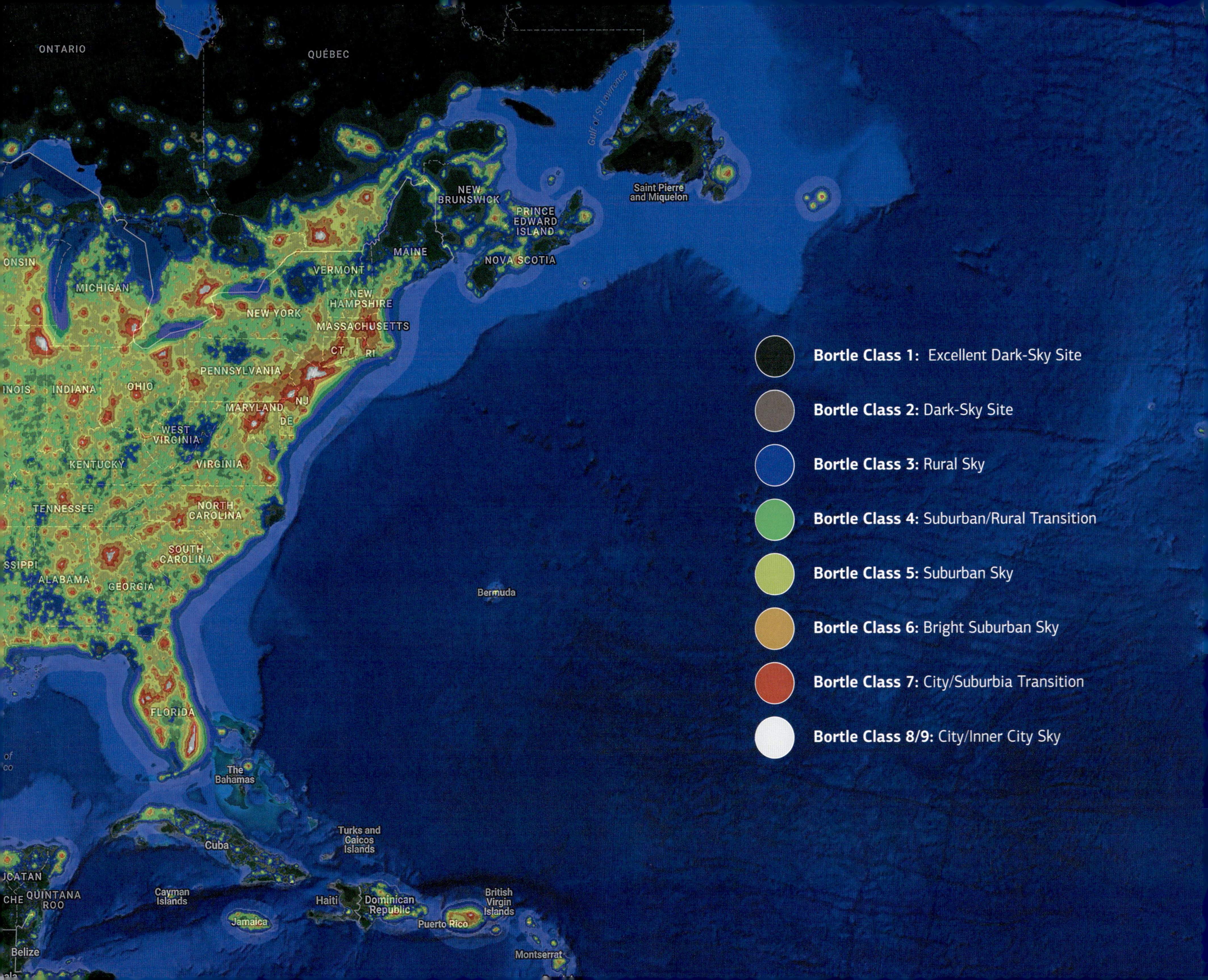

ONTARIO
QUÉBEC
Gulf of St Lawrence
NEW BRUNSWICK
PRINCE EDWARD ISLAND
Saint Pierre and Miquelon
MAINE
NOVA SCOTIA
VERMONT
NEW HAMPSHIRE
MICHIGAN
NEW YORK
MASSACHUSETTS
CT
RI
PENNSYLVANIA
INDIANA
OHIO
MARYLAND
NJ
DE
WEST VIRGINIA
KENTUCKY
VIRGINIA
TENNESSEE
NORTH CAROLINA
SOUTH CAROLINA
ALABAMA
GEORGIA
FLORIDA
Bermuda
The Bahamas
Turks and Caicos Islands
Cuba
Cayman Islands
Haiti
Dominican Republic
British Virgin Islands
Jamaica
Puerto Rico
QUINTANA ROO
Belize
Montserrat
Bortle Class 1: Excellent Dark-Sky Site
Bortle Class 2: Dark-Sky Site
Bortle Class 3: Rural Sky
Bortle Class 4: Suburban/Rural Transition
Bortle Class 5: Suburban Sky
Bortle Class 6: Bright Suburban Sky
Bortle Class 7: City/Suburbia Transition
Bortle Class 8/9: City/Inner City Sky

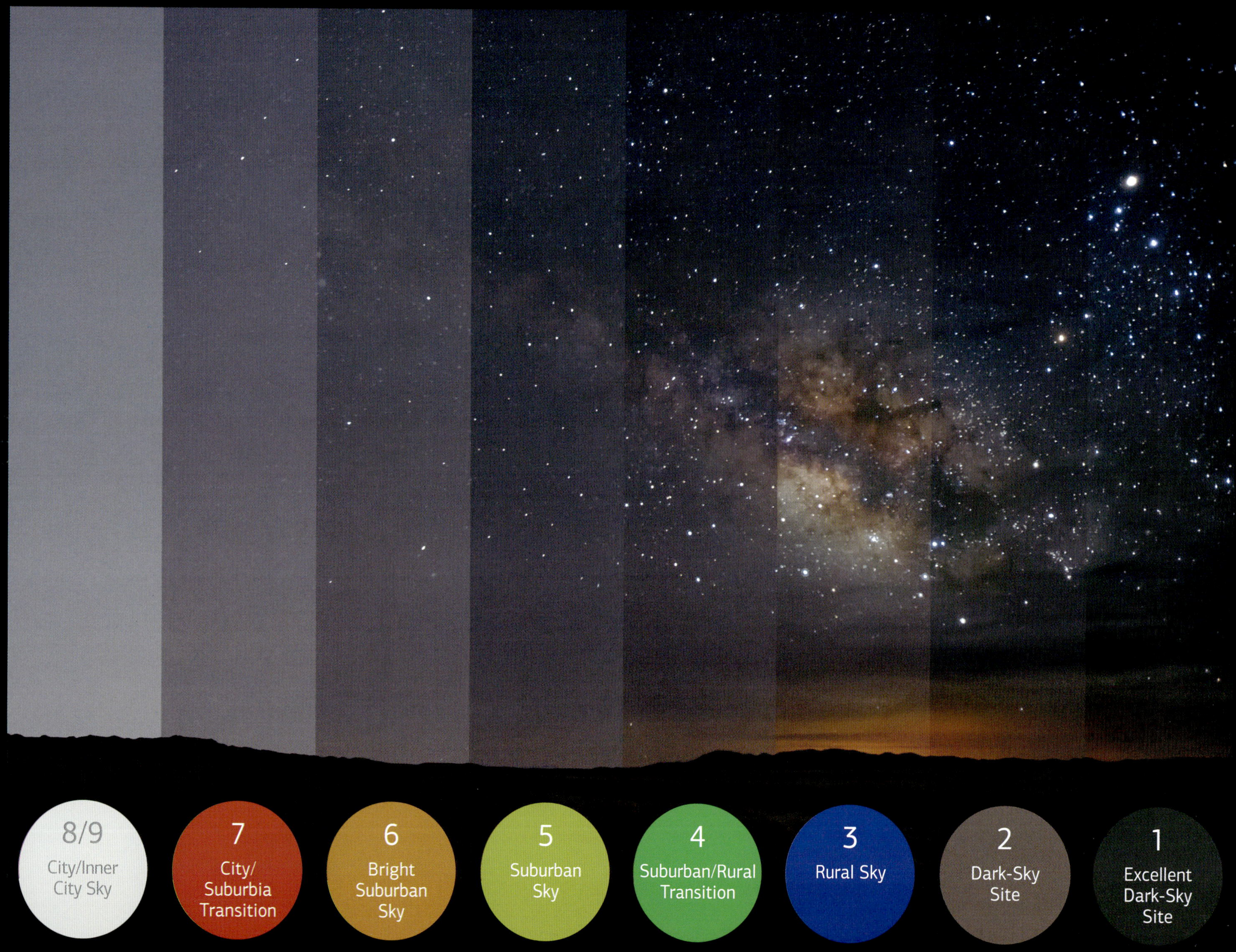
8/9
City/Inner City Sky
7
City/ Suburbia Transition
6
Bright Suburban Sky
5
Suburban Sky
4
Suburban/Rural Transition
3
Rural Sky
2
Dark-Sky Site
1
Excellent Dark-Sky Site

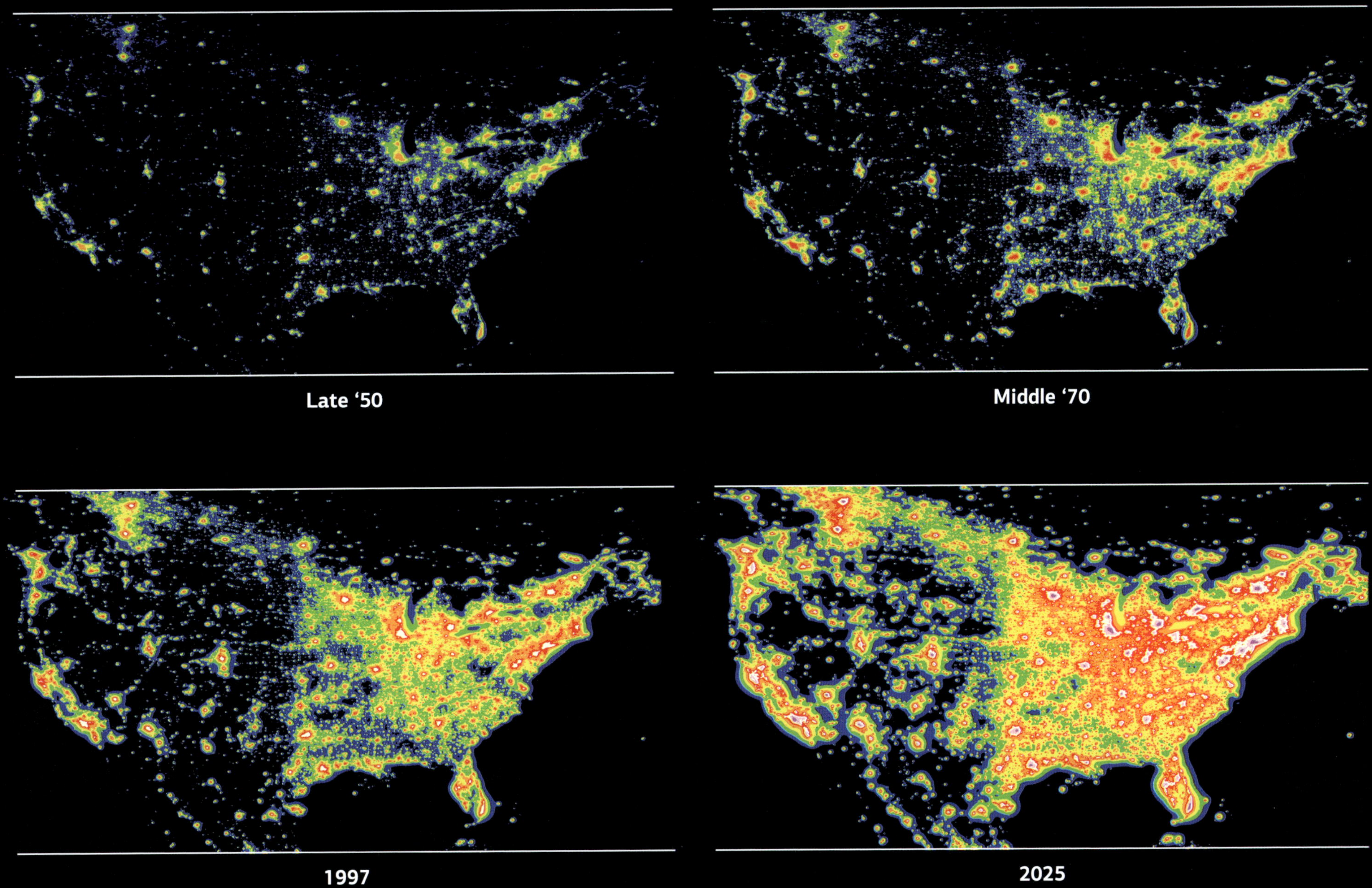

Since the 1950s, light pollution in the United States has grown at an average rate of 6-percent per year, for a total of close to 400-percent growth.

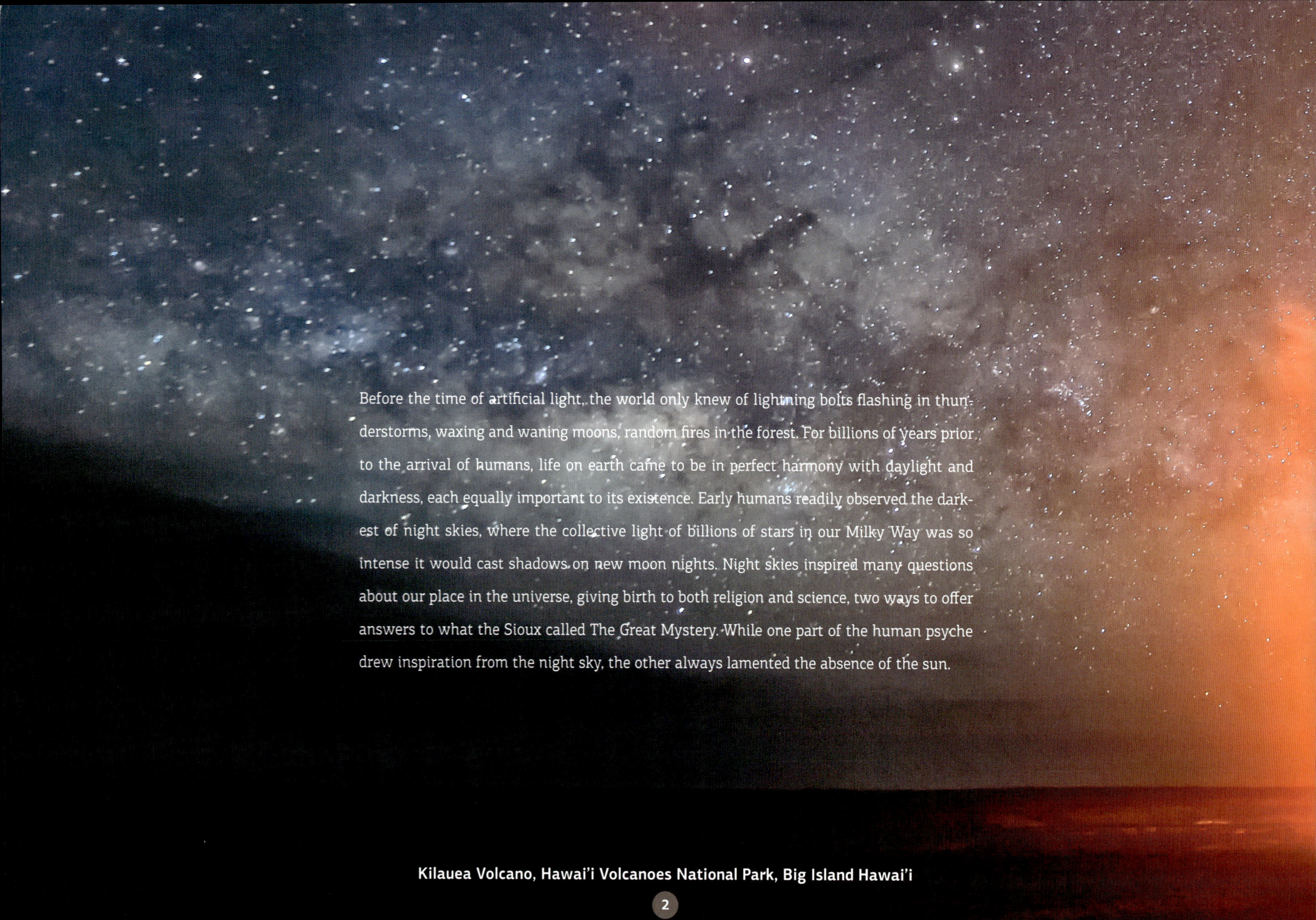

Before the time of artificial light, the world only knew of lightning bolts flashing in thunderstorms, waxing and waning moons, random fires in the forest. For billions of years prior to the arrival of humans, life on earth came to be in perfect harmony with daylight and darkness, each equally important to its existence. Early humans readily observed the darkest of night skies, where the collective light of billions of stars in our Milky Way was so intense it would cast shadows on new moon nights. Night skies inspired many questions about our place in the universe, giving birth to both religion and science, two ways to offer answers to what the Sioux called The Great Mystery. While one part of the human psyche drew inspiration from the night sky, the other always lamented the absence of the sun.

Kilauea Volcano, Hawai'i Volcanoes National Park, Big Island Hawai'i

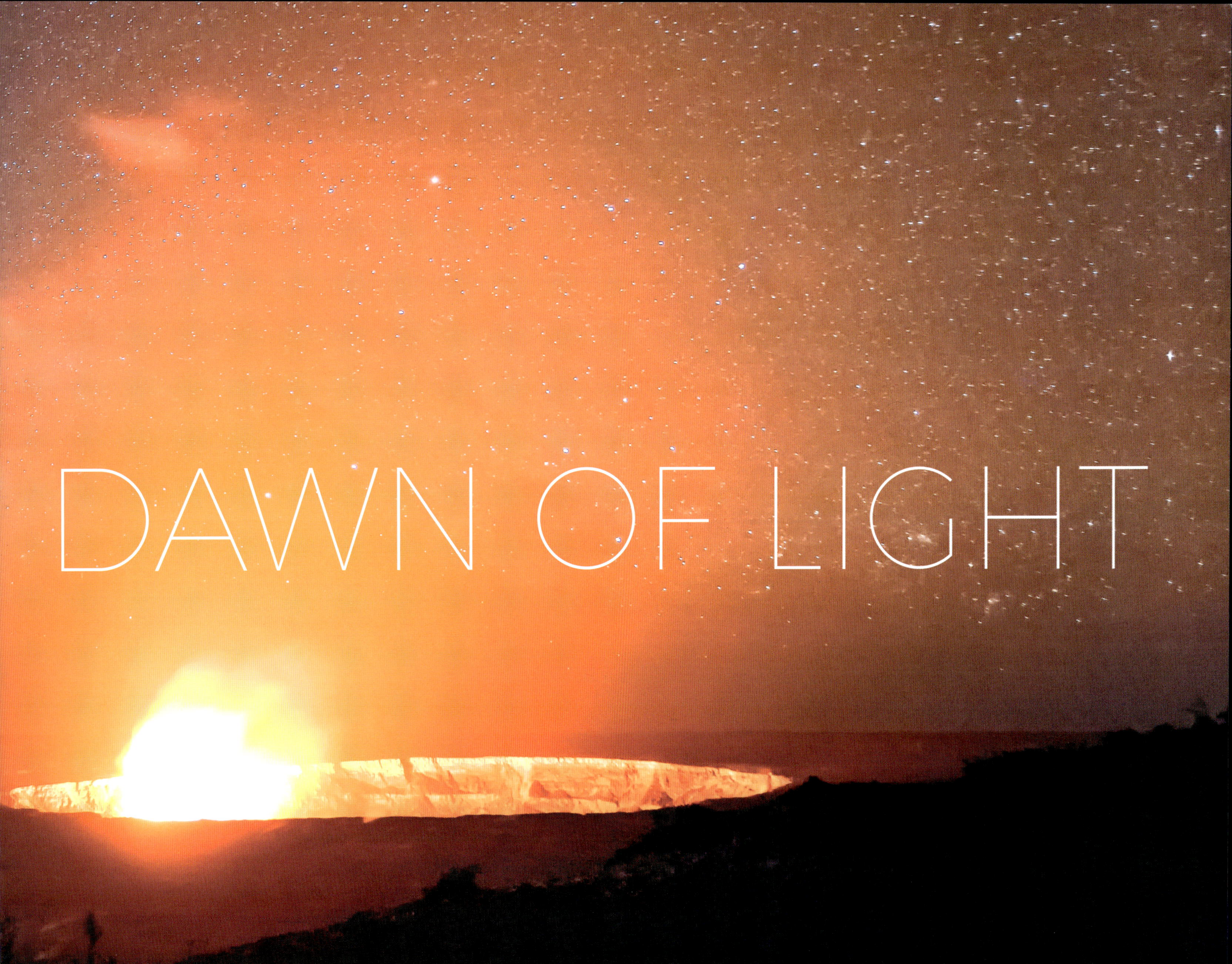
DAWN OF LIGHT

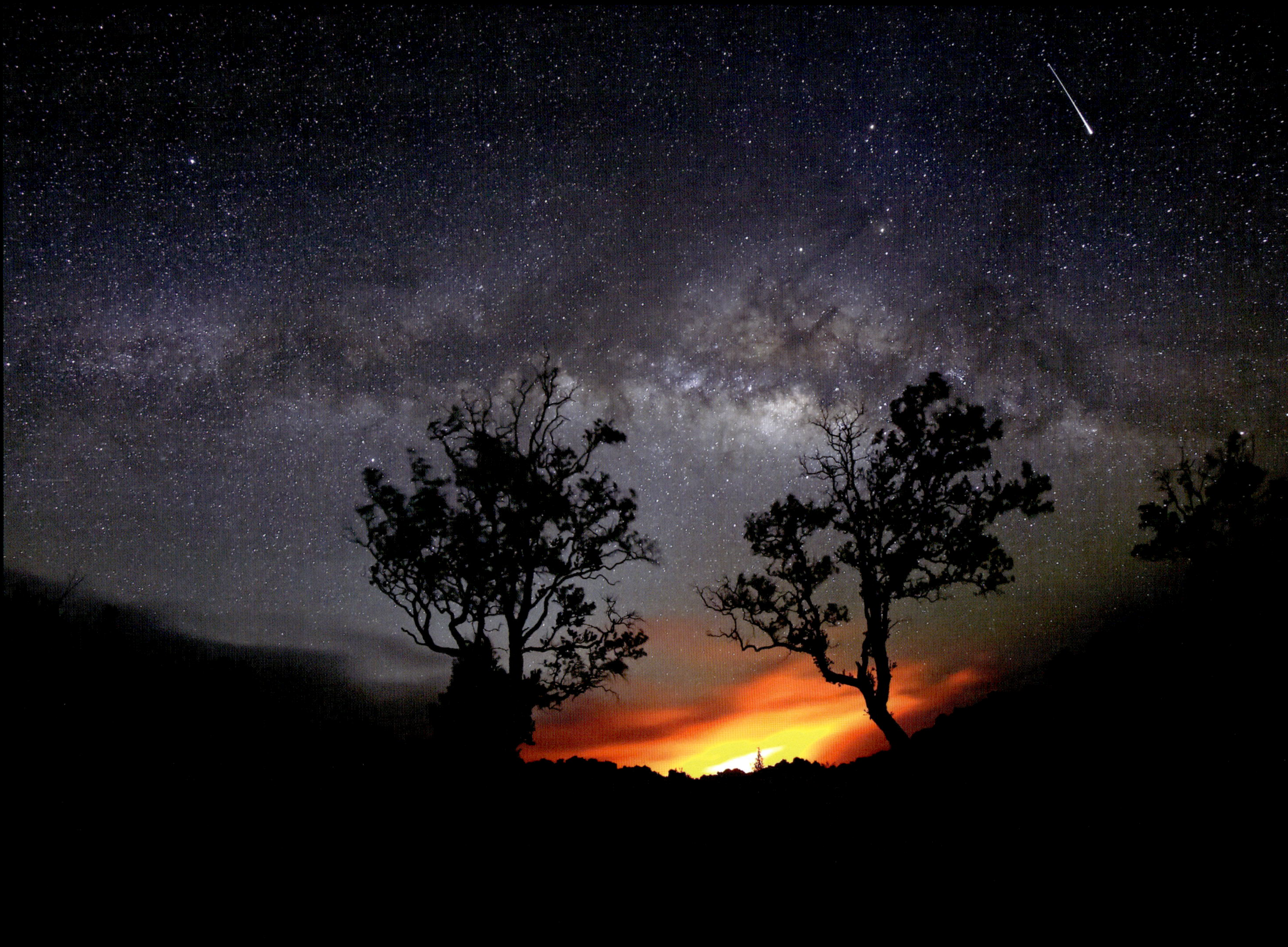

Kilauea Volcano, Hawai'i Volcanoes National Park, Big Island Hawai'i

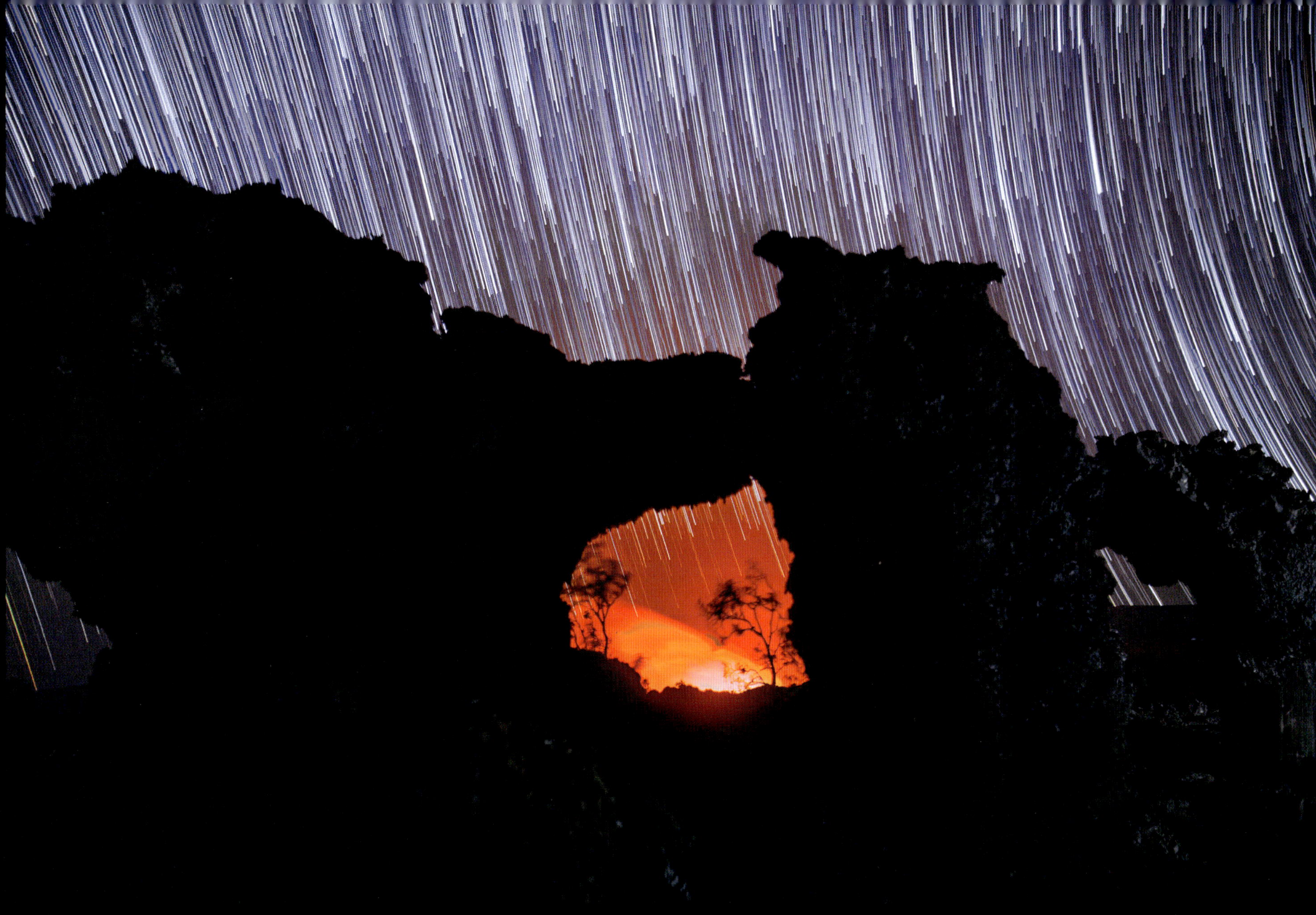

The first presence of natural light pollution on Earth proved to be essential to its creation. Endless volcanic activity--a mixture of lava, ash, and gas that escapes from a magma chamber deep below the surface--has been shaping the planet from the very beginning. This phenomenon continues in Hawai'i, Earth's most volcanically active spot, where new land is created on a daily basis.

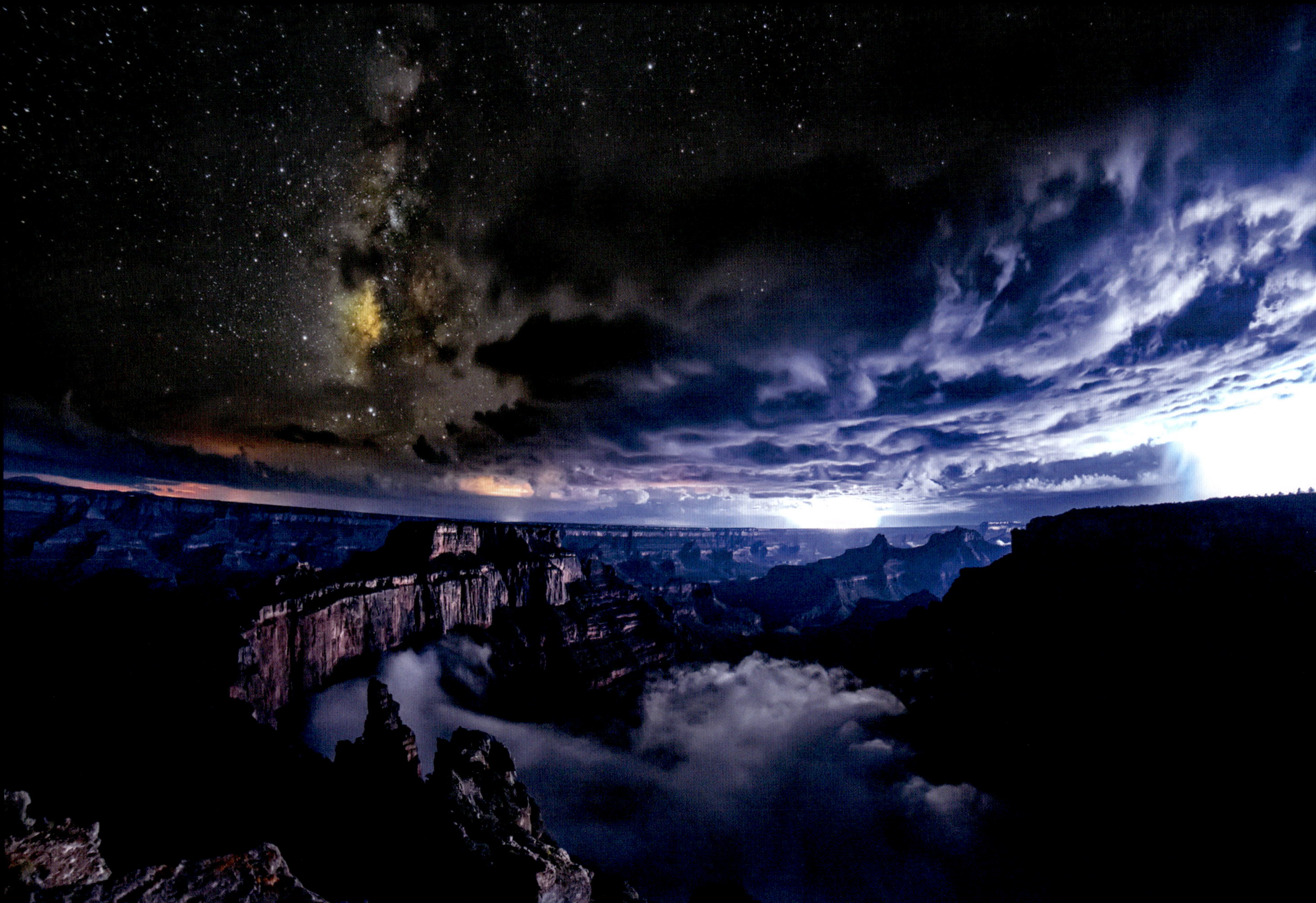

Cape Royal Overlook, North Rim, Grand Canyon National Park, Arizona

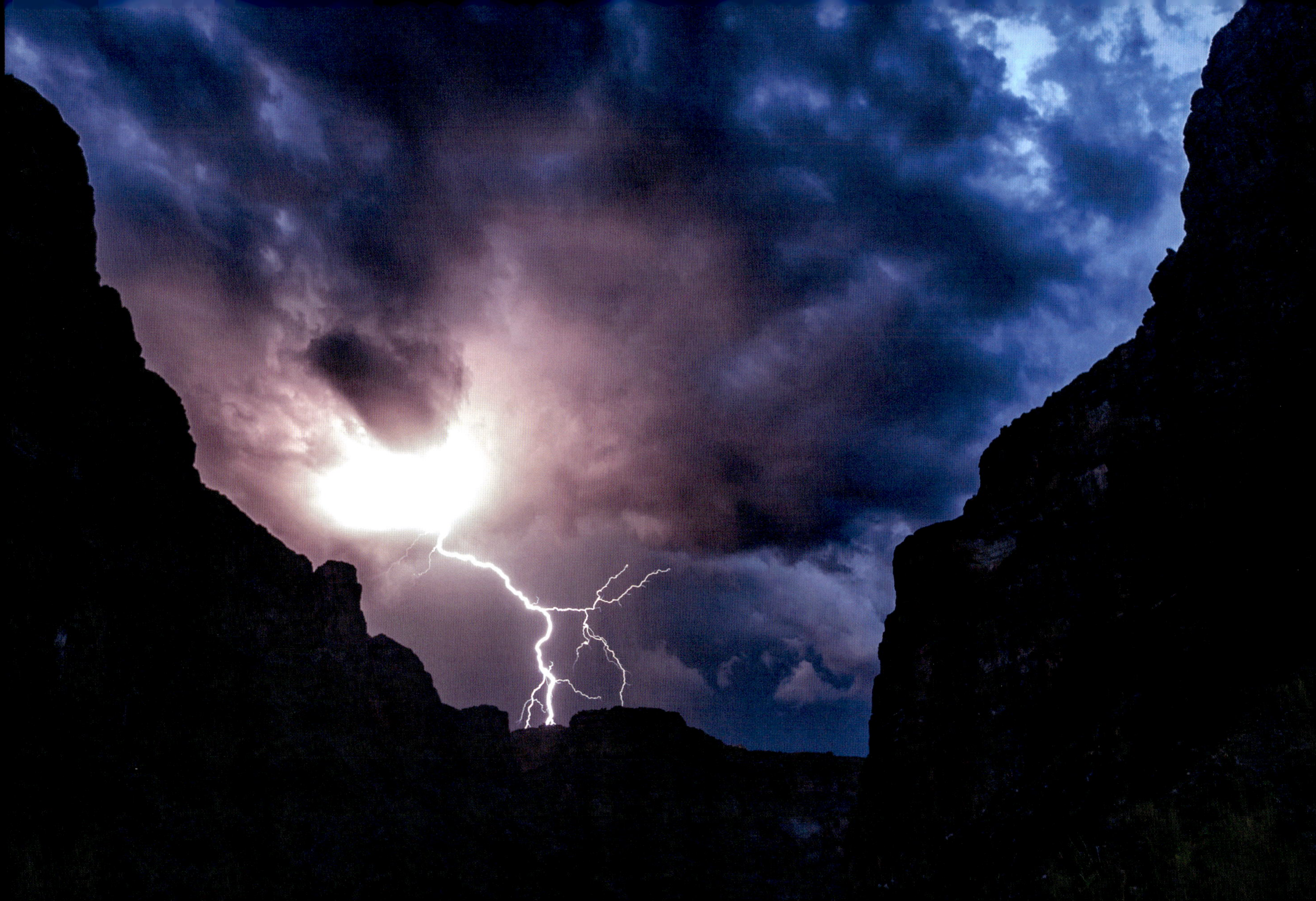

For billions of years, lightning was the brightest light in the night sky. Each flash lasted only a millisecond, providing brief but blinding illuminations, as seen here at the Grand Canyon, where thunderbolts often strike near the edges of the rim in spectacular fashion.

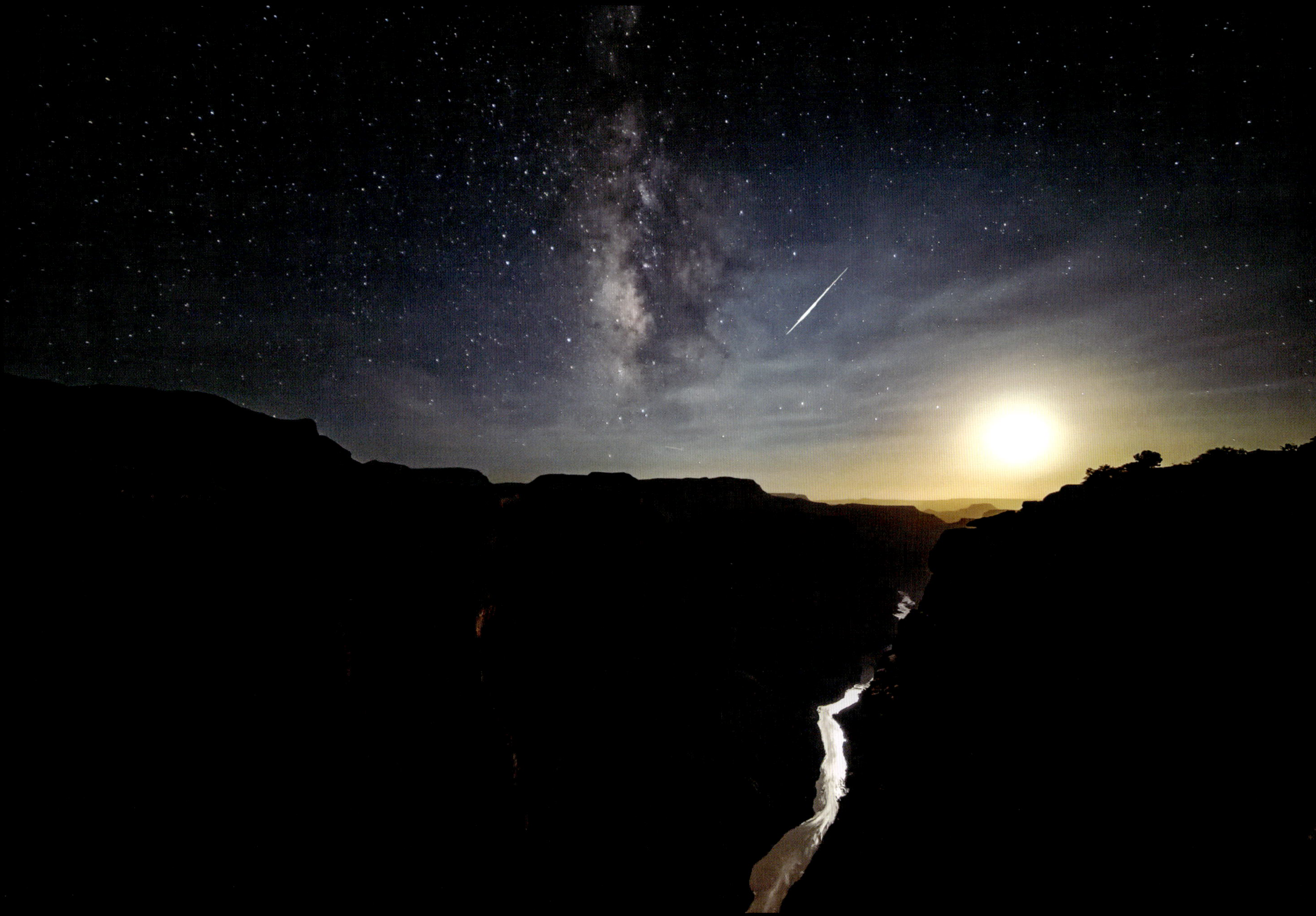

Moonlight too presented the early inhabitants of Earth with daylight-like illumination of the landscape, as seen here during a Grand Canyon moonset.

Toroweap Overlook, Grand Canyon National Park, Arizona

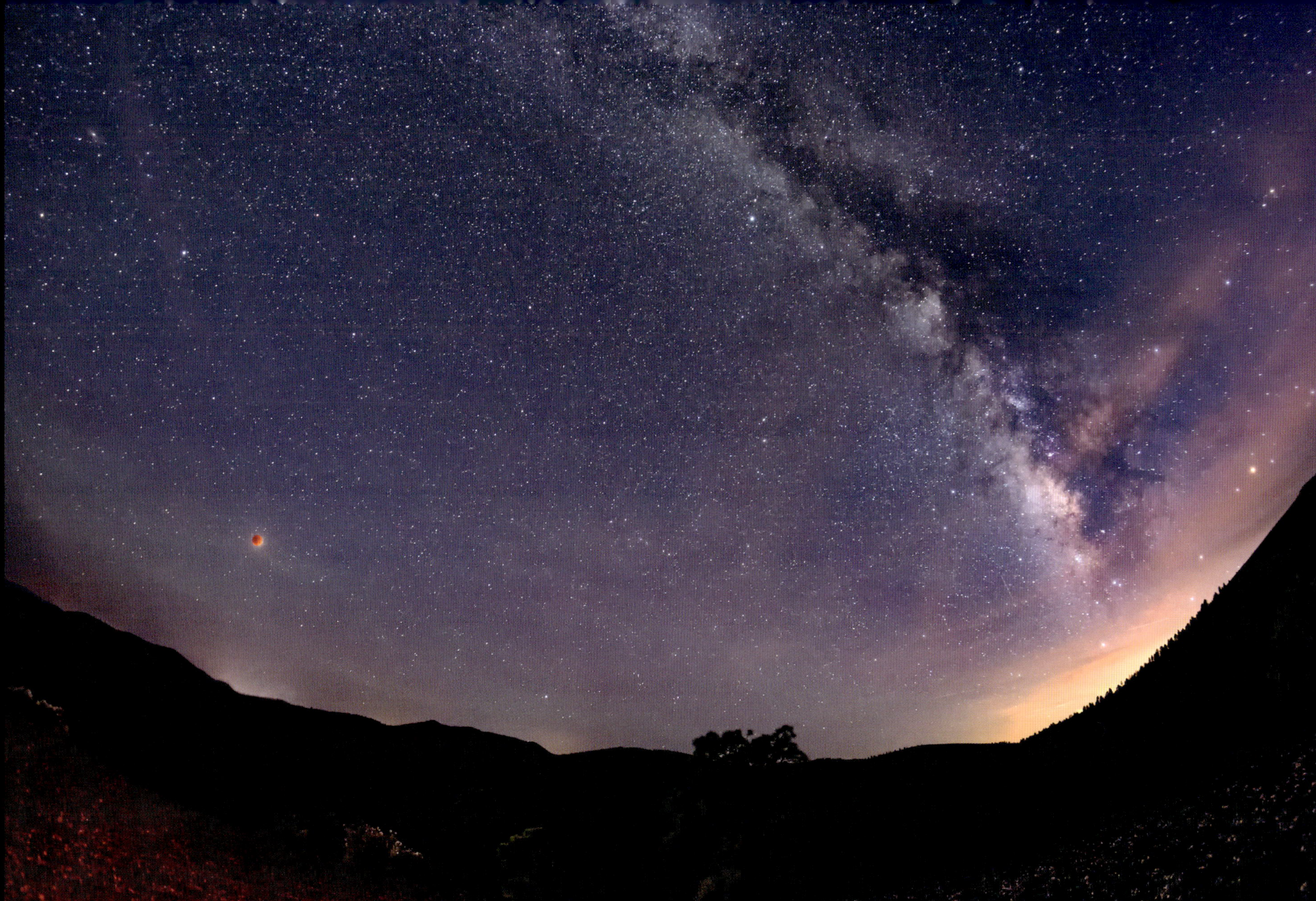

A rare Blood Moon lunar eclipse briefly brings pristine darkness, revealing sights of the heavens only seen during new moon nights.

Cinder Hills Overlook, Sunset Crater National Monument, Arizona

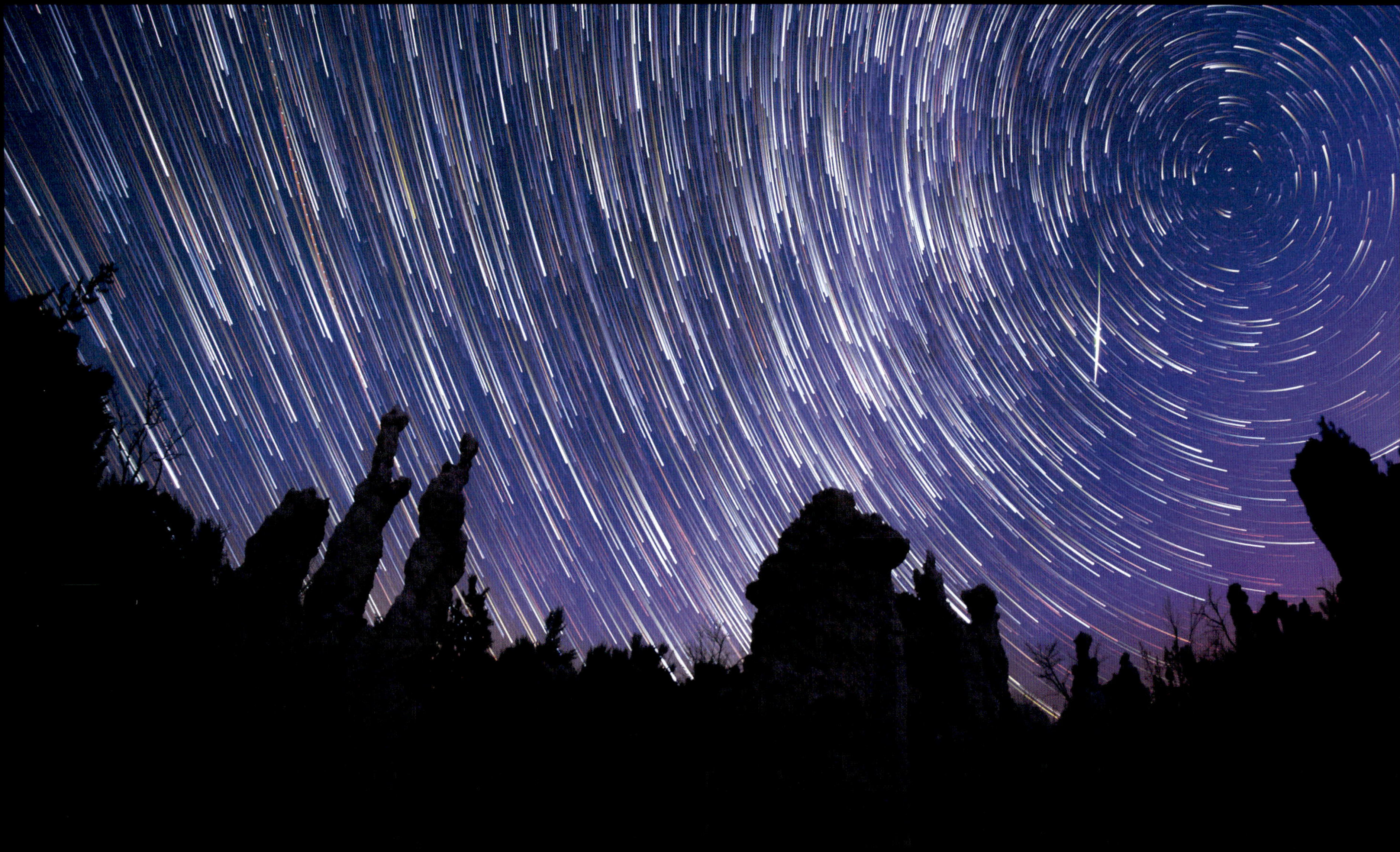

Mono Lake, California

1

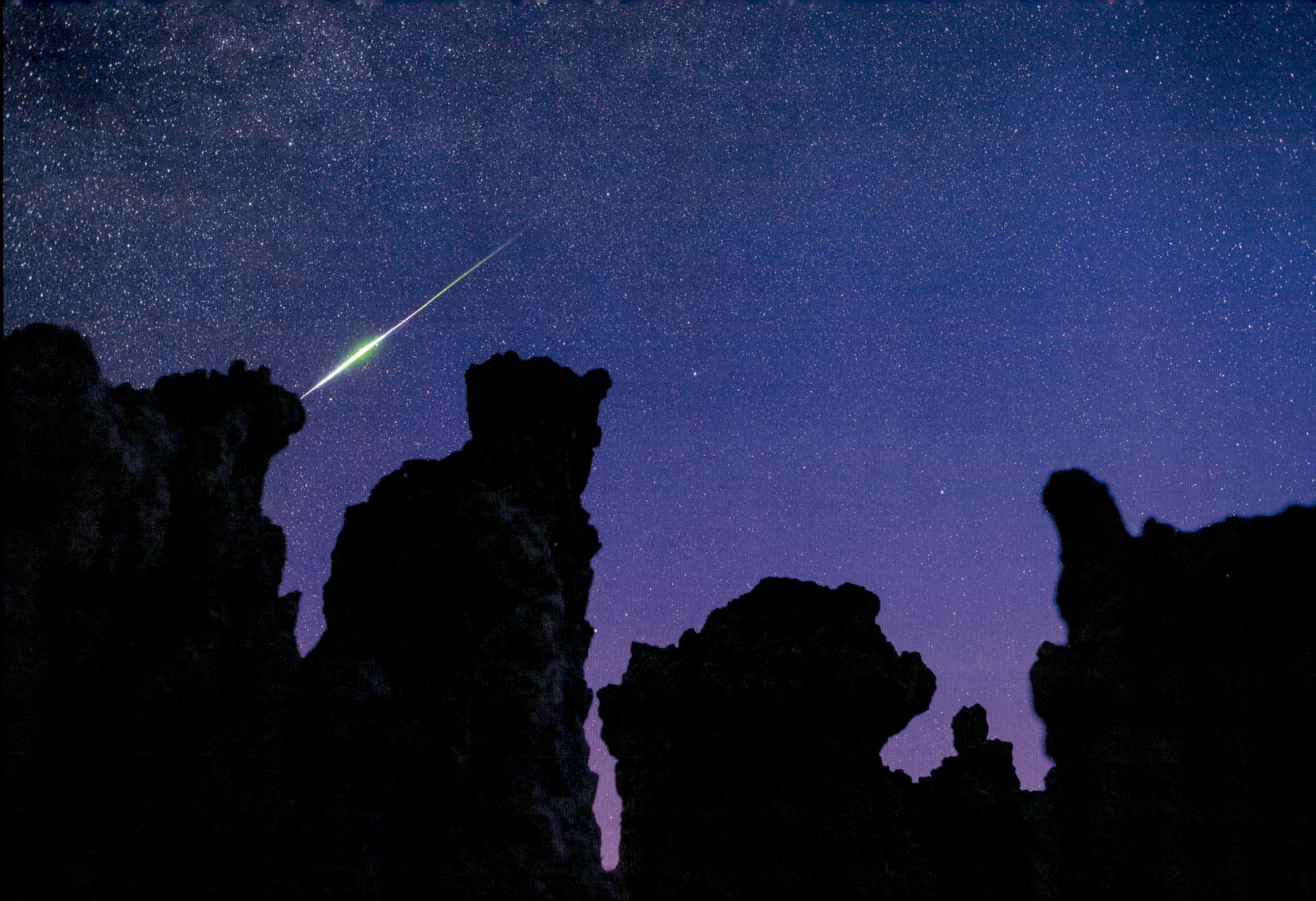

Two other rare, but nonetheless present forms of natural light pollution, were meteors burning in the earth's atmosphere, as well as Aurora Borealis (Northern Lights), both seen here at Owens Valley's Mono Lake, where calcium deposit formations, known as Tufa Formations, have been formed as far back as 100,000 years.

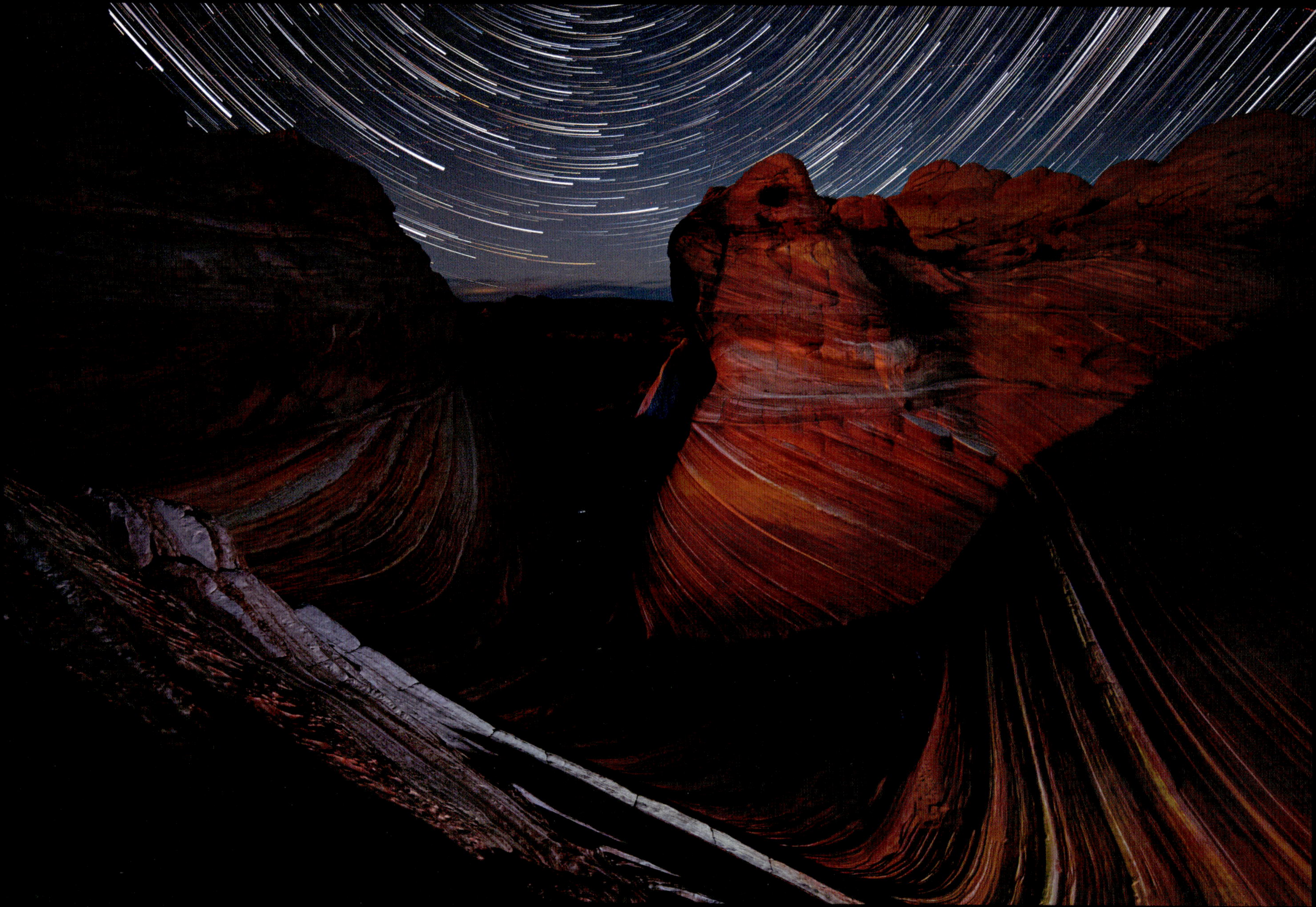

The Wave, Vermilion Cliffs National Monument, Arizona

1

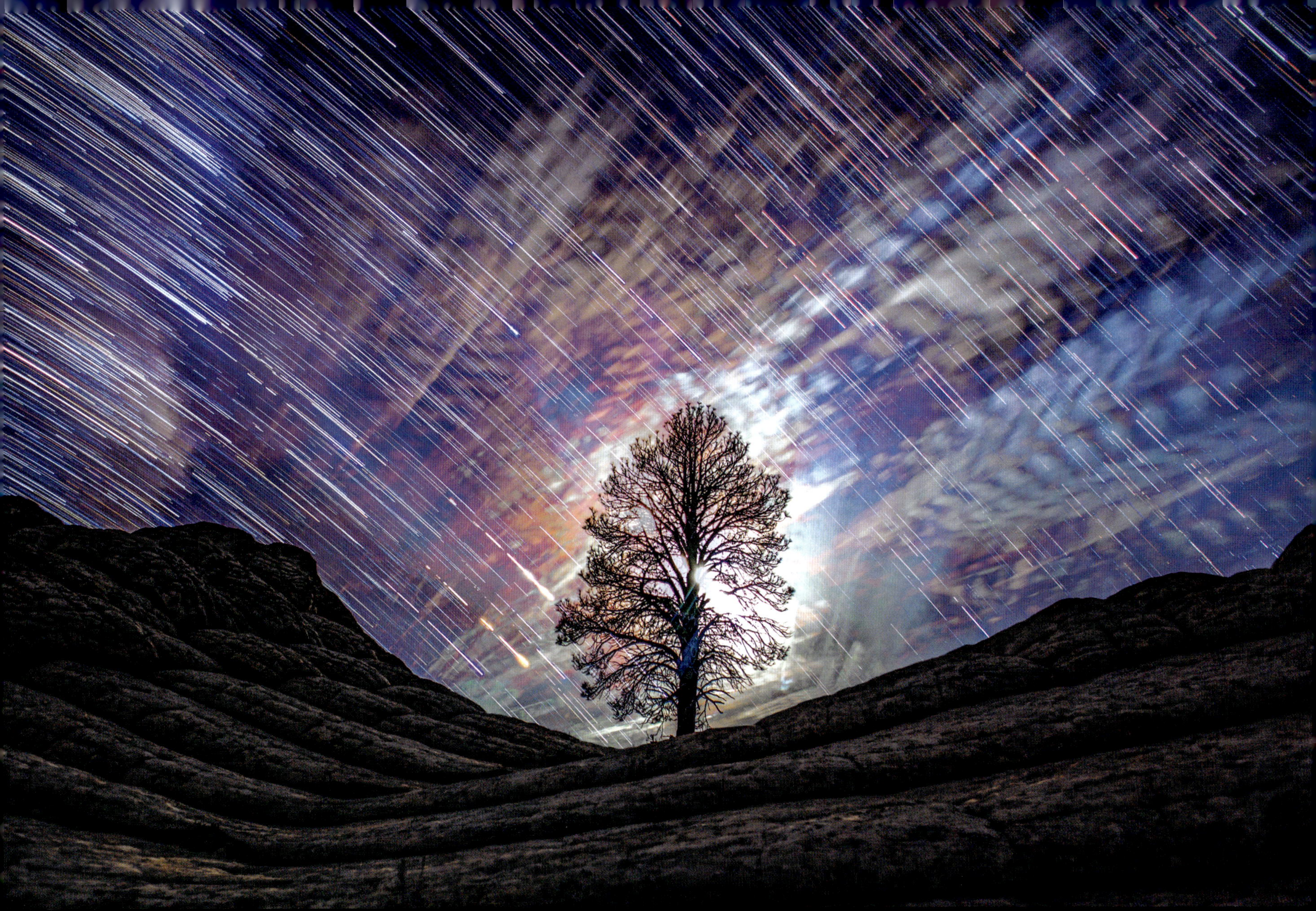

White Pocket, Vermilion Cliffs National Monument, Arizona

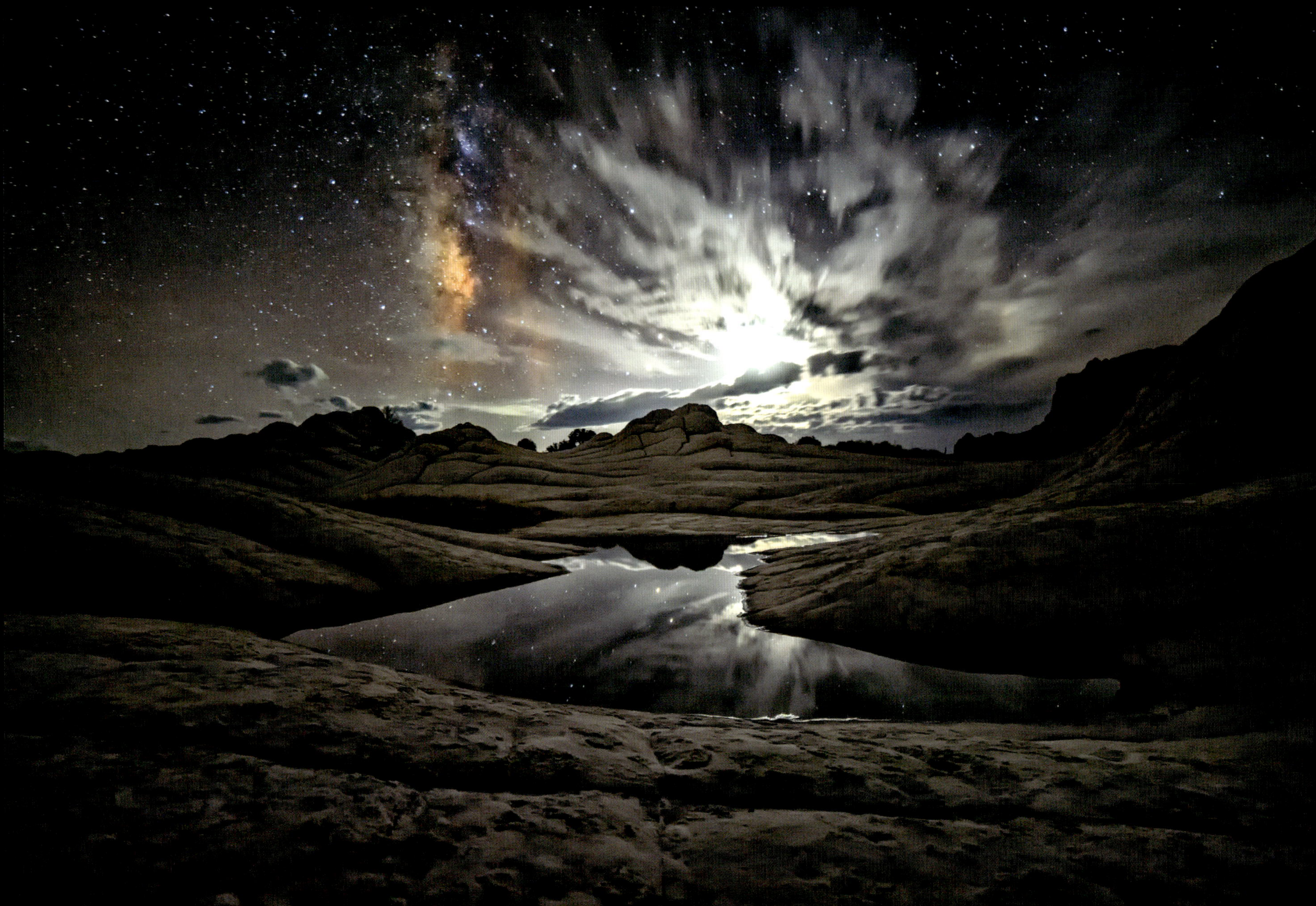

White Pocket, Vermilion Cliffs National Monument, Arizona

1

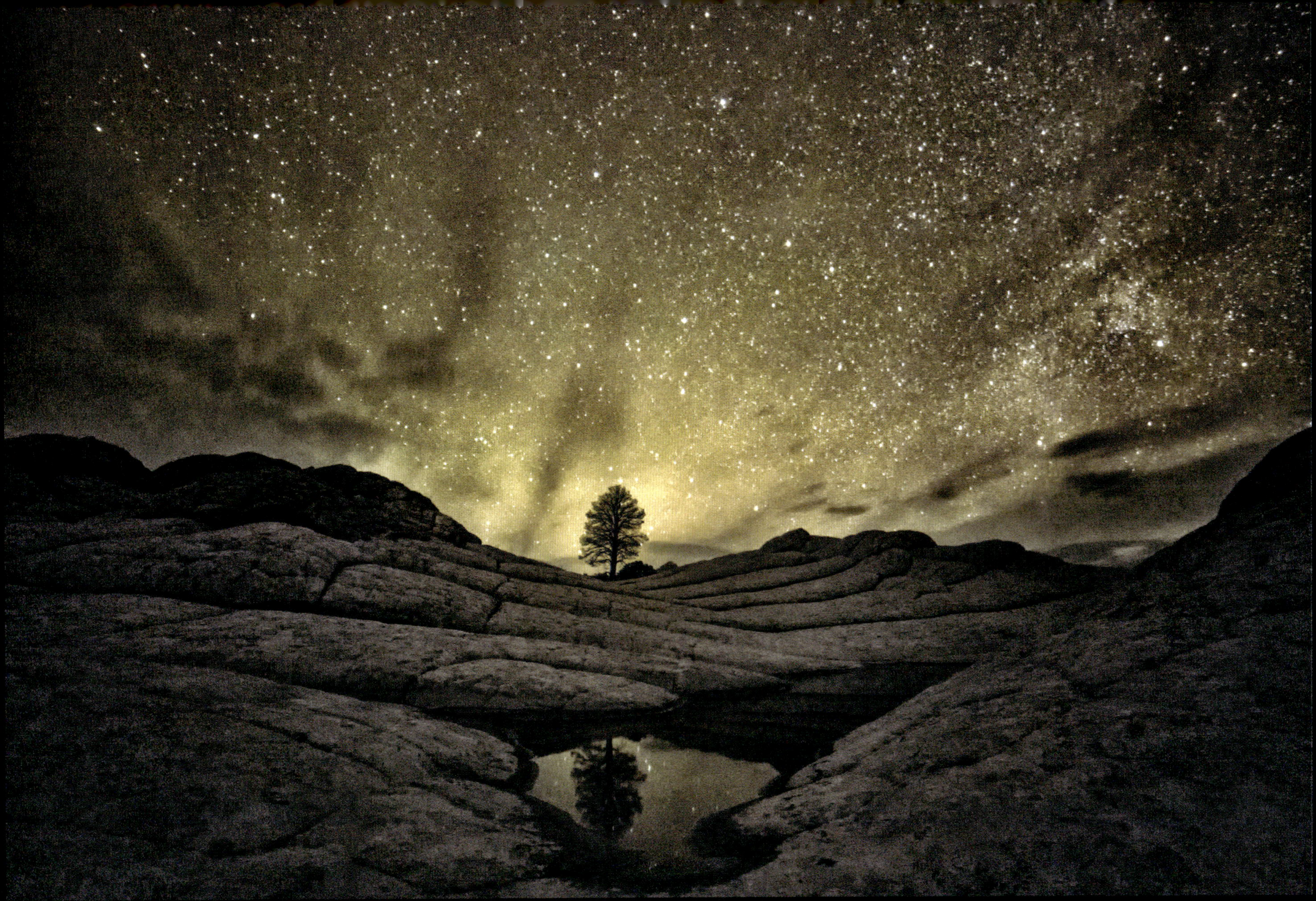

Vermilion Cliffs National Monument is covered in lithified rocks that formed during the Jurassic Period, when this area was completely underwater for millions of years. Near-perfect dark skies here, when the conditions are right, offer a glimpse at the time when dinosaurs roamed the earth.

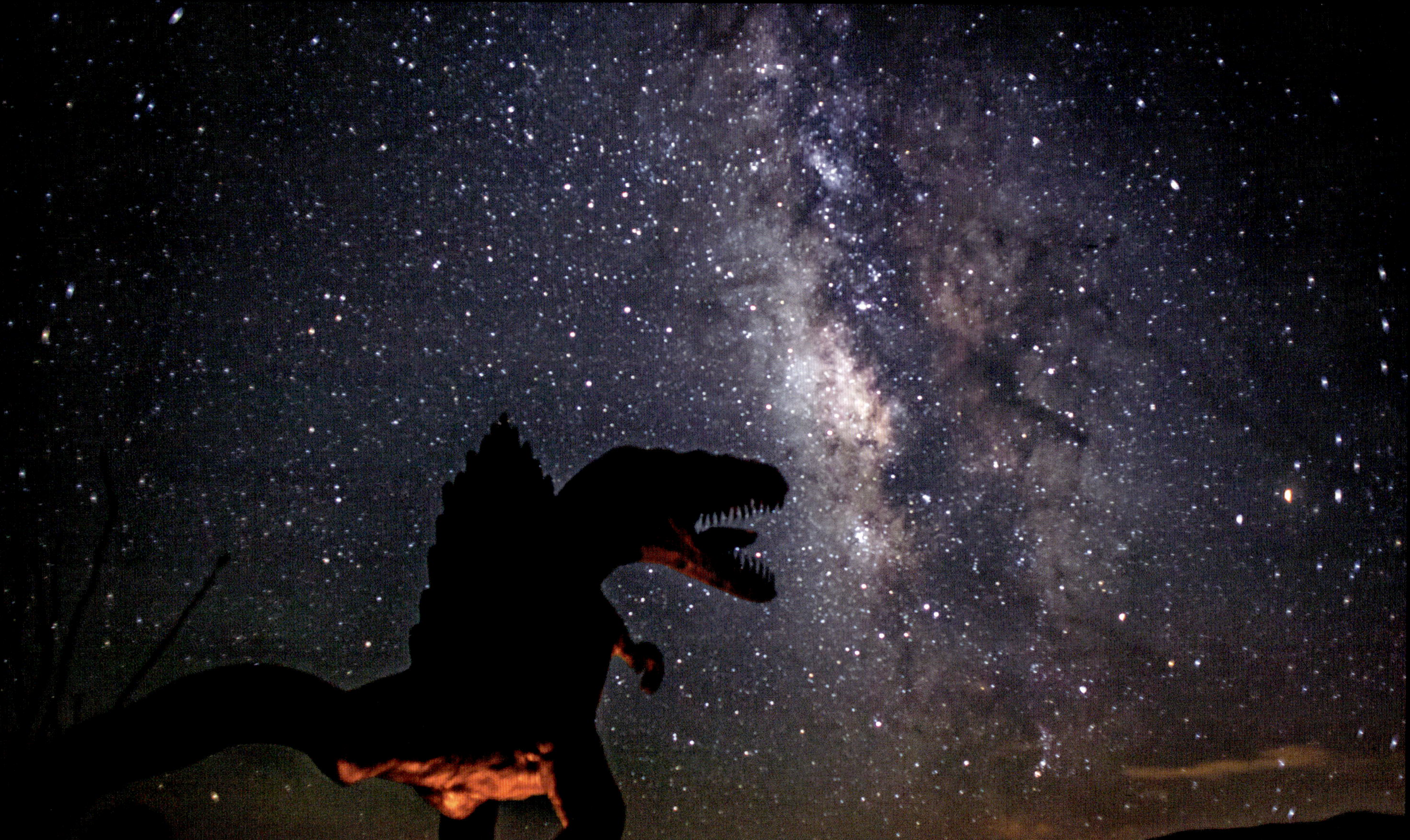

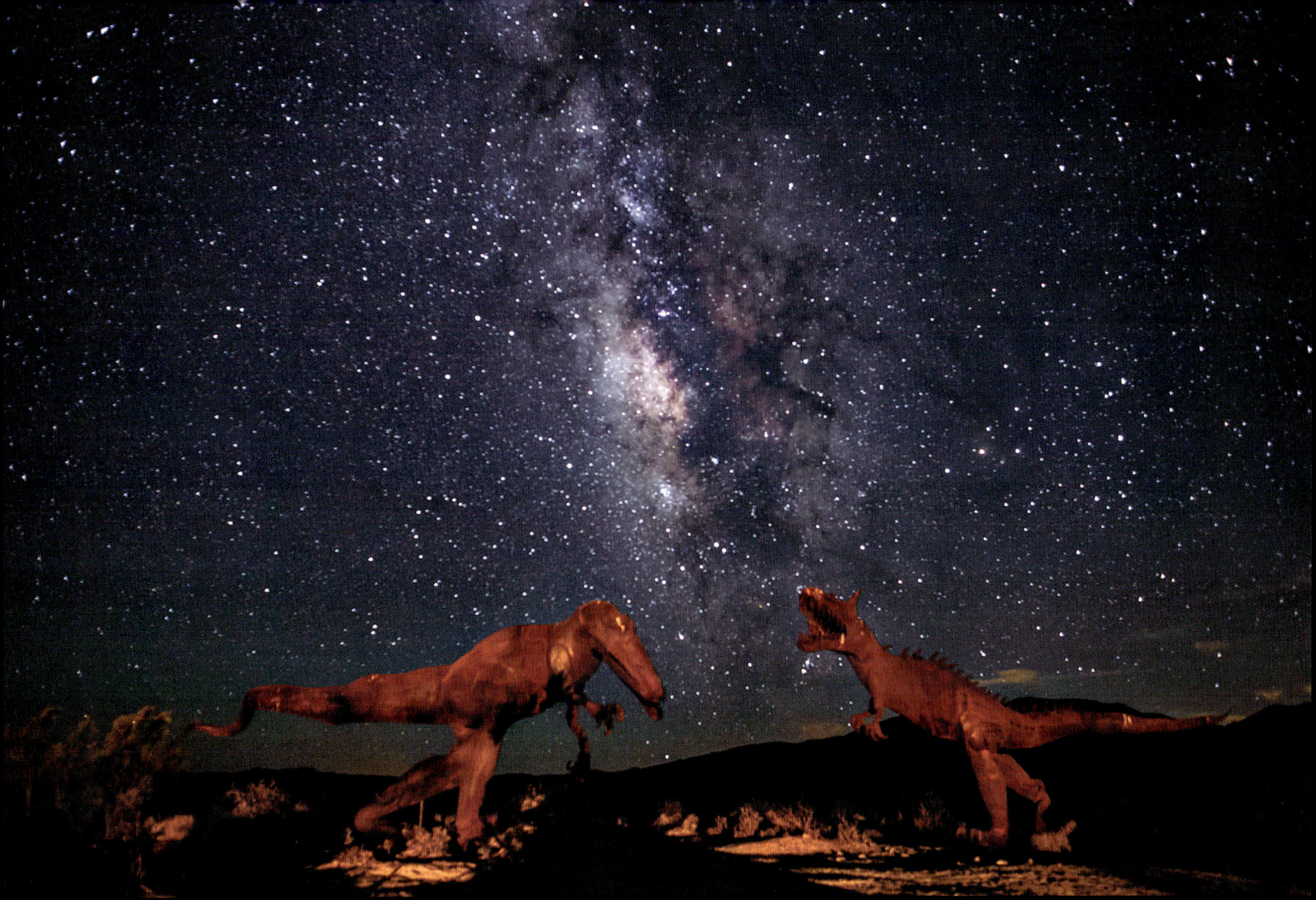

During the darkest nights, in the new moon portion of the lunar cycle, when there was minimal or non-existent reflection of sunlight into the night sky atmosphere, the Milky Way would get so bright it would cast shadows and illuminate the landscape.

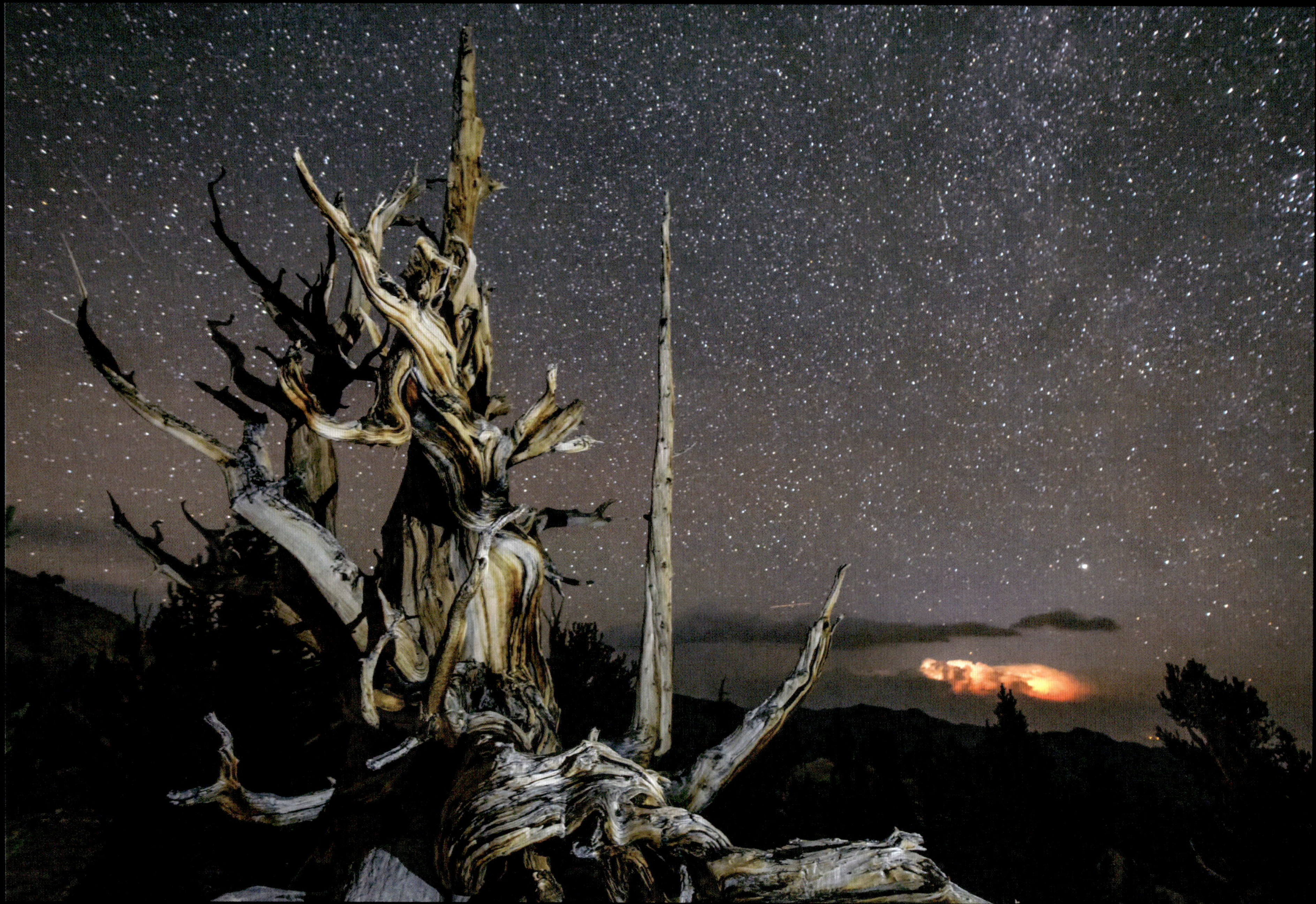

Patriarch Grove, Ancient Bristlecone Pine Forest, California

1

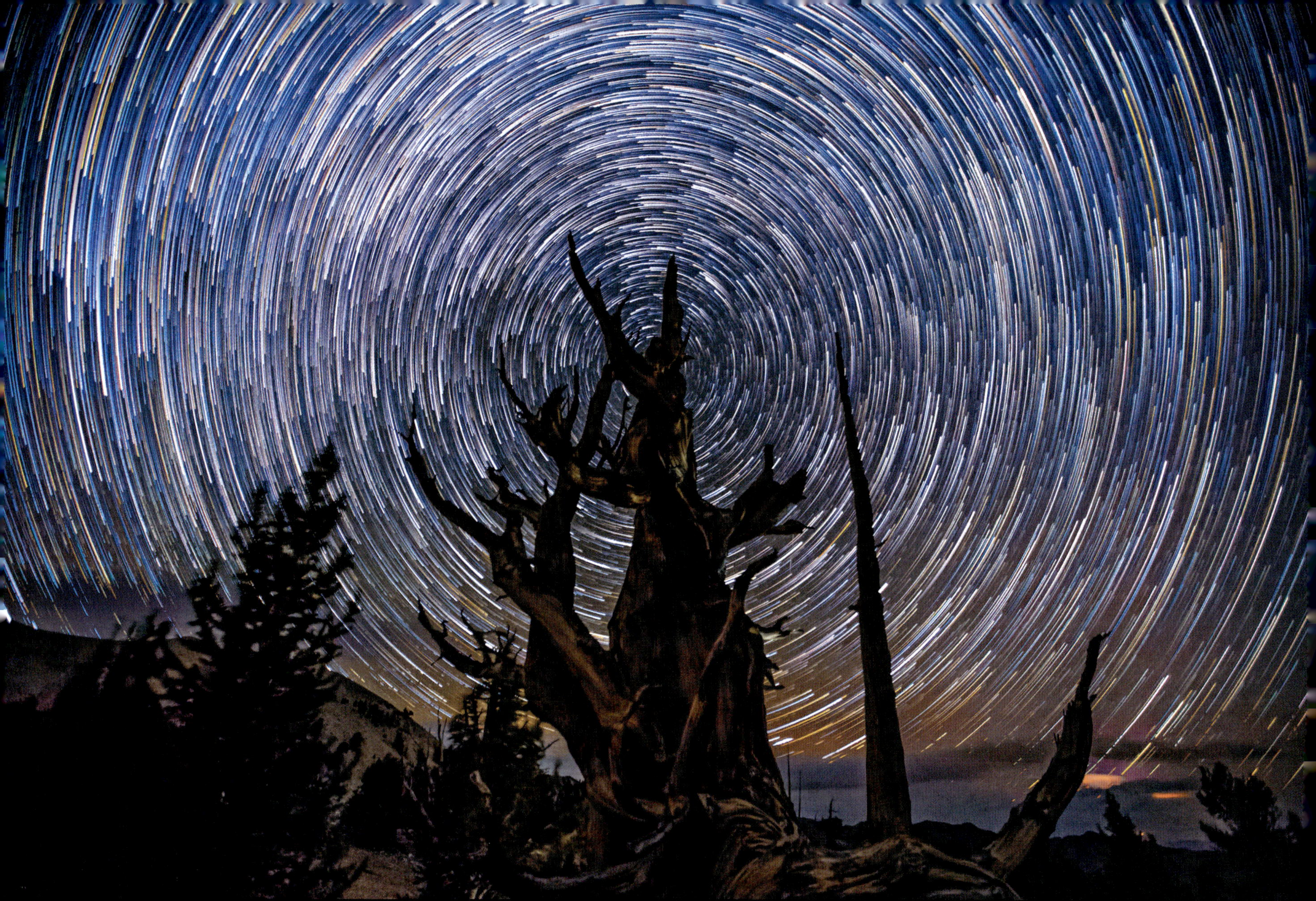

Like the ancient rocks of Vermilion Cliffs, California's Inyo National Forest, the home of 4,500-year-old Ancient Bristlecone Pines, lets us see the sights our earliest ancestors would have enjoyed every night.

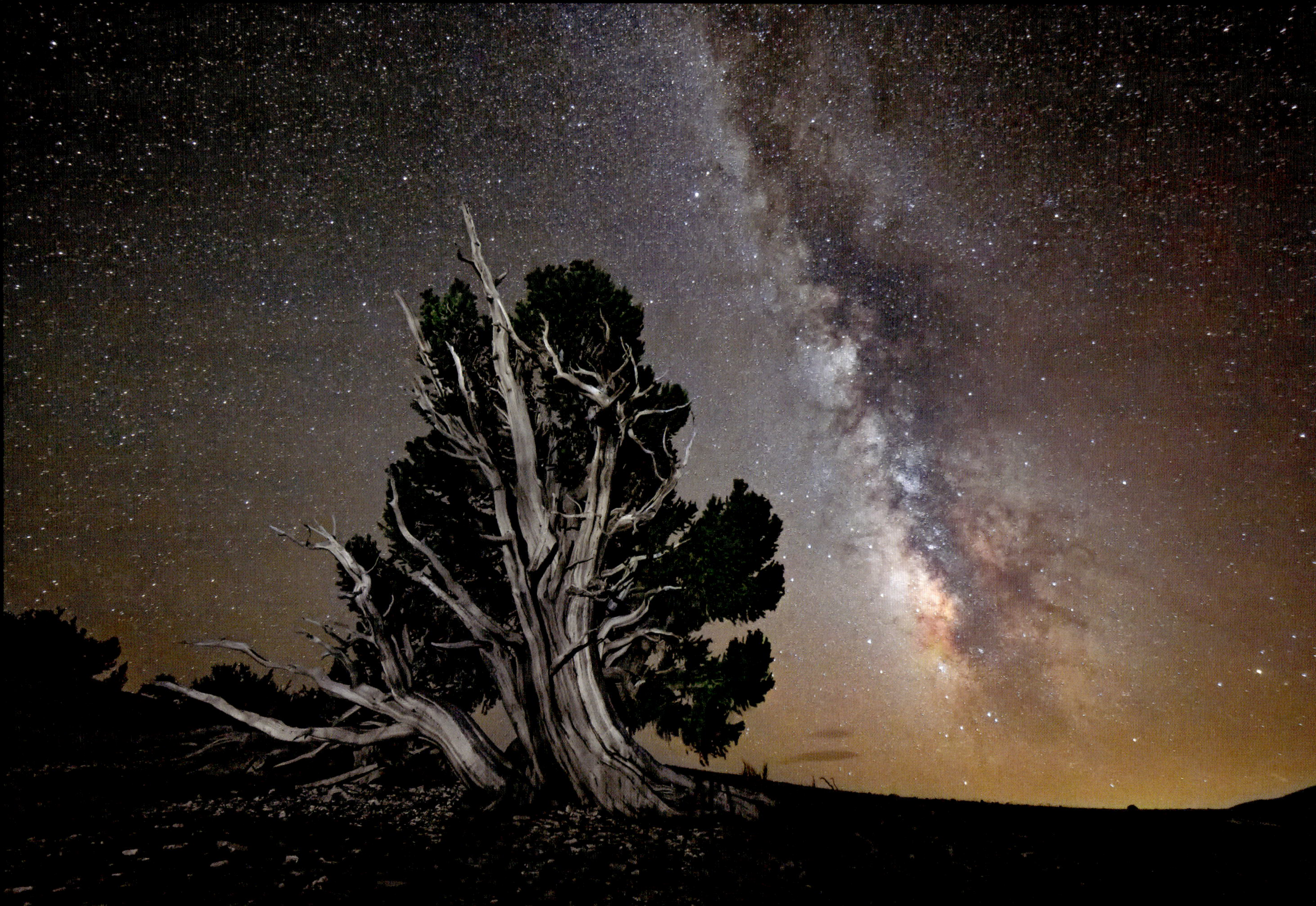

For humans, unlike other animal life, night became a fertile source of both fascination and dread of the unknown.

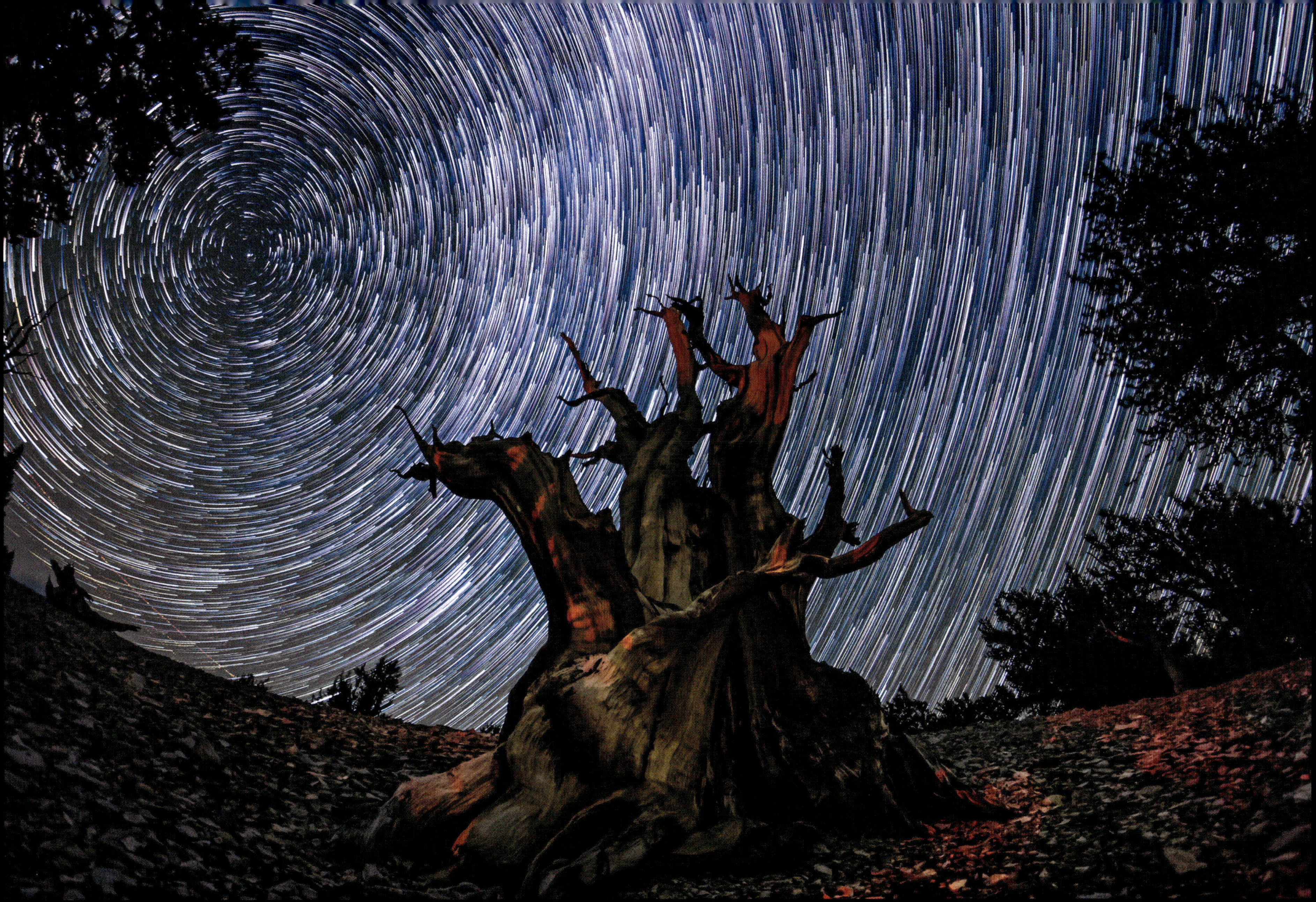

Night sky, the source of endless unsolved mysteries, became the inspiration for some of the oldest known stories and myths.

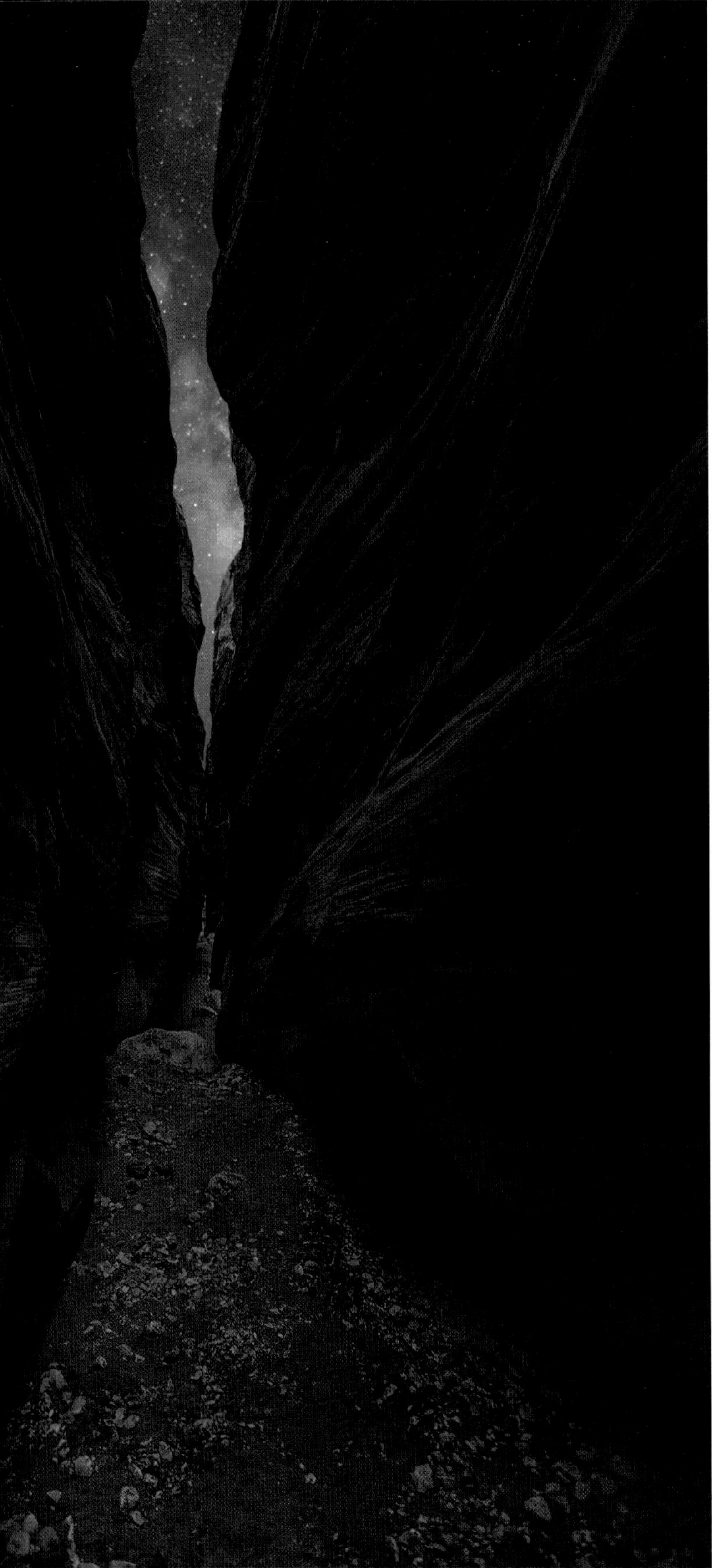

Even after the First Woman created the sun and the moon, the First People of the Fifth World, this current world, demanded more light. In this demand, First Woman saw an opportunity, and made little lights she called "stars."

"I will use these stars to write the laws that are to govern mankind for all time. These laws cannot be written on the water as that is always changing its form, nor can they be written in the sand as the wind would soon erase them, but if they are written in the stars they can be read and remembered forever."

She laid down a blanket on the ground and drew a sky pattern, then placed one of the large stars, Polaris, in the north. "This will never move!" she said, "And it will be known as the Campfire of the North. It will be known as the traveler's guide, and as the lodestar." Then she placed large stars in the other three directions and one in the very center of her sky pattern.

Fire Man began with the North Star and continued climbing the ladder until all the large stars were in the sky, while First Woman placed other stars into groups to form the constellations. It was slow work, as there were many stars and the ladder was very tall.

While all this work was taking place Coyote had been standing close by, impatient as he watched this slow process of placing the constellations. He said to First Woman, "This is taking too long! Why don't you permit me to help? Then we would have this work finished twice as fast!"

First Woman answered, "You always make mistakes and then there is trouble!" But Coyote insisted, saying, "I will do exactly as you say and follow the pattern just as you have placed it on the ground."

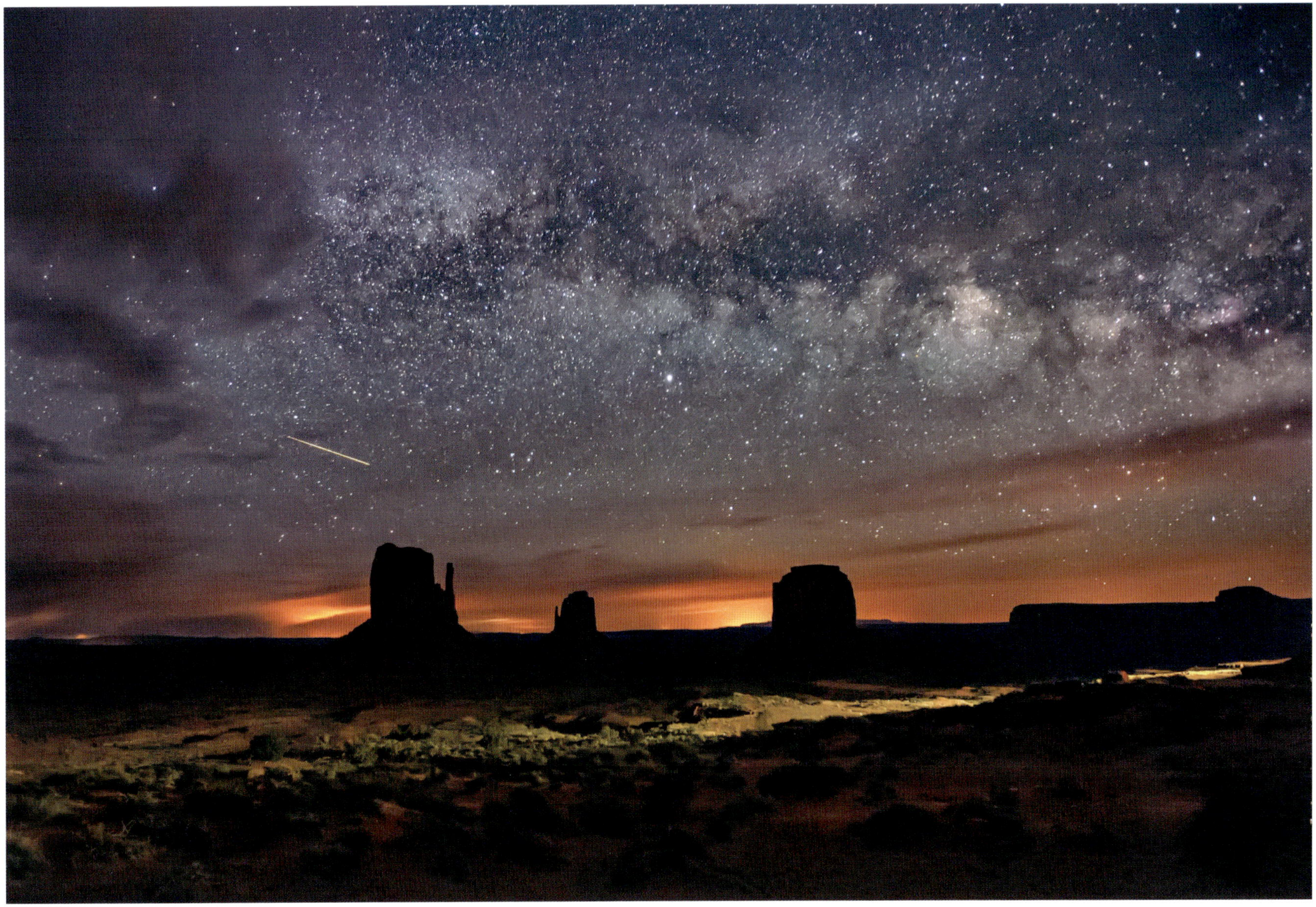

LEFT:
**Waterholes Canyon,
Navajo Nation, Arizona**

3

RIGHT:
**Monument Valley Navajo Tribal Park,
Navajo Nation, Arizona/Utah**

1

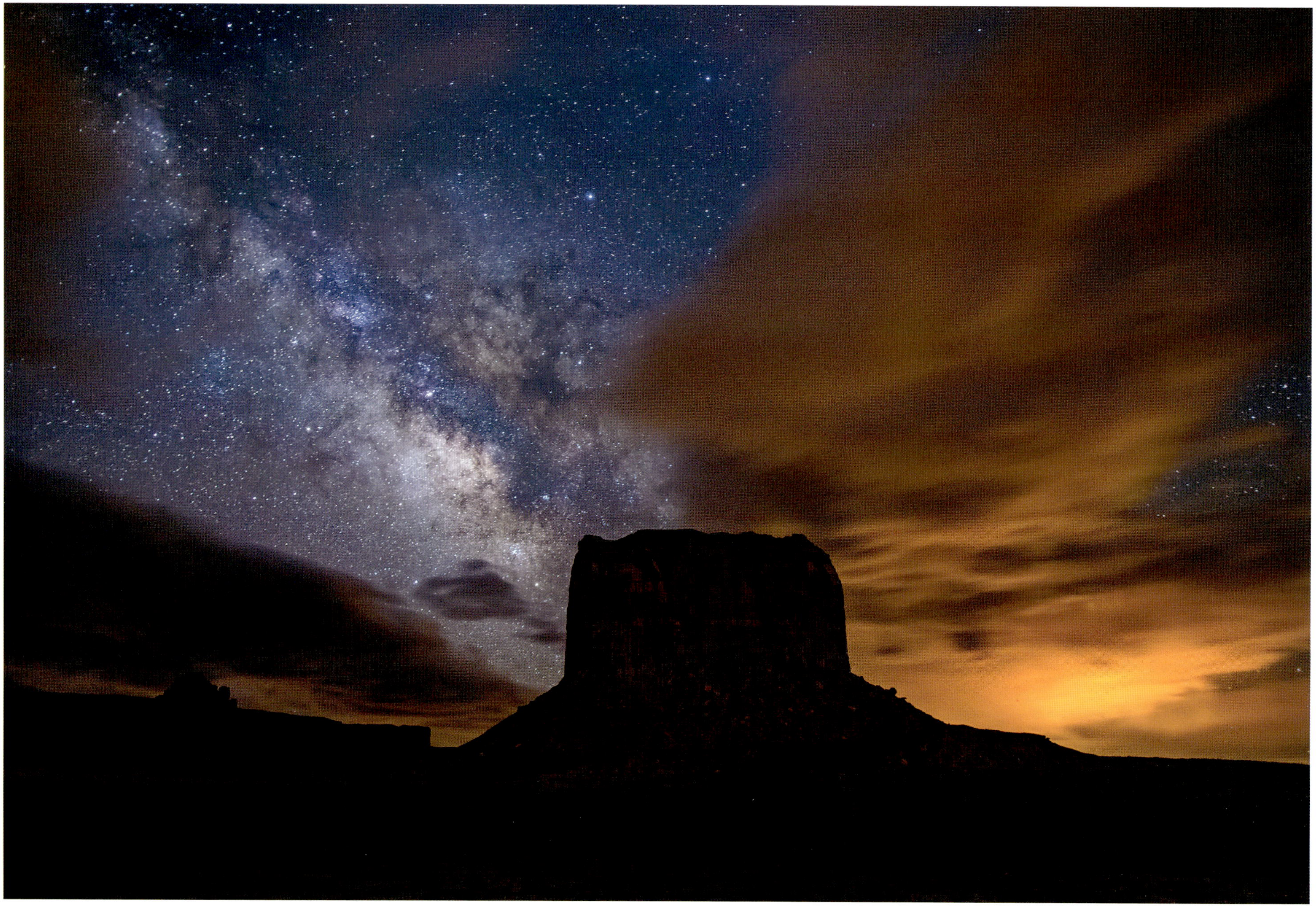

Monument Valley Navajo Tribal Park, Arizona/Utah

1

First Woman was putting two identical stars into her pattern and had named them "the twins." The two lines that marked their paths ran side by side across the sky. She pointed to them and said to Coyote, "Take these two stars and place them somewhat to the west where they will walk hand in hand across the center of the sky."

Then Coyote picked up the two identical stars, the Gemini, one in each hand, and walked to the ladder. He had seen Fire Man climb the ladder with his hands full of stars and thought he could do the same, but when he was halfway up he looked down, and the distance was so great that he became dizzy and almost fell. To make matters worse, Wind Boy came whistling by to see what Coyote was doing, and shook the ladder from east to west. Flustered, Coyote placed the stars in the wrong place.

"Now look what you have done!" First Woman cried. "Those two were supposed to establish peace and friendship among all peoples of the earth. Now they will cause enmity, strife, and dissension that will plague mankind forever. You shall carry no more stars to the sky!"

Upset, he howled: "I'll show you a beautiful sky," then stepped up to the blanket of stars on the ground. He grasped it by two corners and swung it into the air so the stars swept across the sky in a great arc that reached from horizon to horizon. This formed Yikáísdáhí, "the one that comes before the dawn": Milky Way, and the pathway for the spirits traveling between heaven and earth.

- Navajo Legend as transcribed by Franc Johnson Newcomb

**Antelope Canyon Navajo Tribal Park,
Navajo Nation, Arizona**

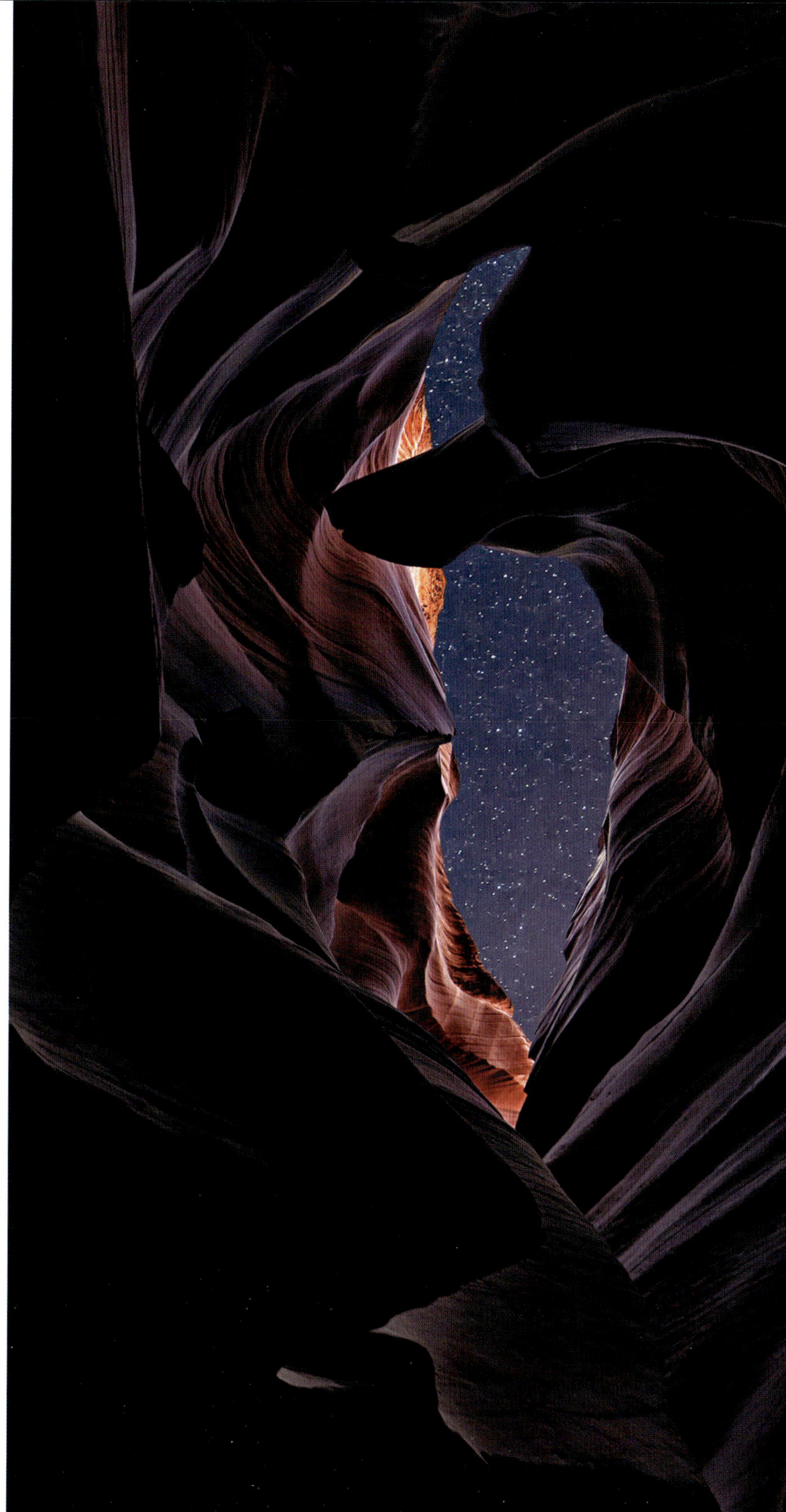

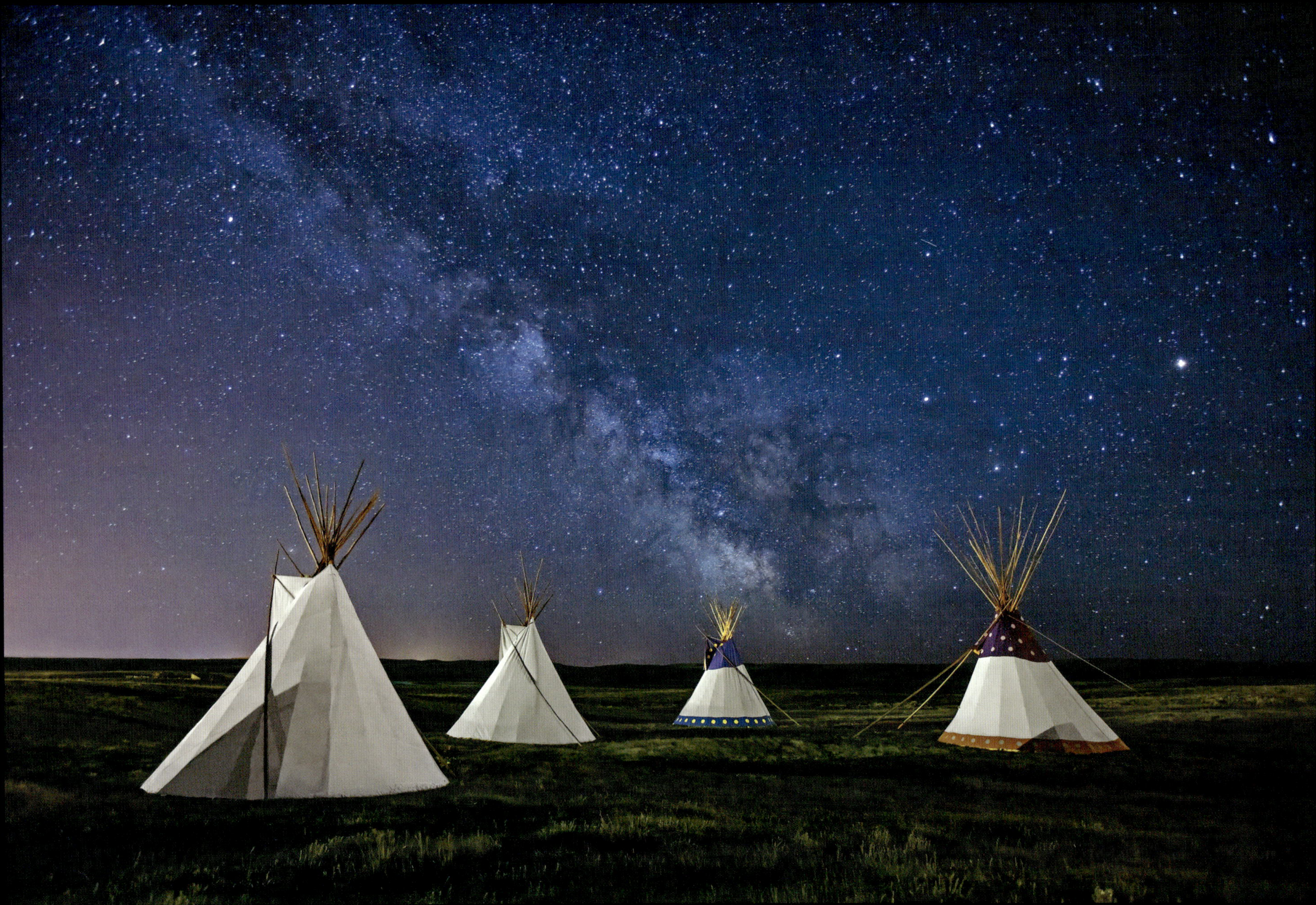

Lodgepole Tipi Village, Blackfeet Indian Reservation, Montana

3

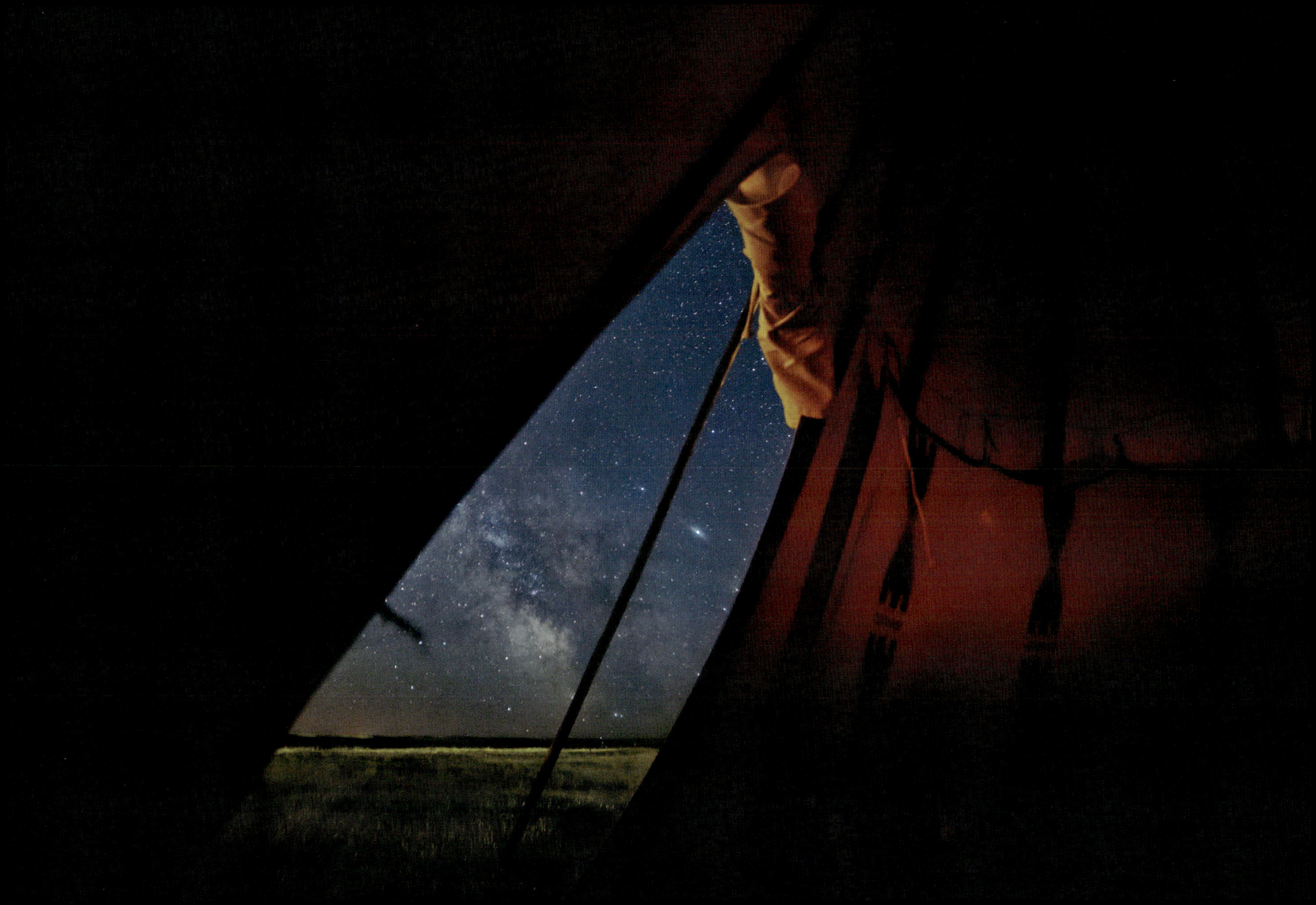

For many centuries, indigenous tribes such as the Blackfeet could witness pure dark skies each night from their homes, developing rich mythologies that stemmed from their observations of the heavens. Stories told and songs performed around campfires gave forms to the sights above.

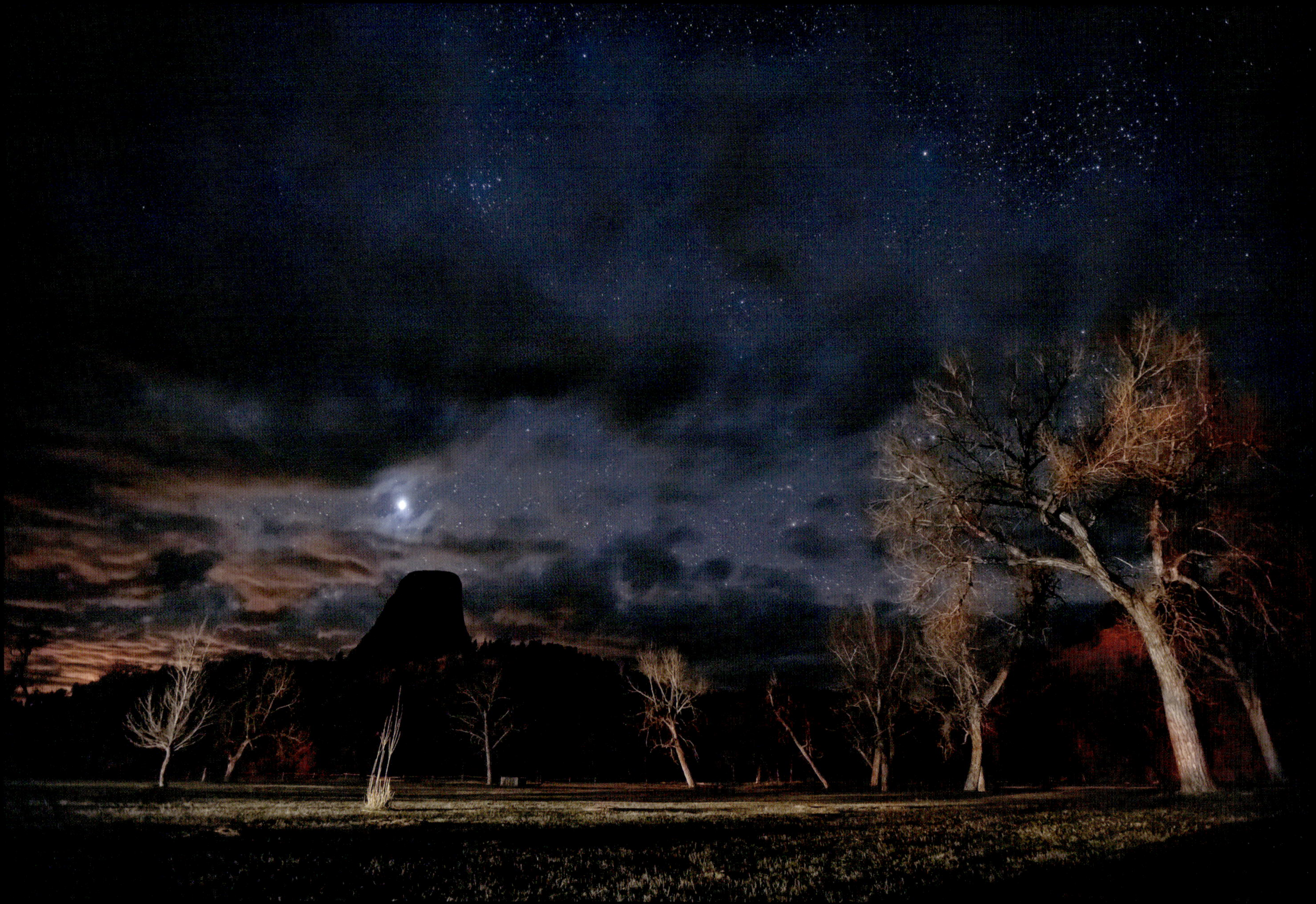

Devil's Tower National Monument, Wyoming

1

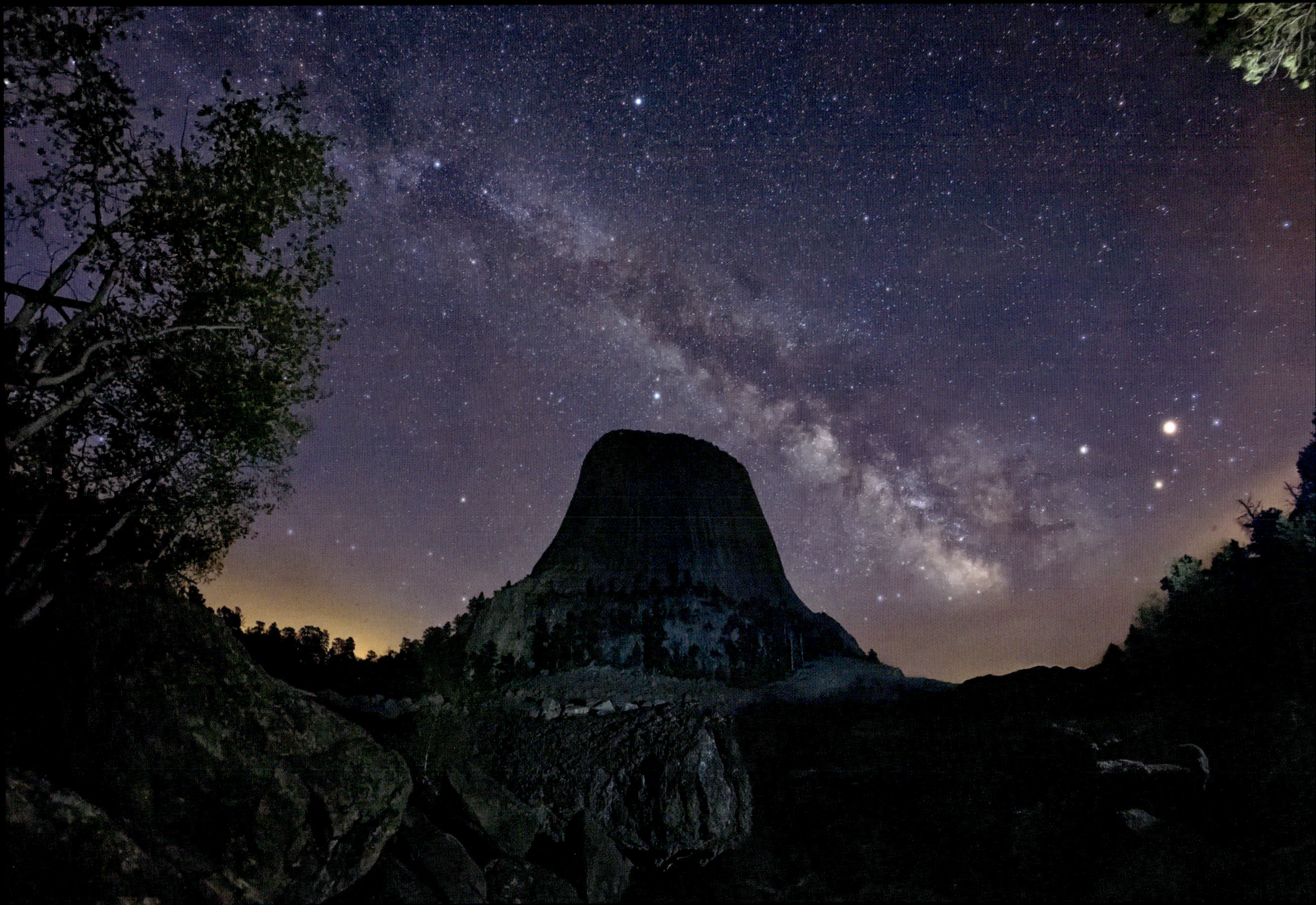

One day, seven little girls were playing away from their village. Suddenly, large black bears chased after them. Knowing they couldn't escape, the girls jumped on a low rock and began to pray: Rock take pity on us, please save us! The rock began to grow upwards, pushing the girls higher and higher. The bears jumped to reach the girls, but only scratched the rock, broke their claws, and fell on the ground. The rock rose higher and higher, all the way to the sky, and the girls became the constellation Pleiades, The Seven Sisters.

- Kiowa Legend of what is today known as Devil's Tower

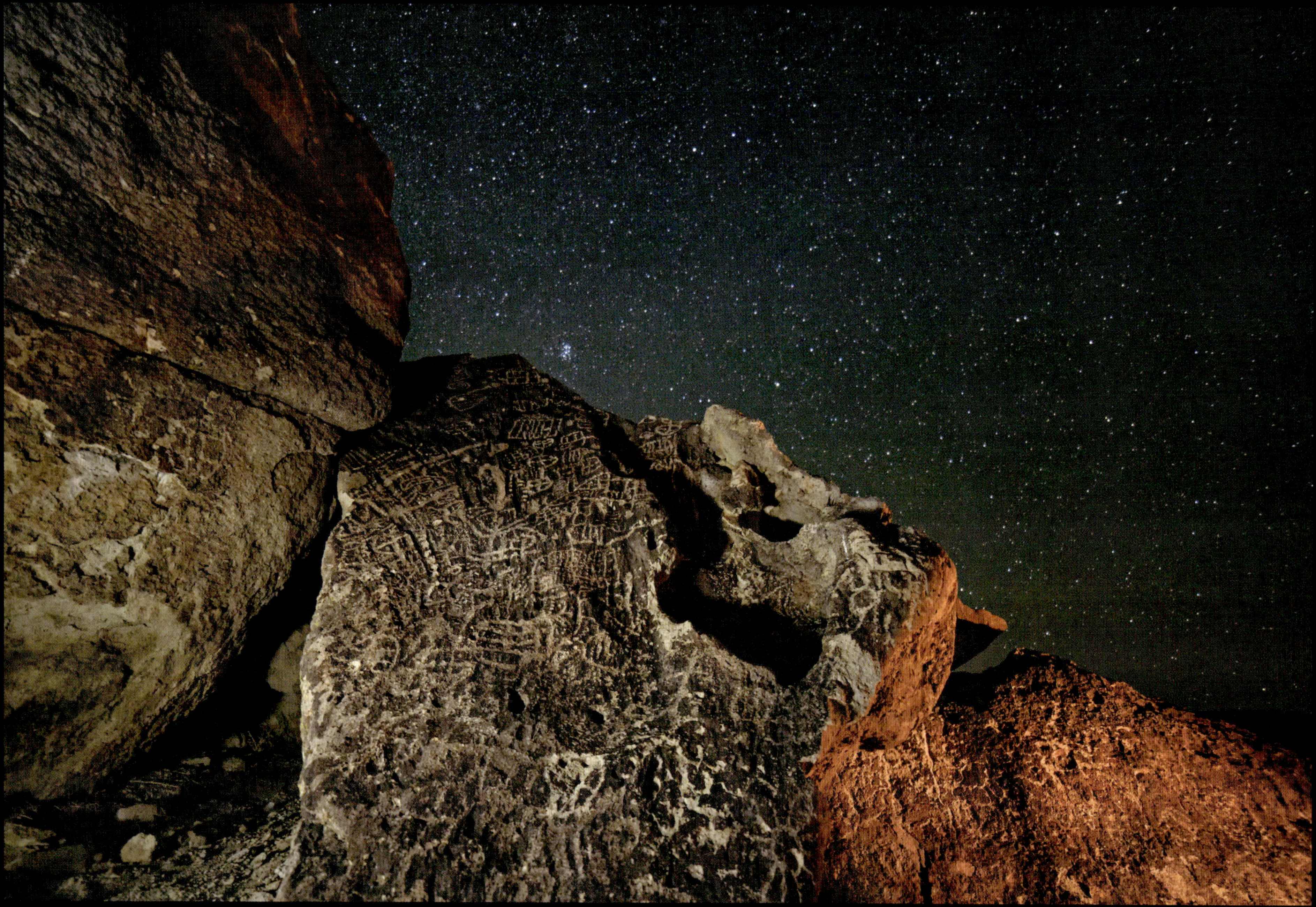

Ancient Paiute Petroglyphs, Bishop, California

1

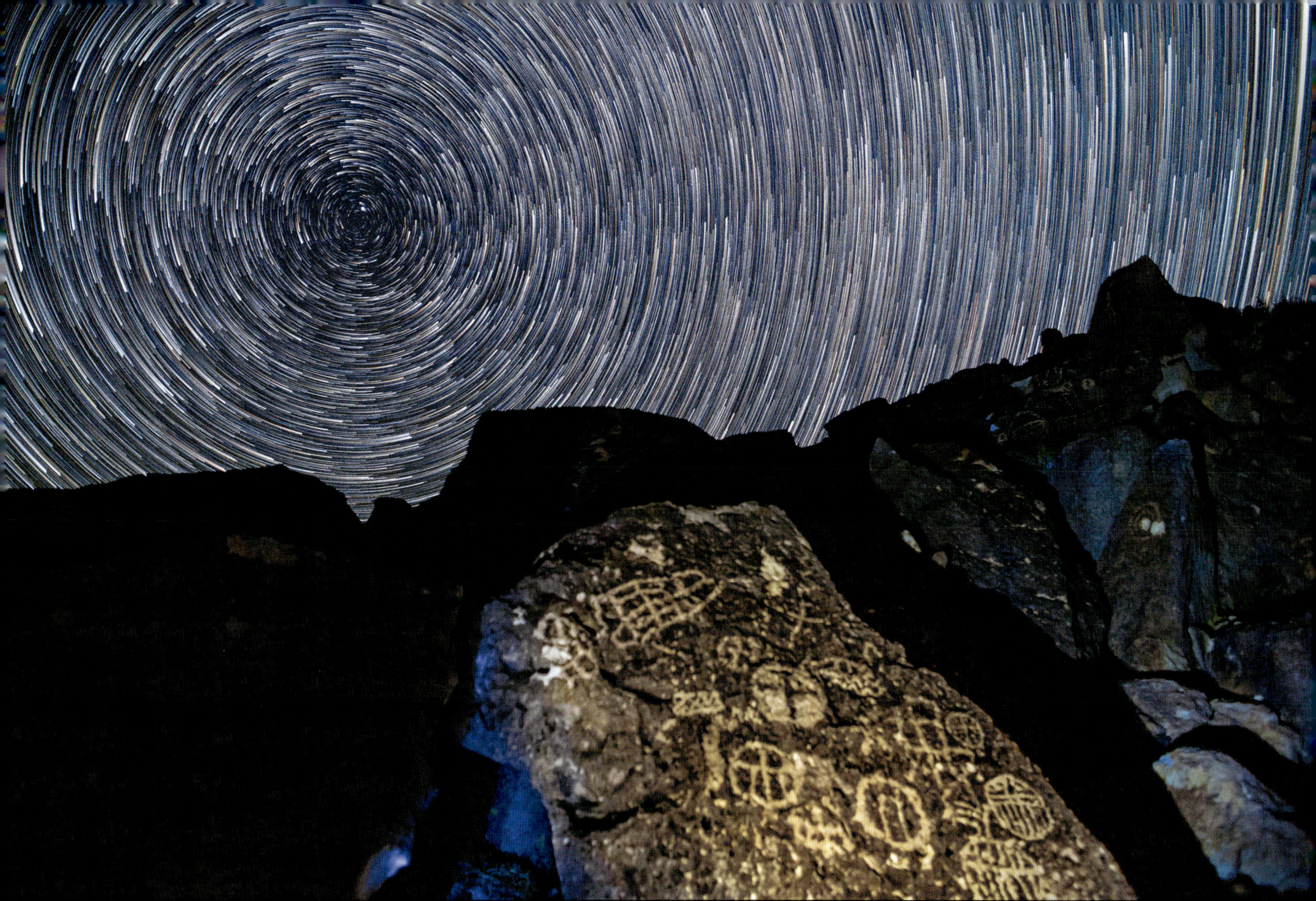

While many passed their celestial mythologies through stories, song, and dance, other tribes began writing their ancient astronomical myths on rocks, leaving behind night sky interpretations etched as petroglyphs, such as those scattered around Bishop, California, by the Paiute tribe

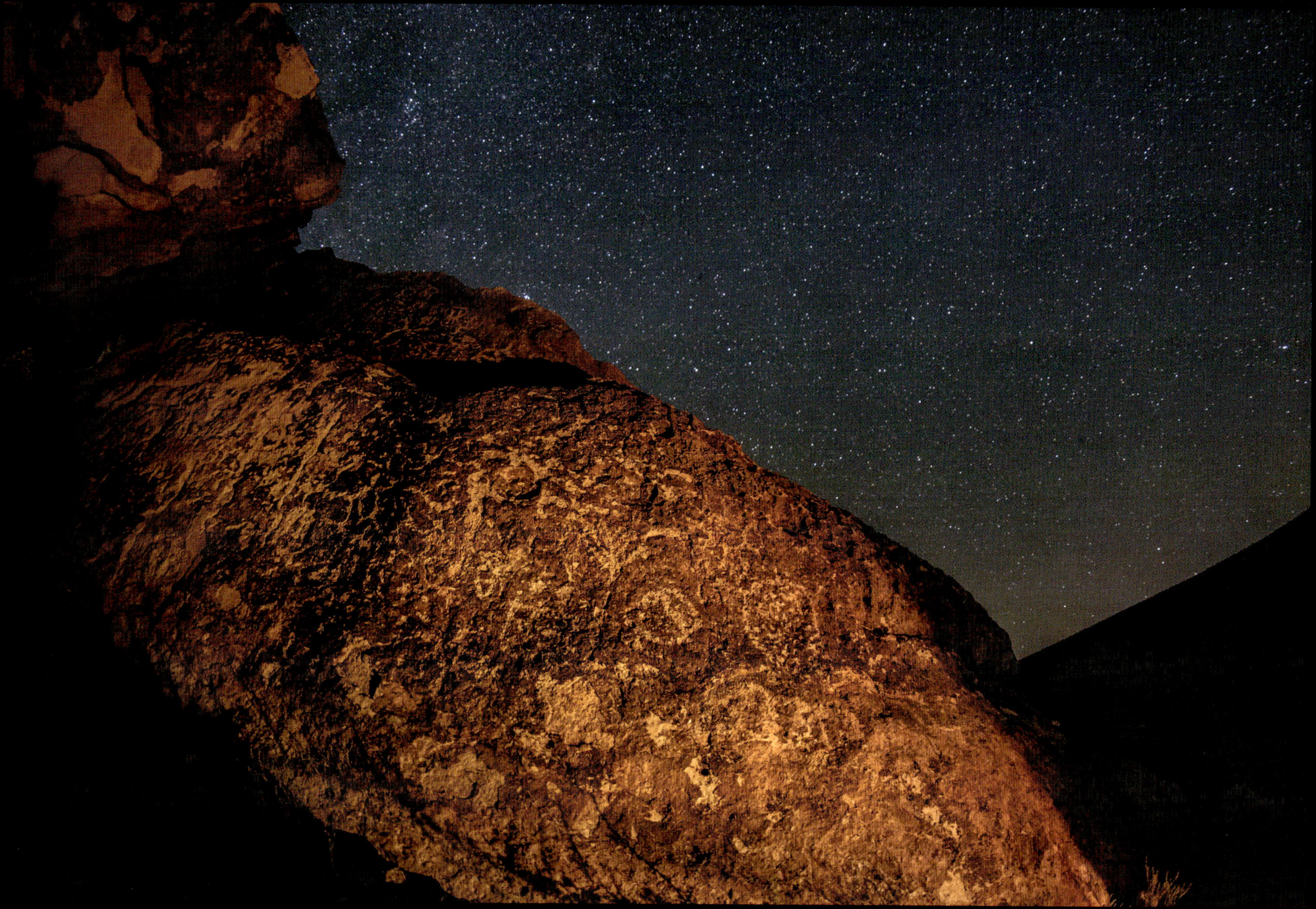

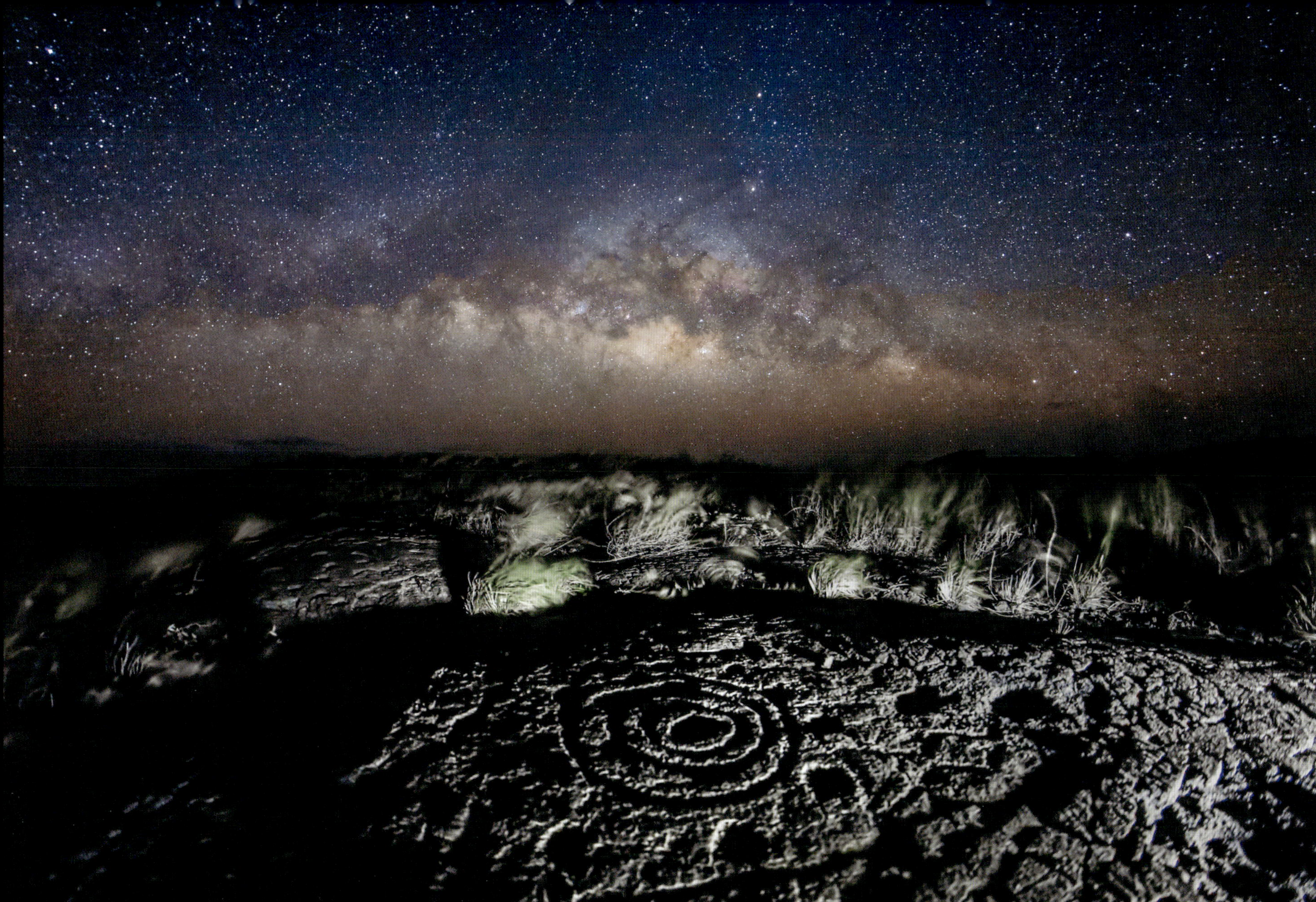

Pu`u Loa Petroglyphs, Hawai'i Volcanoes National Park, Big Island Hawai'i

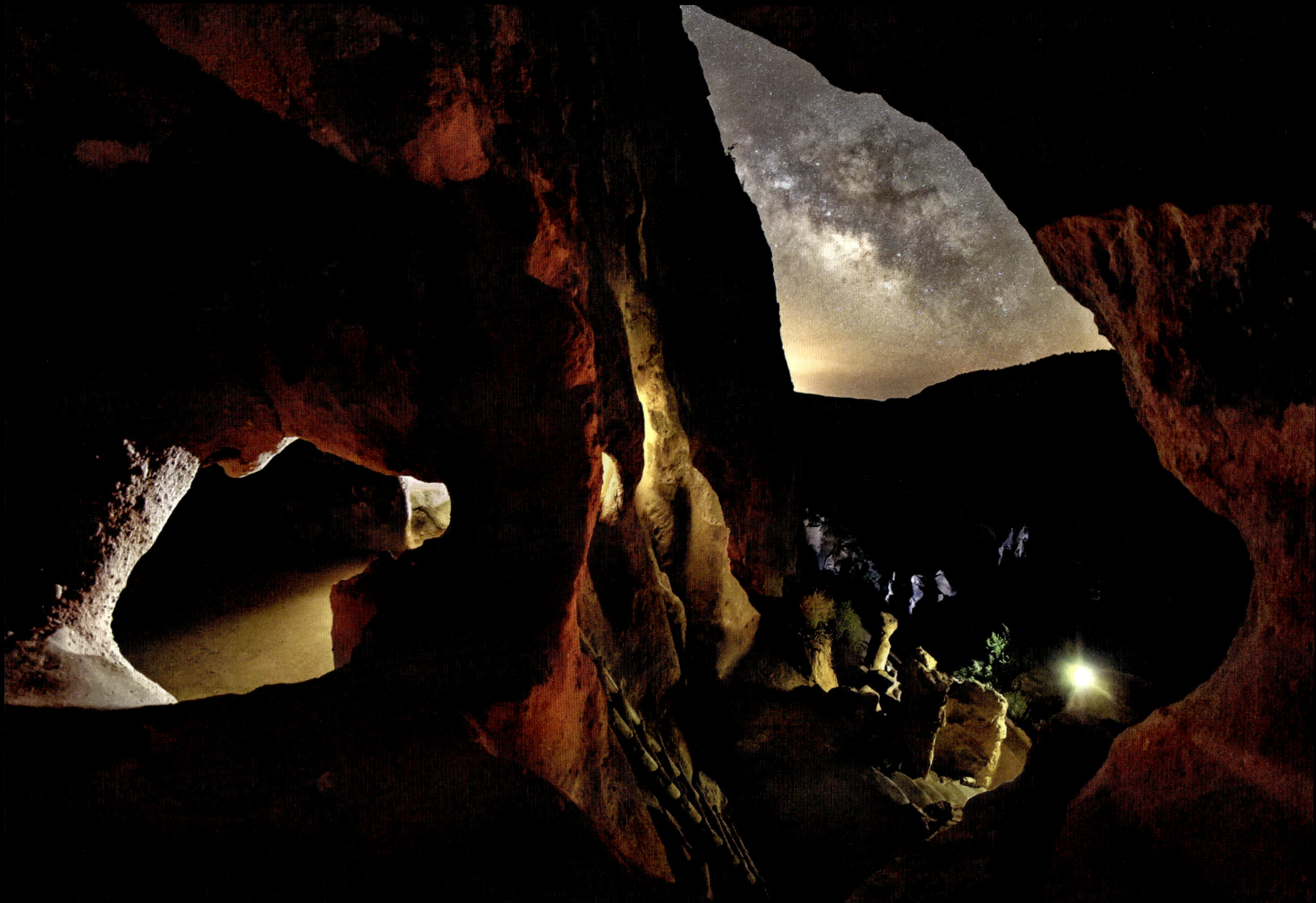

Ancient Puebloan Cave Dwellings, Bandelier National Monument, New Mexico

4

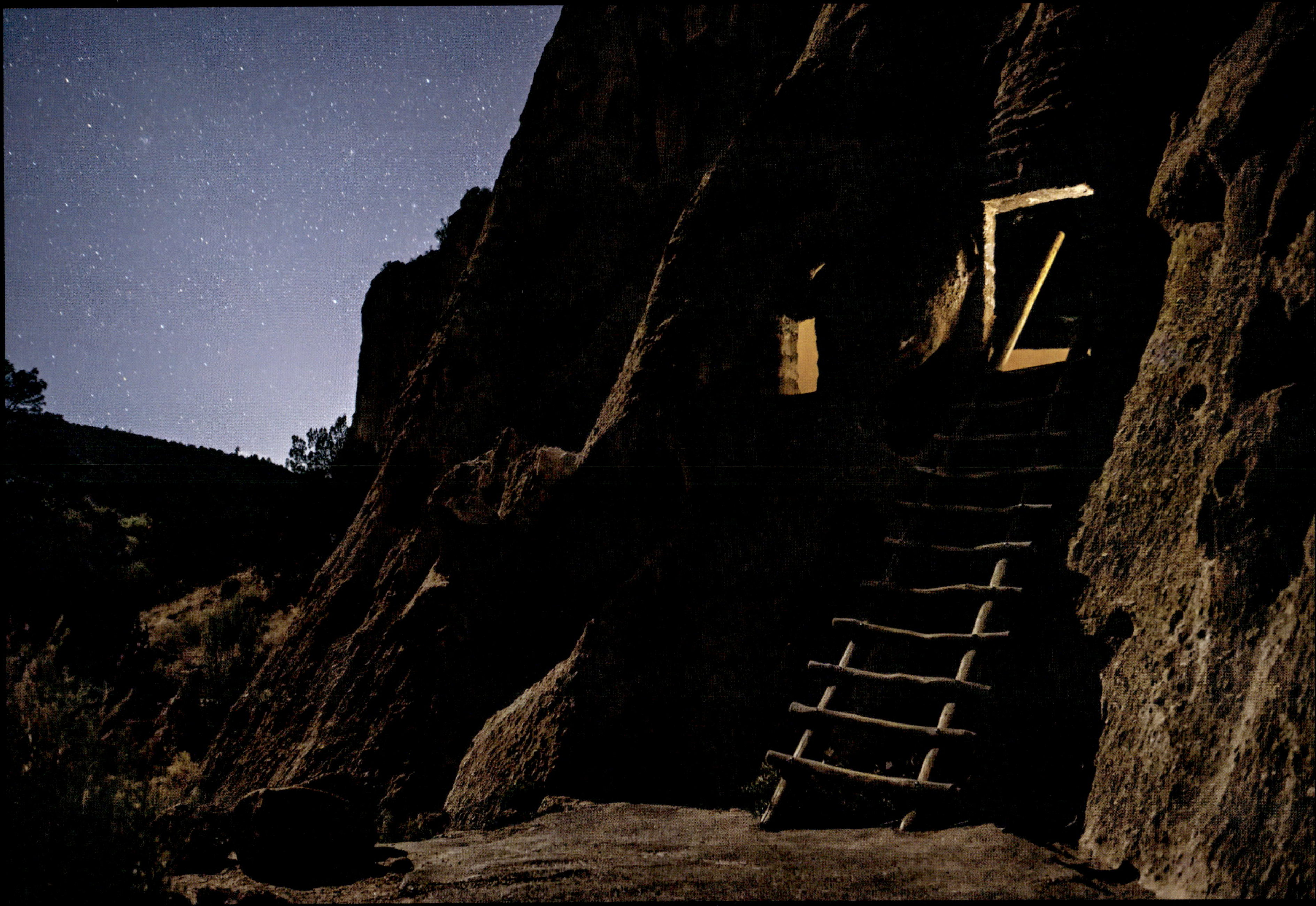

Cave dwellings, such as those at Bandelier, give us a glimpse of cultures that began the transition from hunter-gatherer to agrarian cultures, and with that, to advanced technologies.

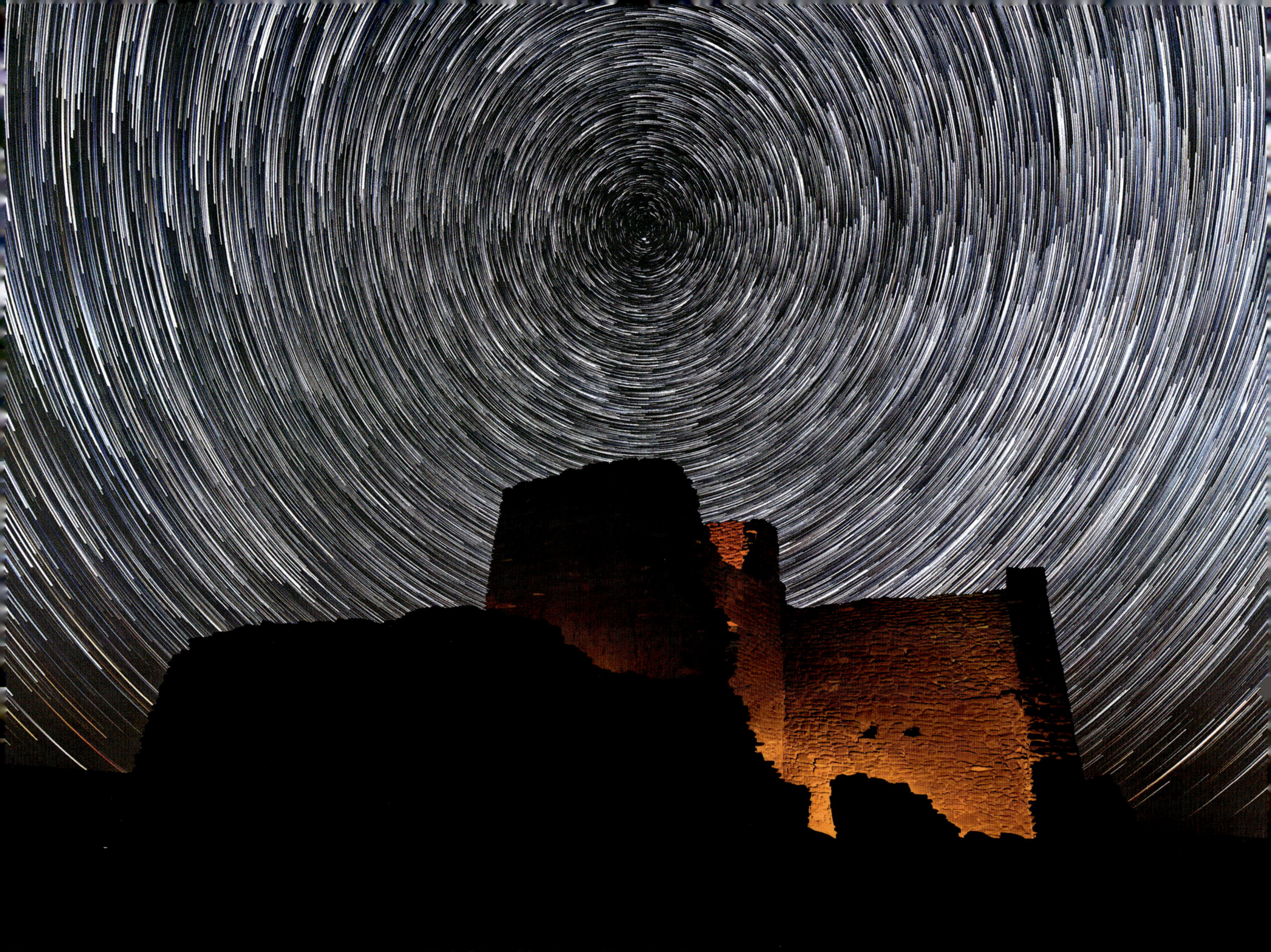

Wukoki Pueblo, Wupatki National Monument, Arizona

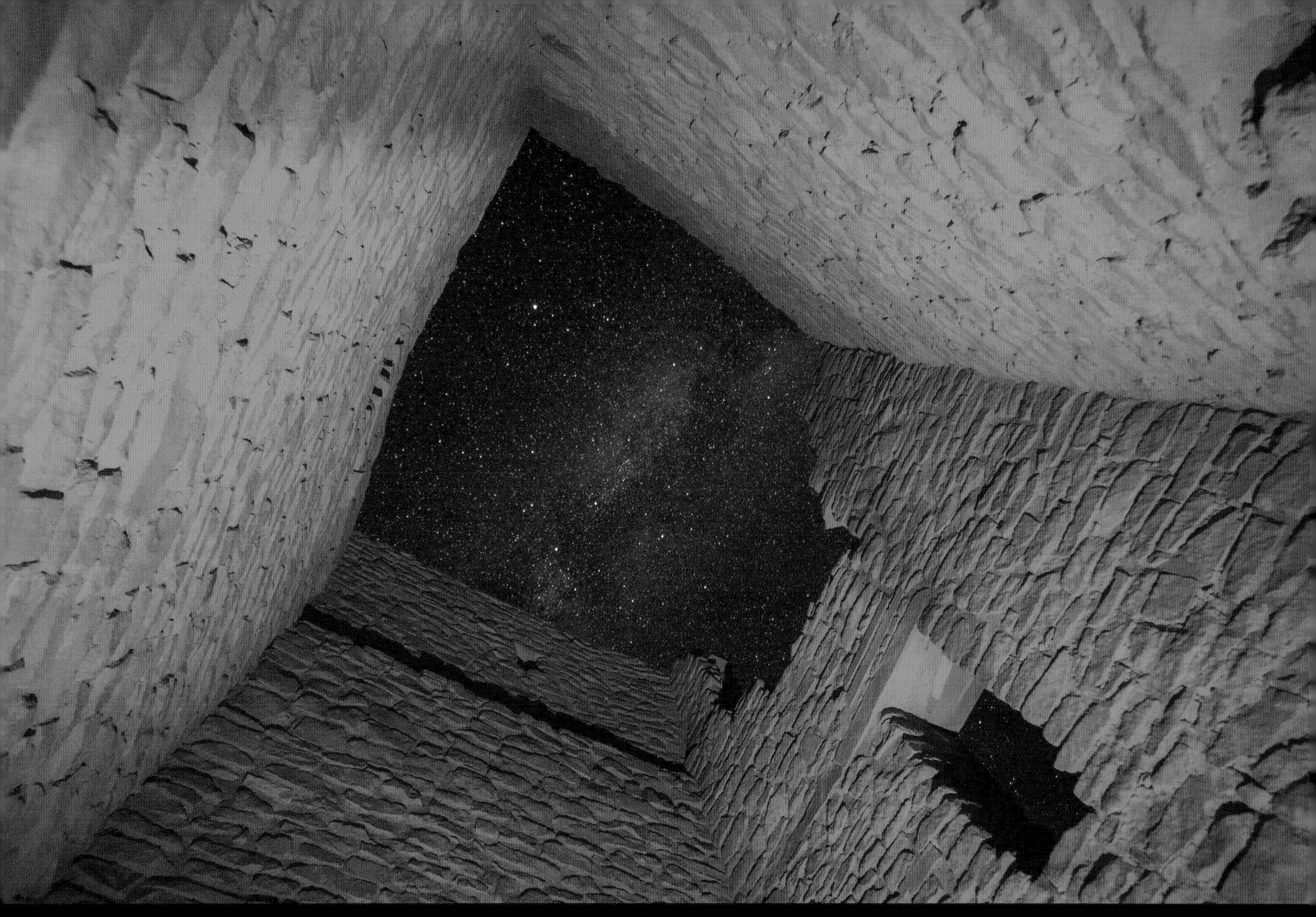

These technological advancements included buildings that were constructed not only for habitation but also for observation of solar, lunar, and celestial cycles and events. These buildings were forms of early calendars, and became intimately connected with religious ceremonies and cultural practices.

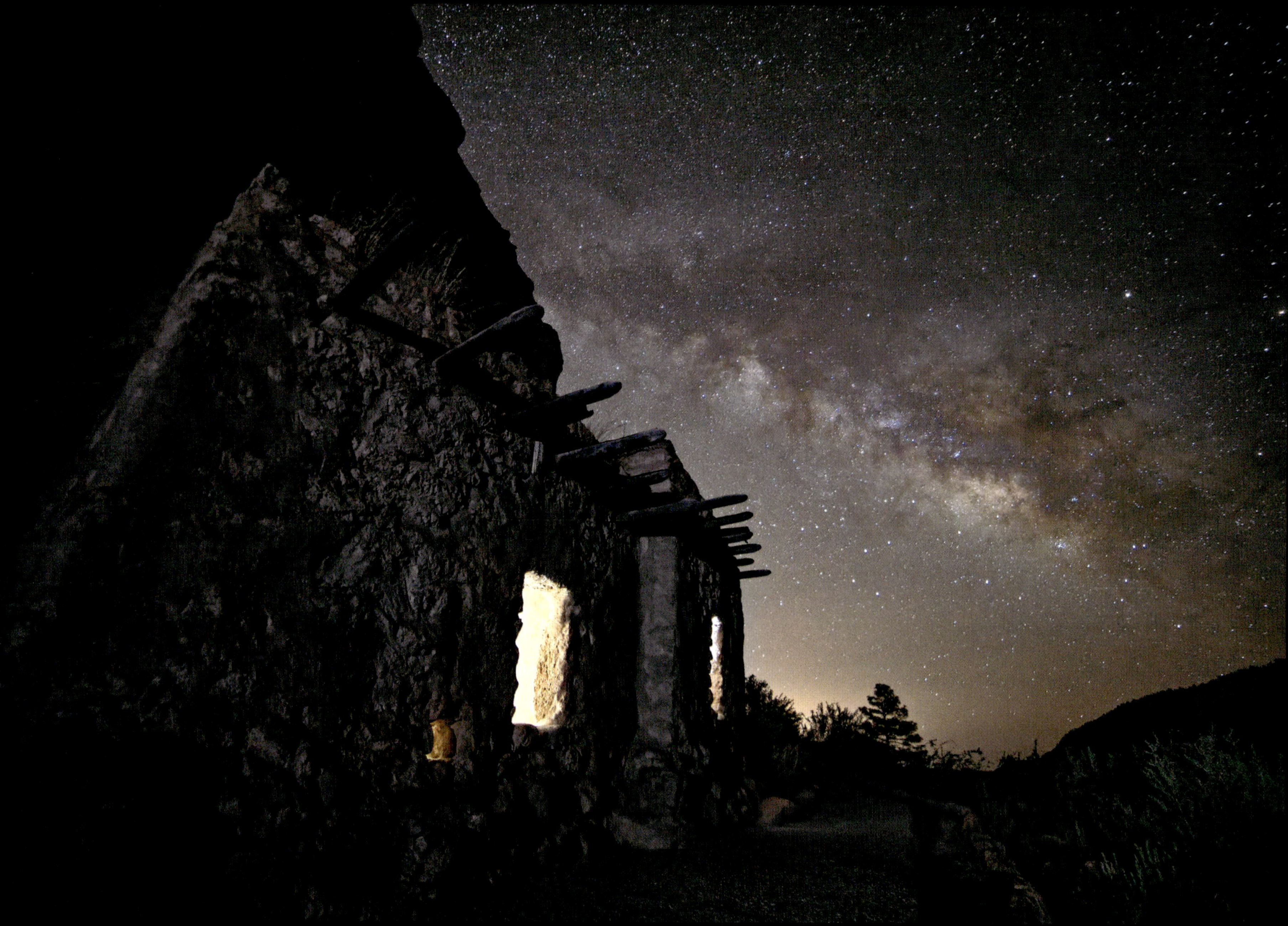

Talus House, Bandelier National Monument, New Mexico

4

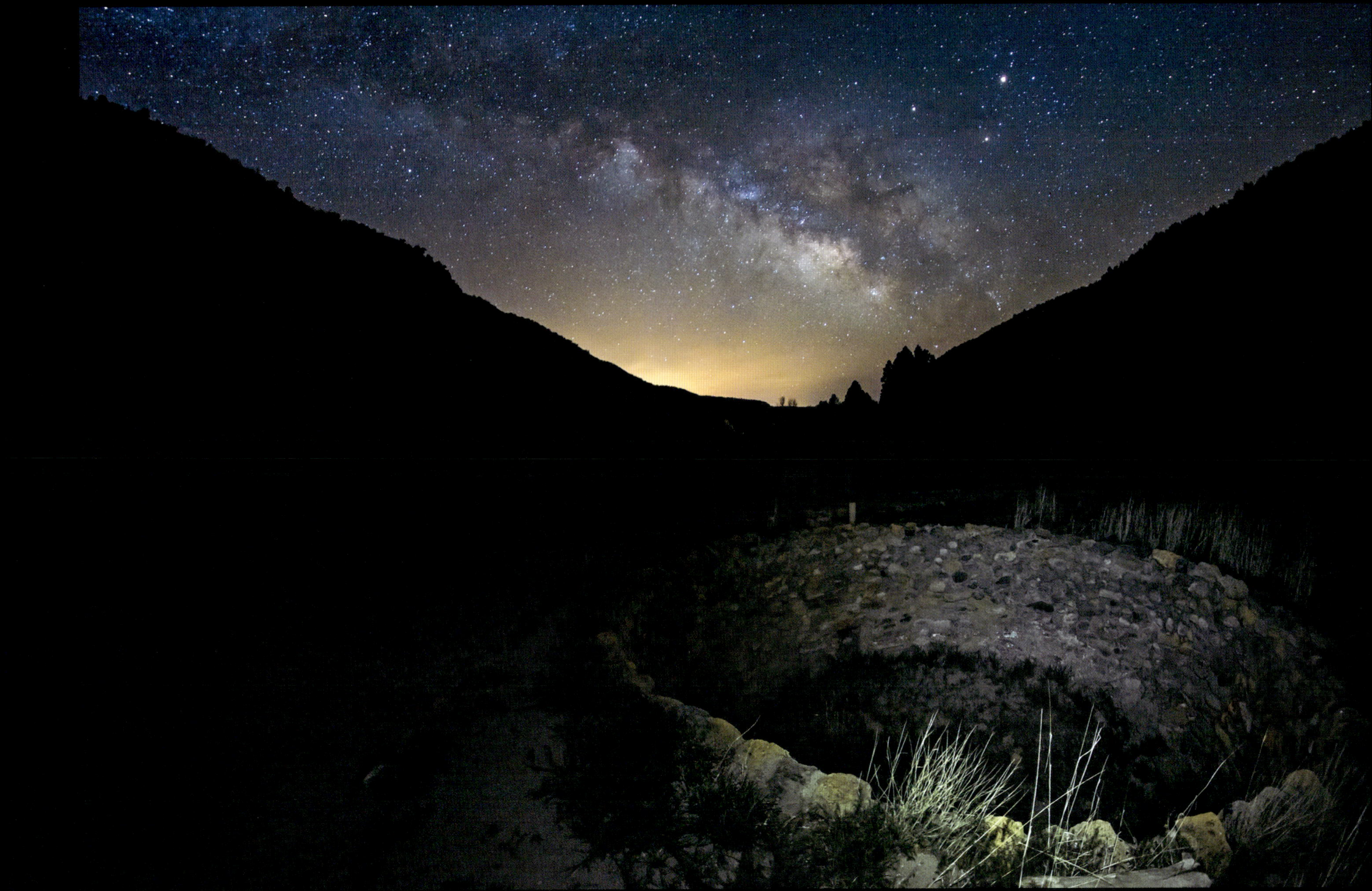

Some argue that buildings such as Ancient Puebloan kivas, places of spiritual gatherings, would feature openings in the ceiling aimed at specific celestial events and night sky observation, as well as solstices. These practices of building to celestial events have come to be known as Archaeoastronomy.

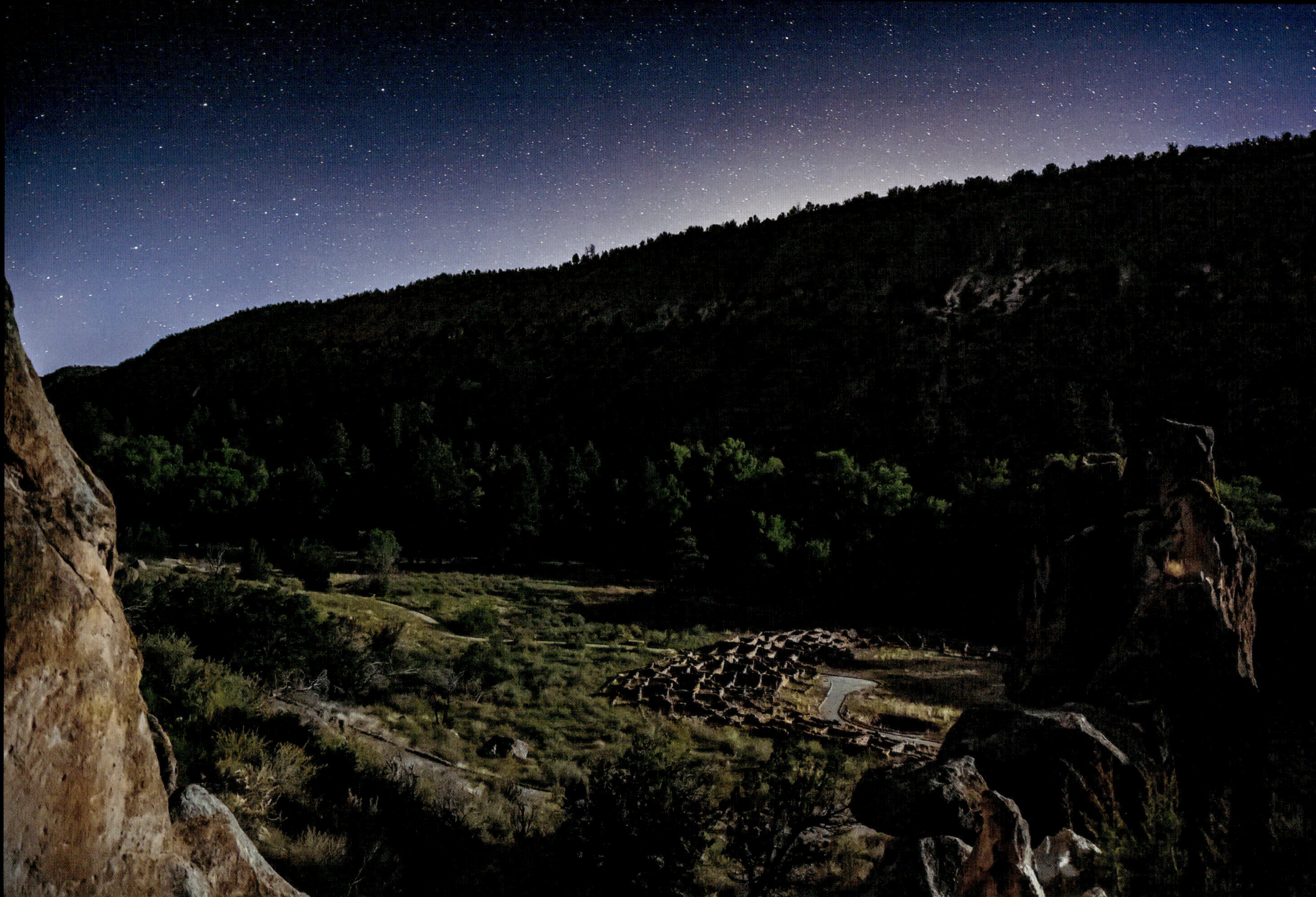

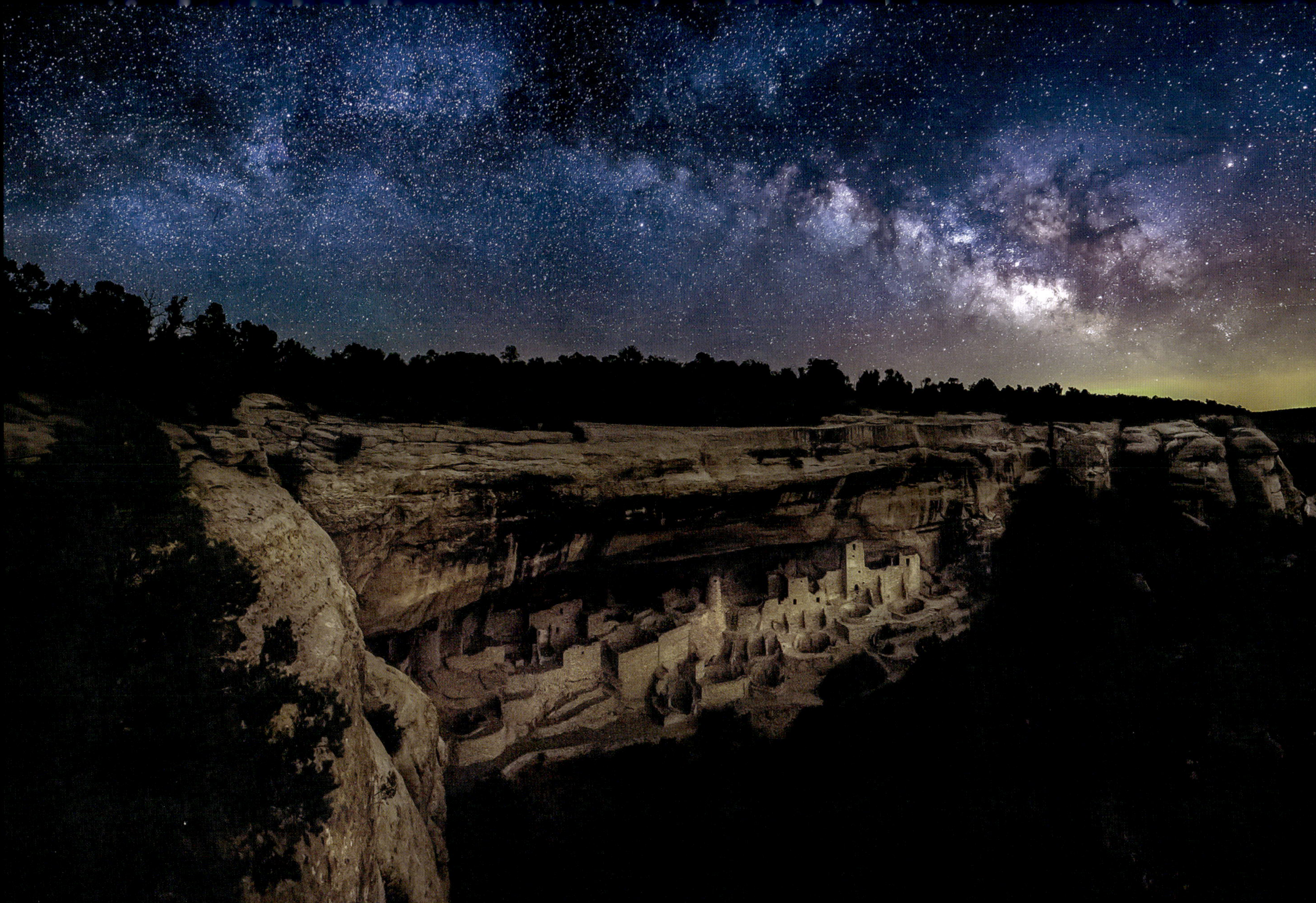

In places such as Mesa Verde, buildings like Sun Temple are thought to have been built solely as astronomical observatories.

Cliff Palace, Mesa Verde National Monument, Colorado

1

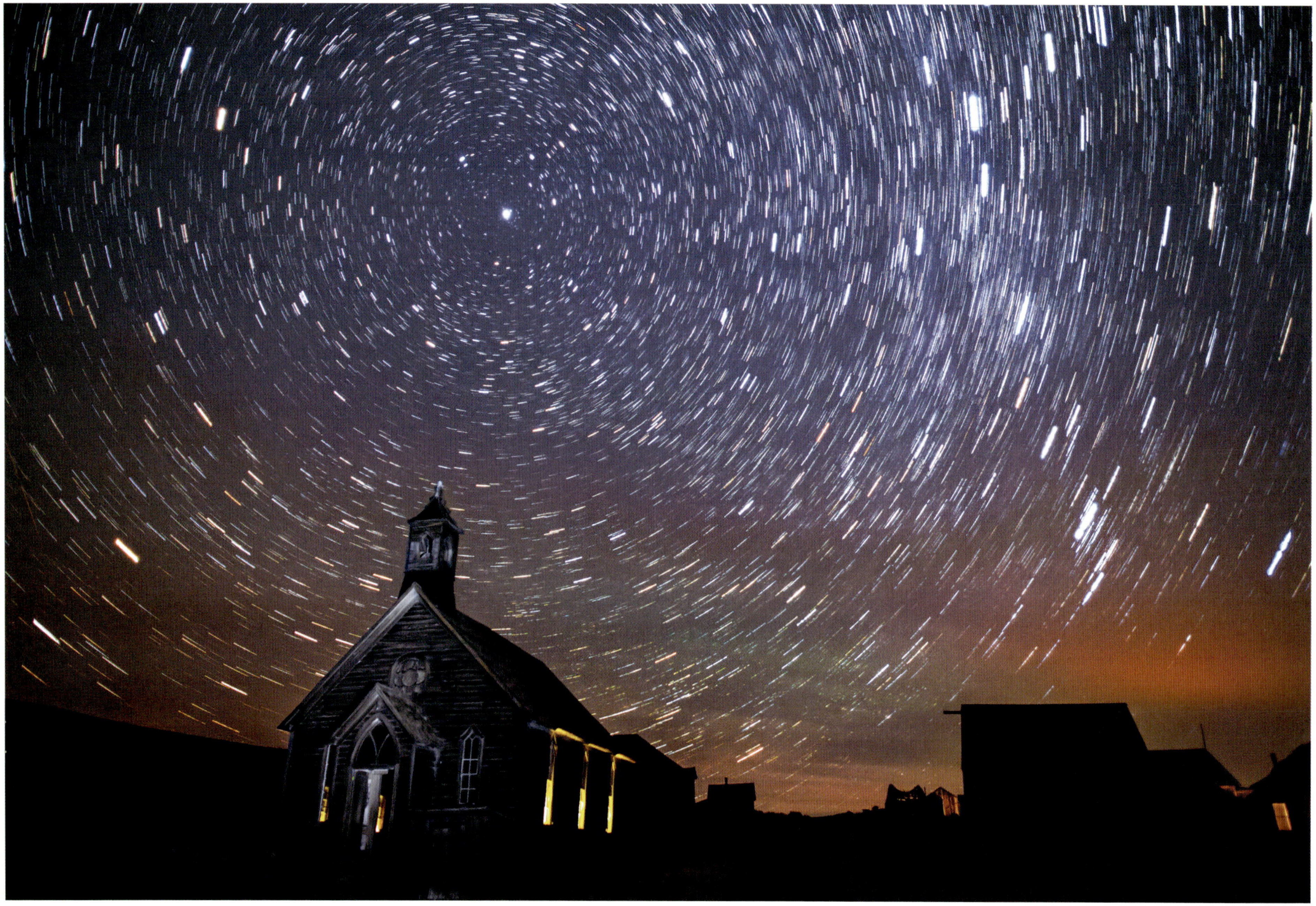

Methodist Church, Bodie State Historic Park, California

1

Traditional ways of life and methods of relating to the universe that had endured for thousands of years came to a swift end as a new culture arrived on North America's shores. European colonizers, bringing a different culture and social order, used the stars to navigate the seas, placing far less importance on the night sky as a social and religious symbol. Nonetheless, observations of the night sky had made their impact on the newly arrived Europeans' dominant Christian religion. Of the few references to the stars in The Bible, the Book of Matthew's mention of the Star of Bethlehem, which signified the coming of Jesus Christ, remains the most notable:

After Jesus was born in Bethlehem of Judea in the days of Herod the king, behold, wise men from the East came to Jerusalem, saying, "Where is He who has been born King of the Jews? For we have seen His star in the East and have come to worship Him." When Herod the king heard this, he was troubled, and all Jerusalem with him. And when he had gathered all the chief priests and scribes of the people together, he inquired of them where the Christ was to be born.

There are various theories as to what astronomic phenomenon the "Star of Bethlehem" may have been referring to—a large comet, a planetary conjunction, a heliacal rising, or possibly others.

Roadside Graves,
Rice Ghost Town, California

1

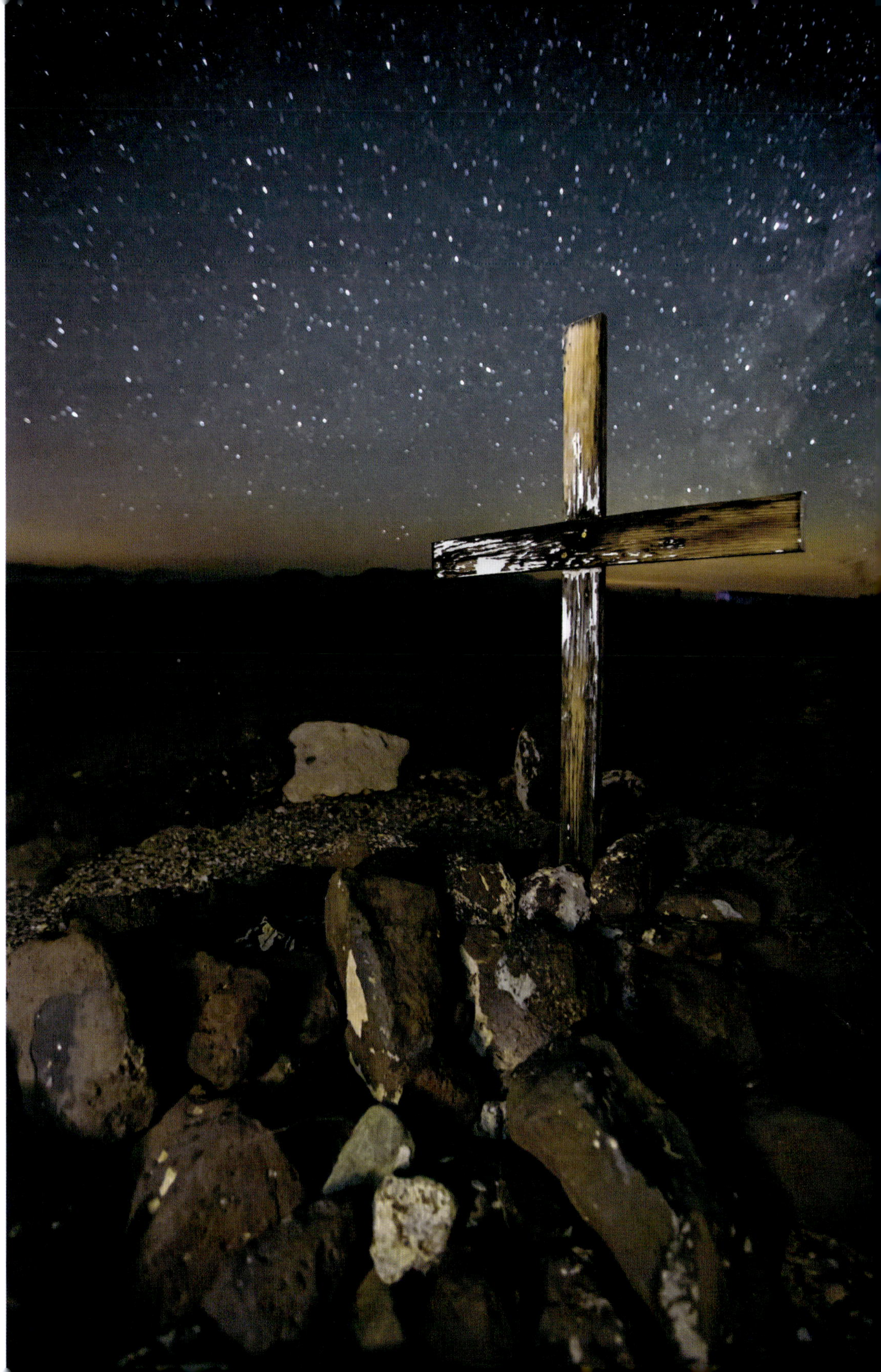

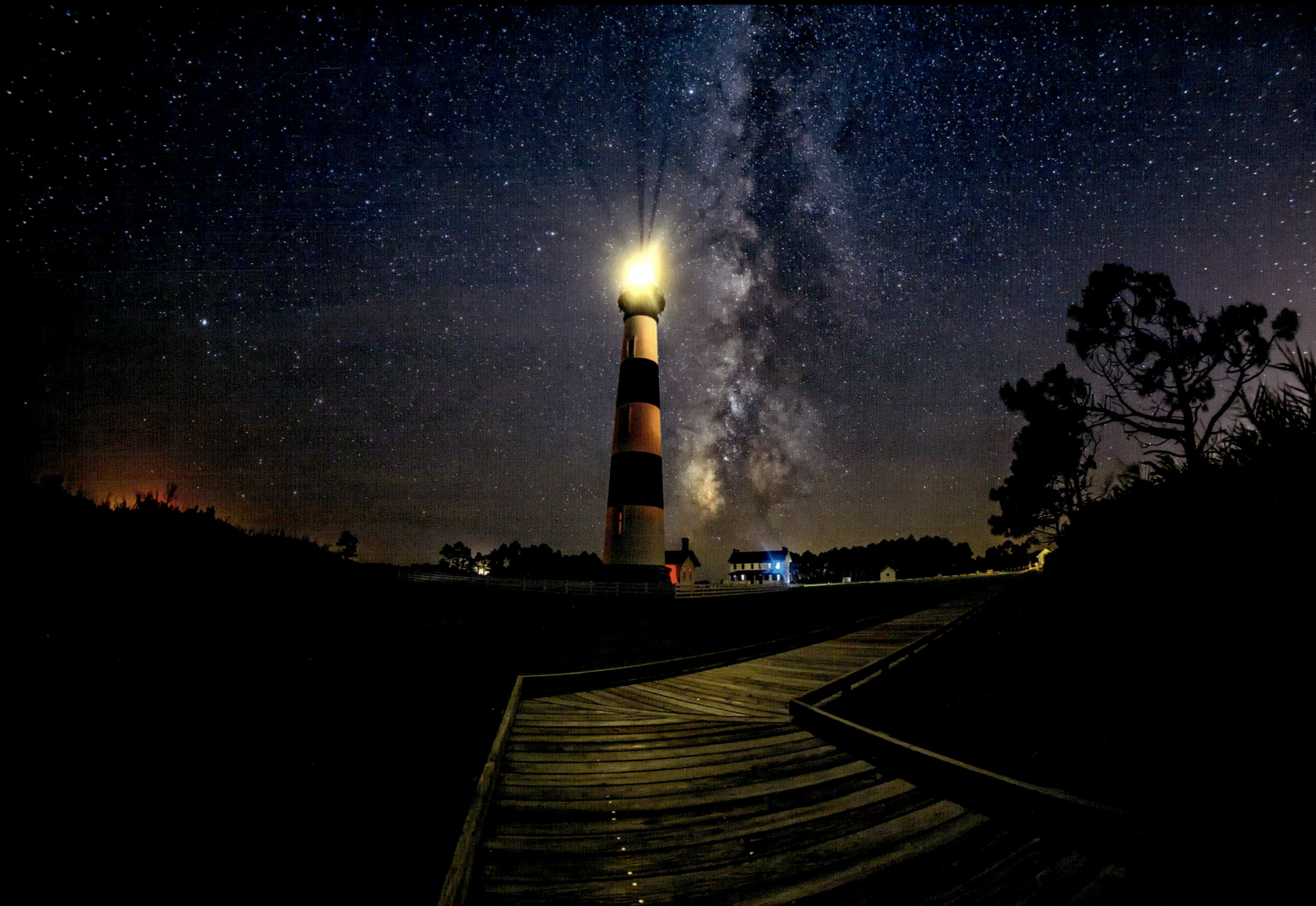

With the arrival of European colonizers to the Americas came new light technologies such as lighthouses, which quickly littered the entire Eastern seaboard.

Bodie Light, Cape Hatteras National Seashore, North Carolina

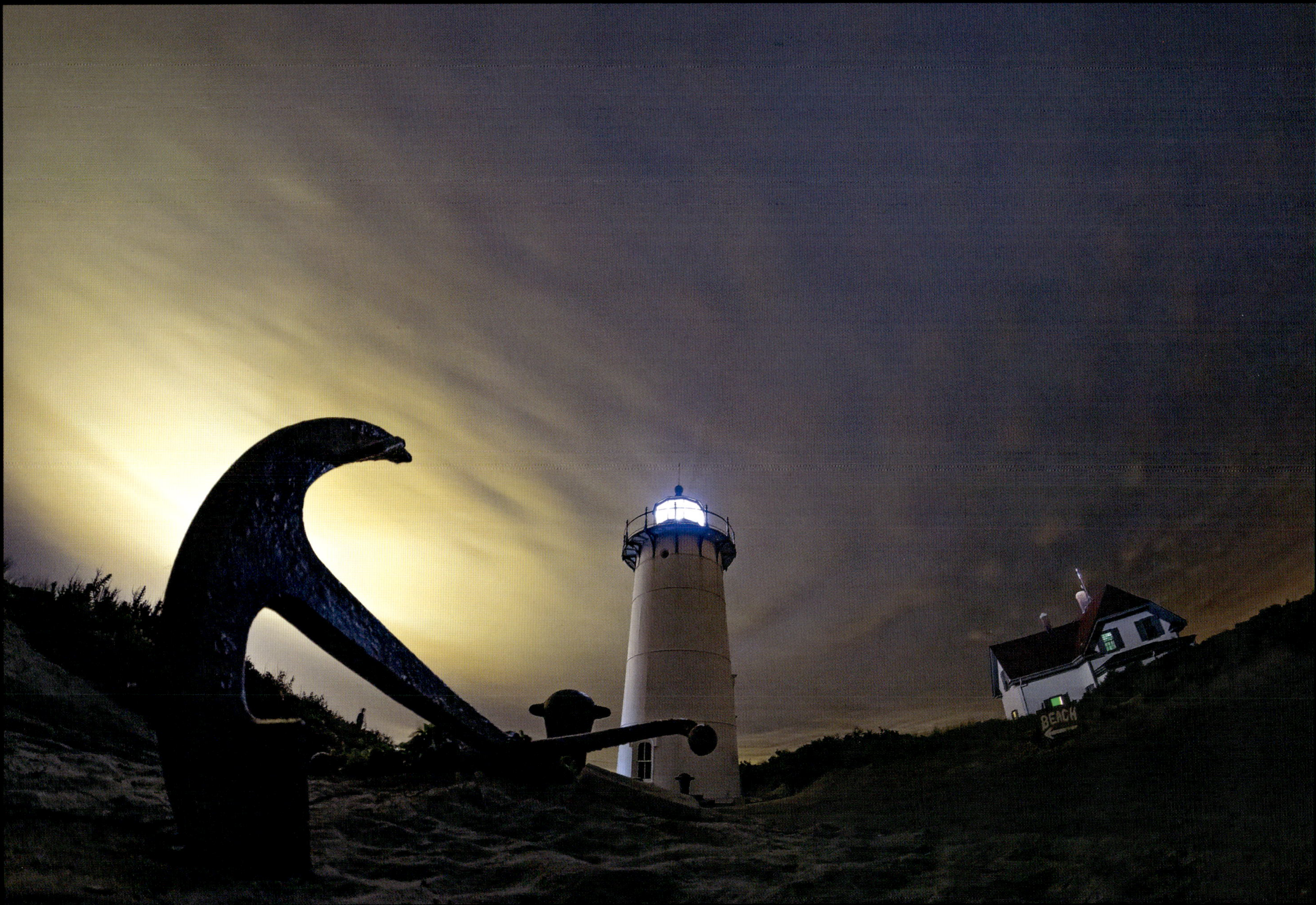

These lighthouses, like most new structures at the time, were built by African slaves, who began adding new chapters to American celestial mythologies.

Race Point Light, Cape Cod National Seashore, Massachusetts

Escaping slaves had to find their way north, where New York and Massachusetts had strong abolitionist societies and benevolent groups—both black and white—that would help them escape. In Canada, the northernmost destination, slavery had been outlawed.

One of the best clues that slaves could use to find the north was the North Star. Polaris, referred to as the North Star, never appears to change its position, as it happens to fall right on Earth's turning axis.

To locate the North Star, people have used a cluster of stars that appear to look like a dipper—a cup with a very long handle. Today this region in space is known as Ursa Major, or as some have called it, The Big Dipper.

Slaves referred to this group of stars as the Drinking Gourd. They sometimes used hollowed-out gourds to dip into and drink water. The gourds looked just like long-handled cups. Two stars on the cup's edge always point to the North Star. Many escaping slaves—including Harriet Tubman, James Pennington, and Josiah Henson—may have followed "the Drinking Gourd" to freedom.

Fort Jefferson, Dry Tortugas National Park, Florida

1

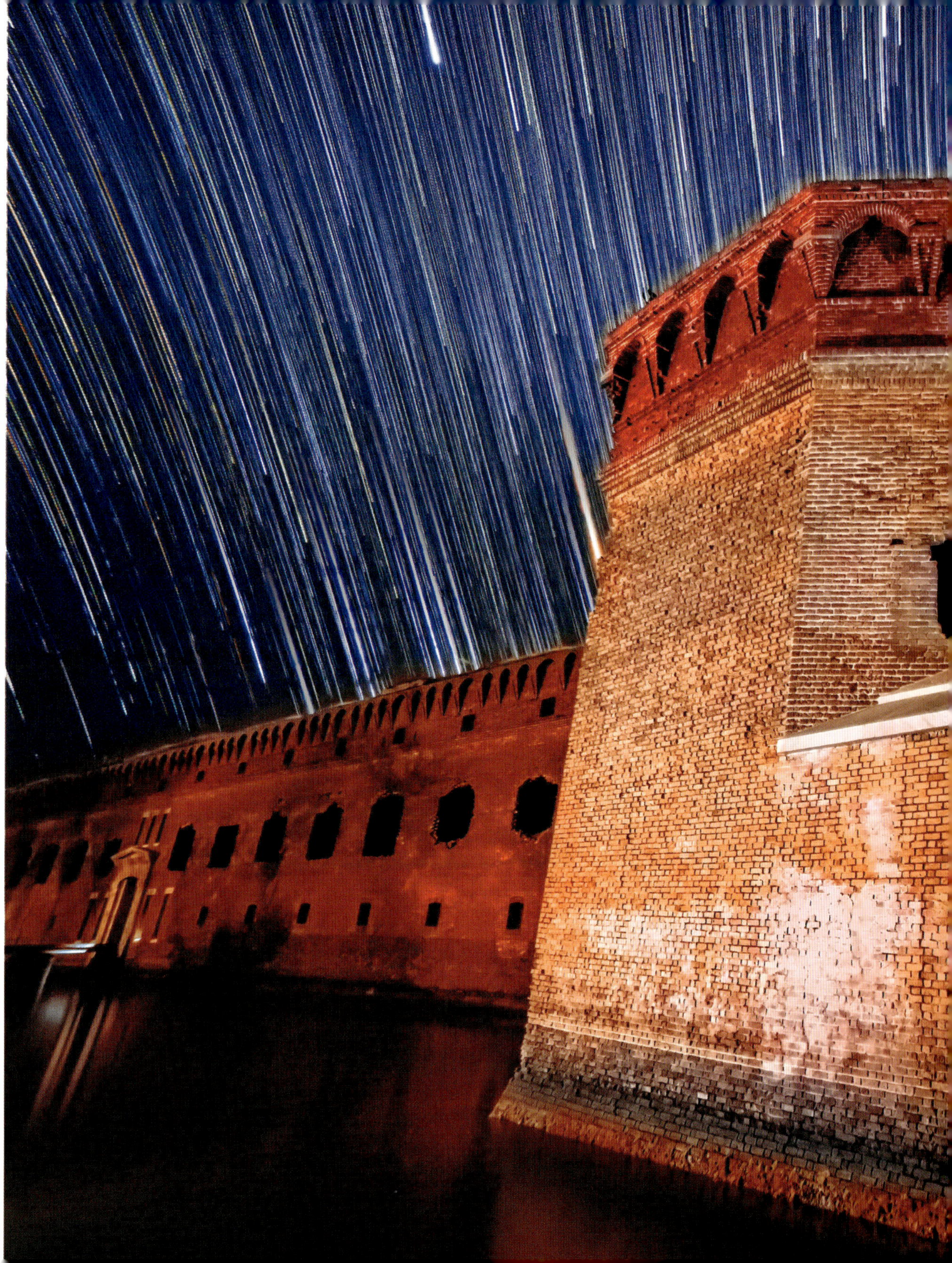

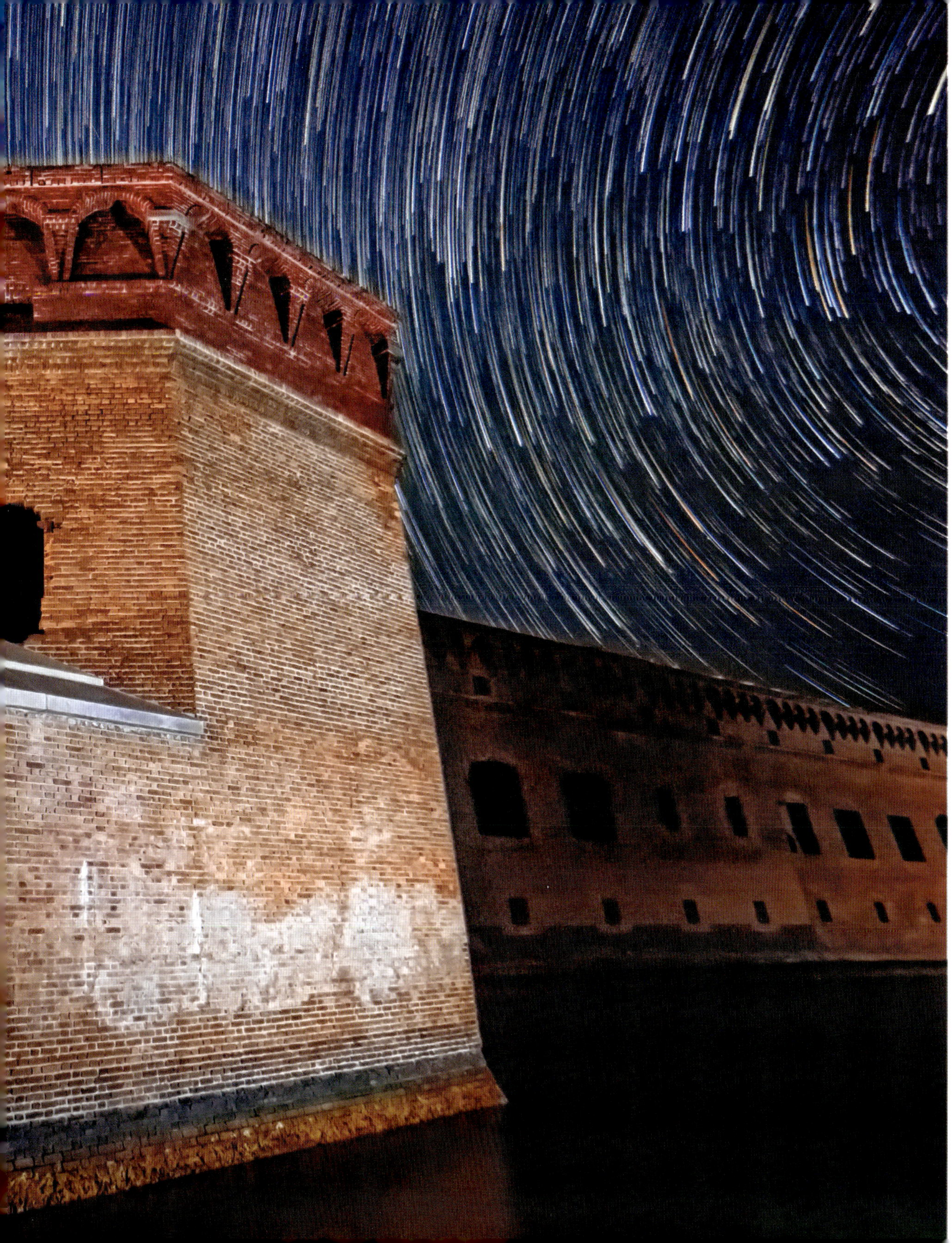

Follow the Drinkin' Gourd

When the Sun comes back
And the first quail calls
Follow the drinking gourd,
For the old man is a-waiting for to carry you to freedom
If you follow the drinking gourd
The riverbank makes a very good road.
The dead trees will show you the way.
Left foot, peg foot, traveling on,
Follow the drinking gourd.
The river ends between two hills
Follow the drinking gourd.
There's another river on the other side
Follow the drinking gourd.
When the great big river meets the little river
Follow the drinking gourd.
For the old man is a-waiting for to carry to freedom
If you follow the drinking gourd.

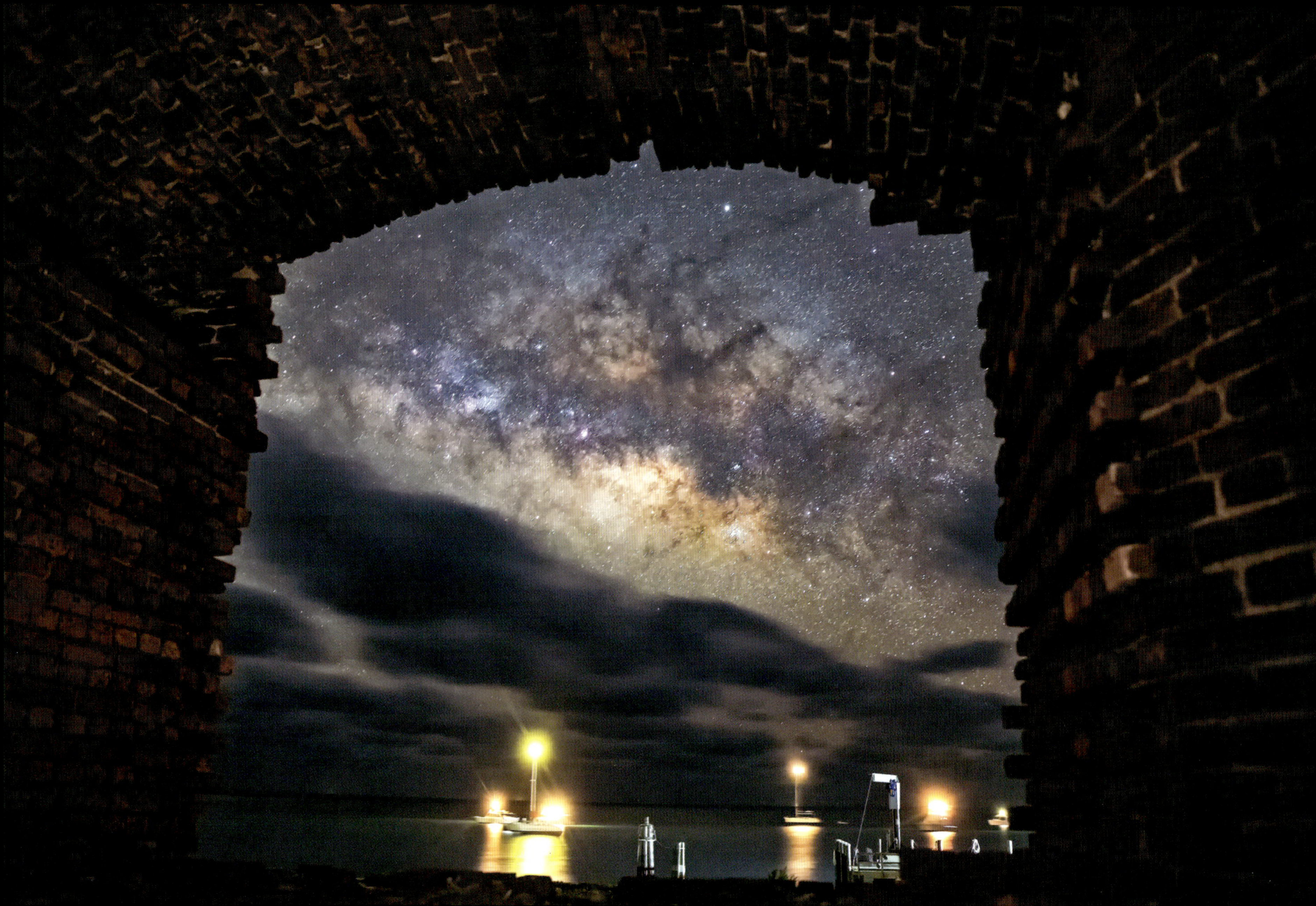

Fort Jefferson, Dry Tortugas National Park, Florida

1

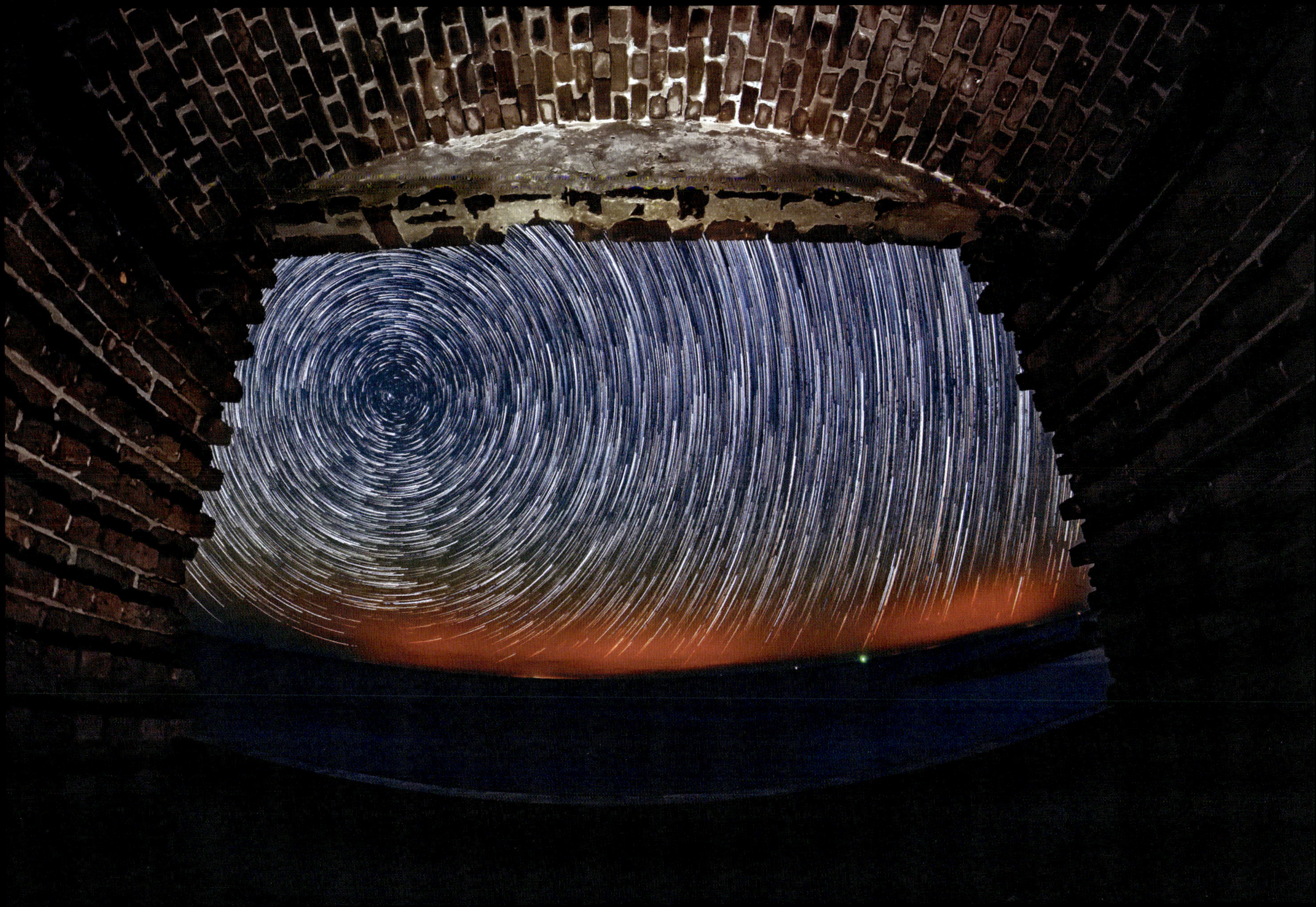

Polaris, also known as the North Star, is seen here from a window on Fort Jefferson's north wall. Fort Jefferson, like many forts at the time, was constructed mainly by slaves, who looked towards their freedom as they gazed at the northern night sky.

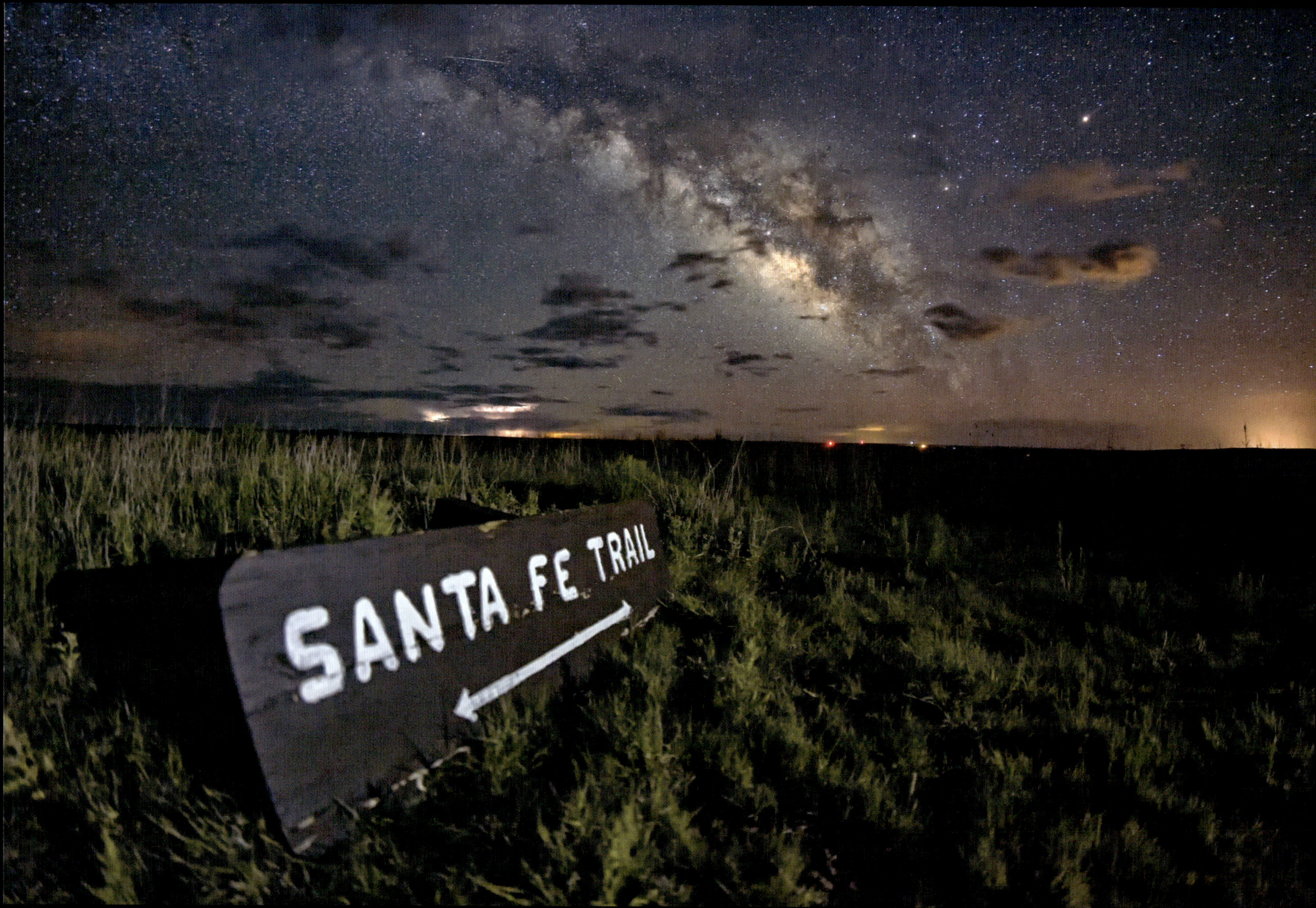

Fort Union National Monument, New Mexico

1

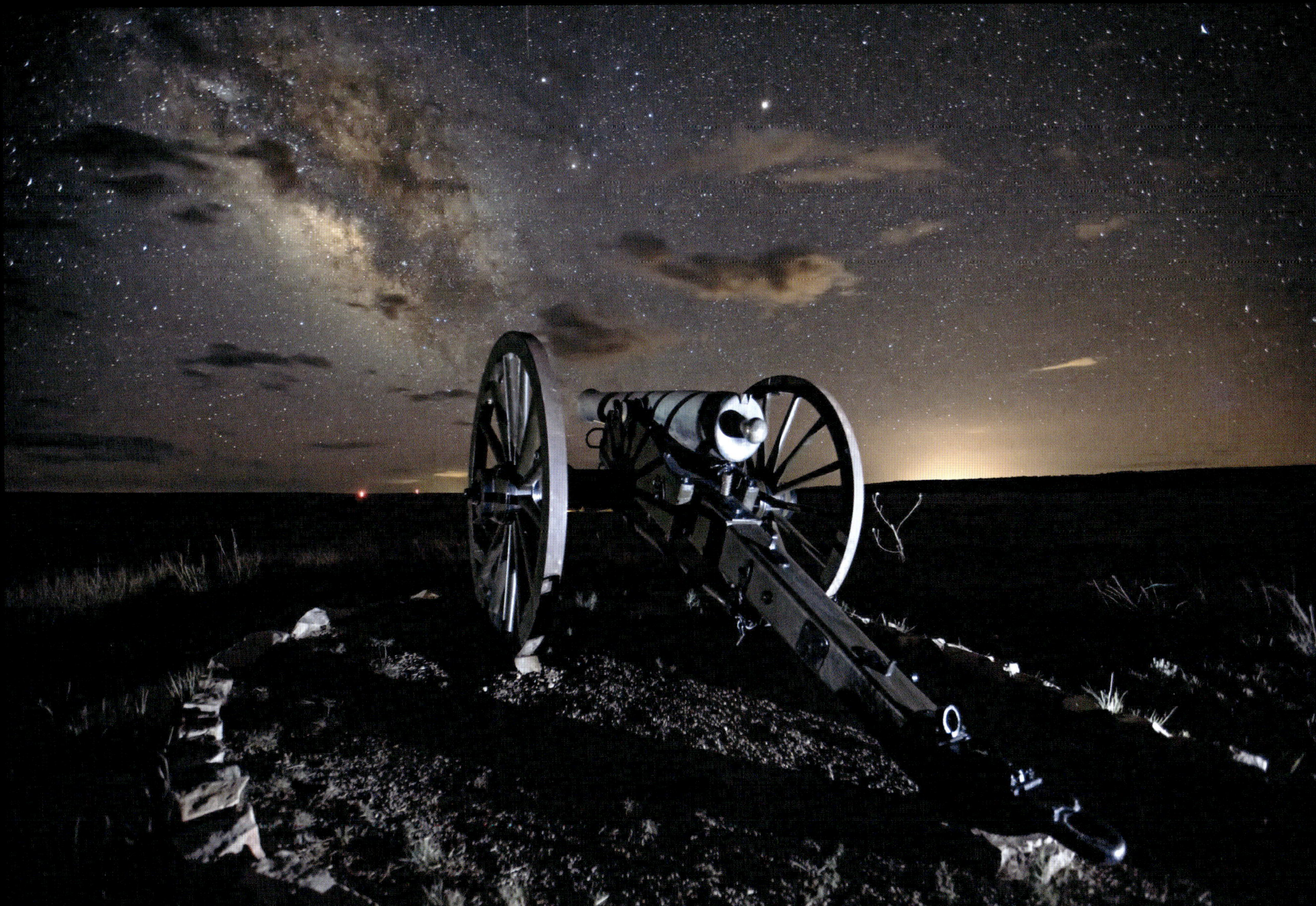

Westward expansion led to the formation of forts along trade routes, such as Fort Union, where soldiers enjoyed perfect night skies at sundown. The night sky, an integral influence on so many Native American mythologies, made its way into the stories and songs of American soldiers as well. One such song, the bugle call *"Butterfield's Lullaby"*—or *"Taps,"* as it is more commonly known today—signified "lights out" or "day is done," and was played each night at sunset.

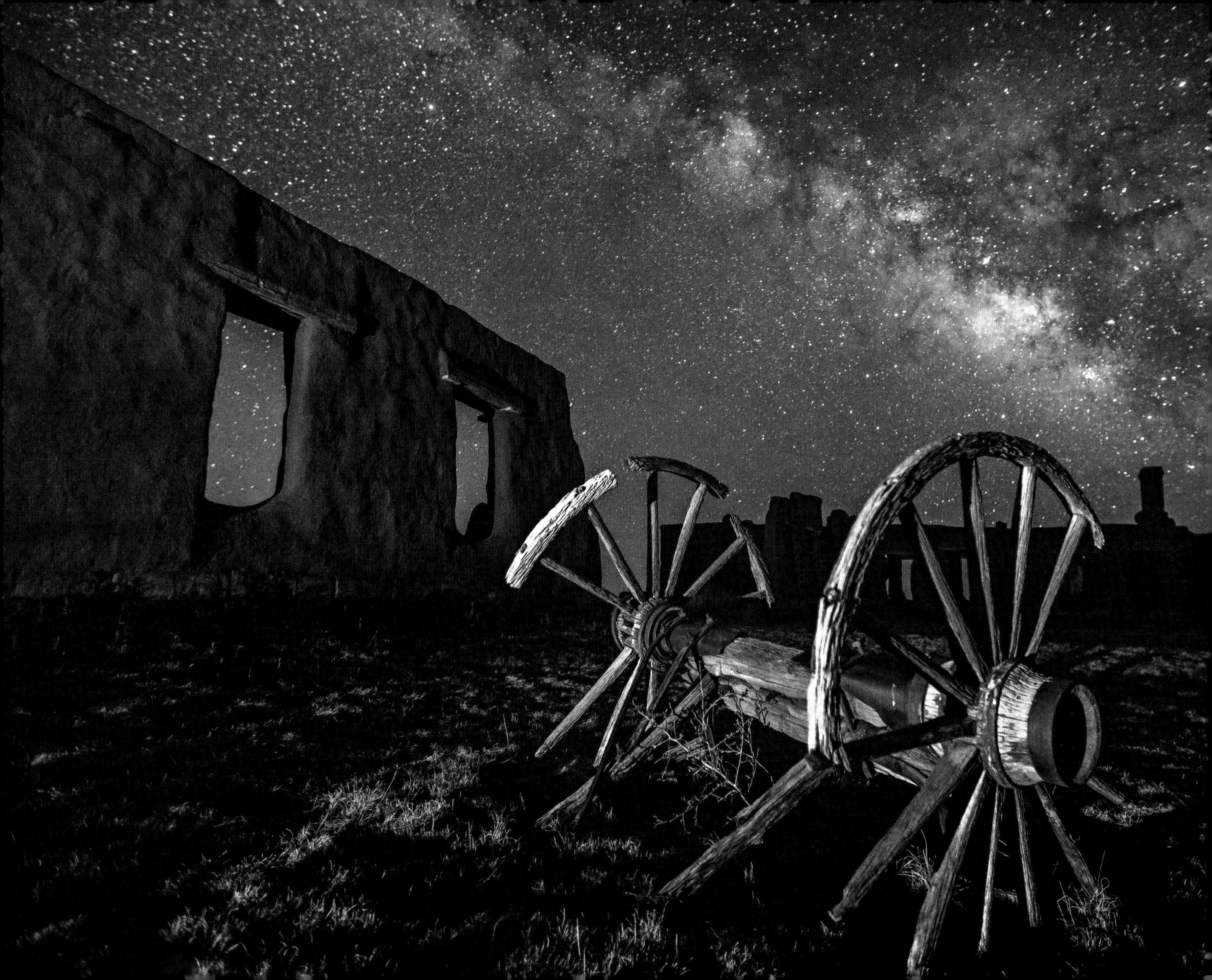

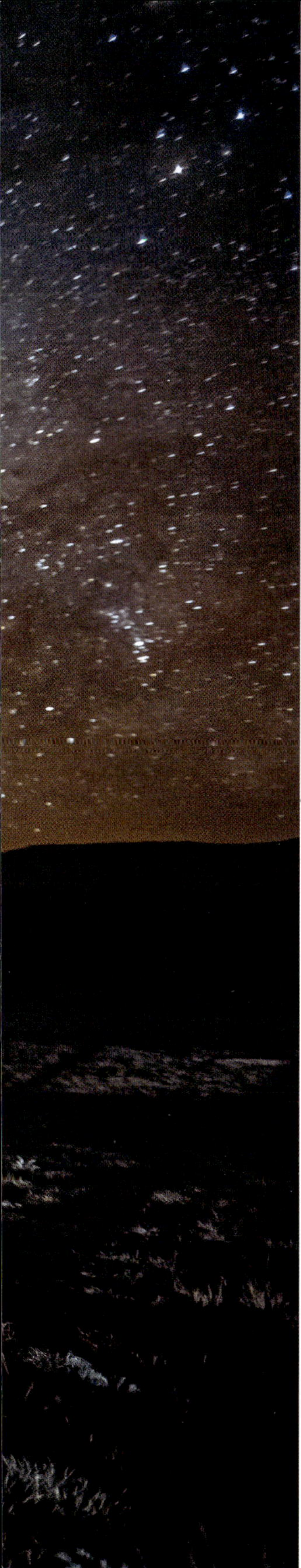

Butterfield's Lullaby

by Horace Lorenzo Trim

Day is done, gone the sun,
From the lake, from the hills, from the sky;
All is well, safely rest, God is nigh.
Fading light, dims the sight,
And a star gems the sky, gleaming bright.
From afar, drawing nigh, falls the night.
Thanks and praise, for our days,
'Neath the sun, 'neath the stars, 'neath the sky;
As we go, this we know, God is nigh.
Sun has set, shadows come,
Time has fled, Scouts must go to their beds
Always true to the promise that they made.
While the light fades from sight,
And the stars gleaming rays softly send,
To thy hands we our souls, Lord, commend.

Fort Union National Monument, New Mexico

1

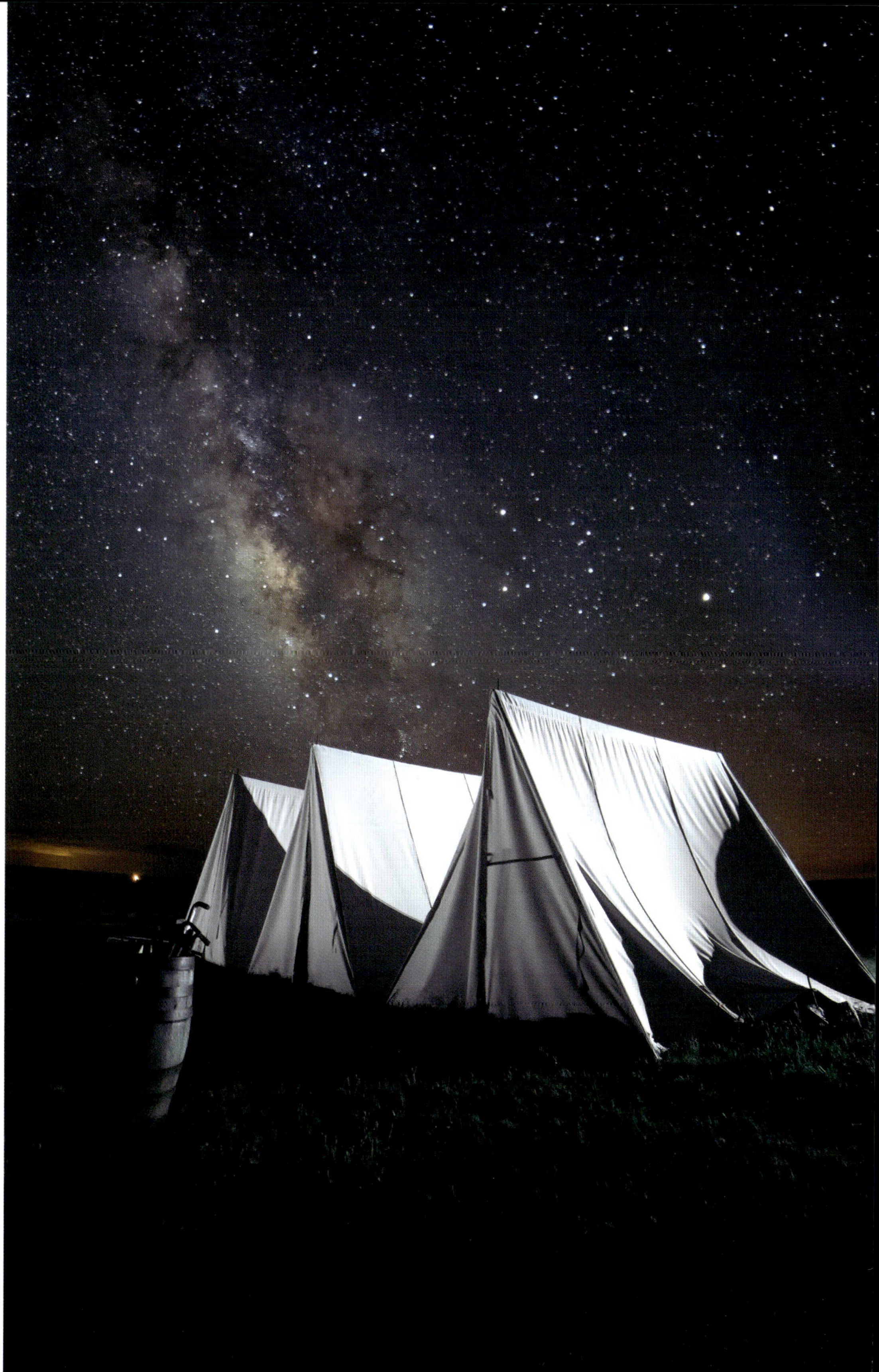

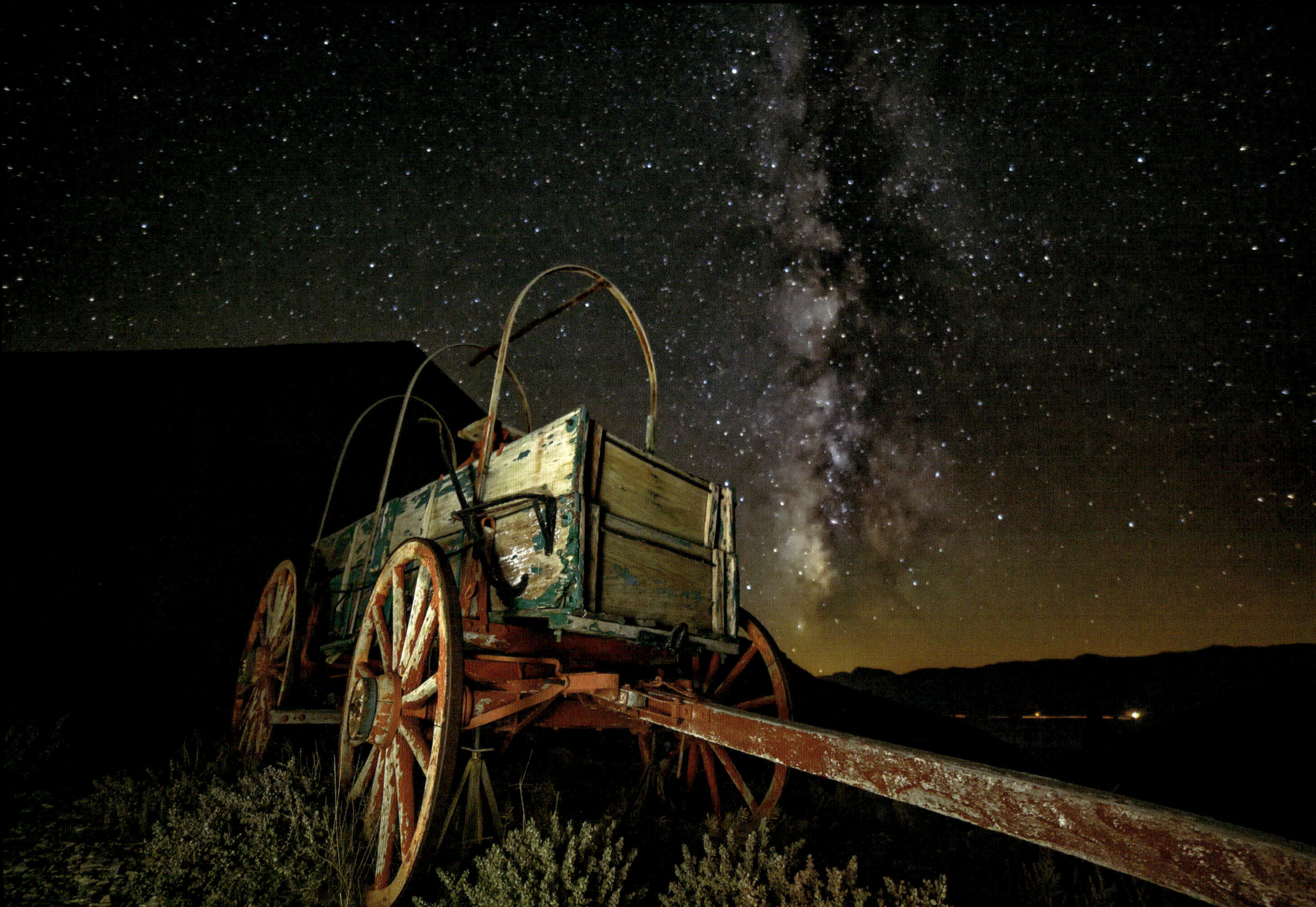

Late-1800s carriages like this one served as primary modes of transportation for those making their way out West looking for riches. It is graced here by a night sky nearly identical to ones that settlers would have seen during their journeys.

Cerro Gordo Ghost Town, California

1

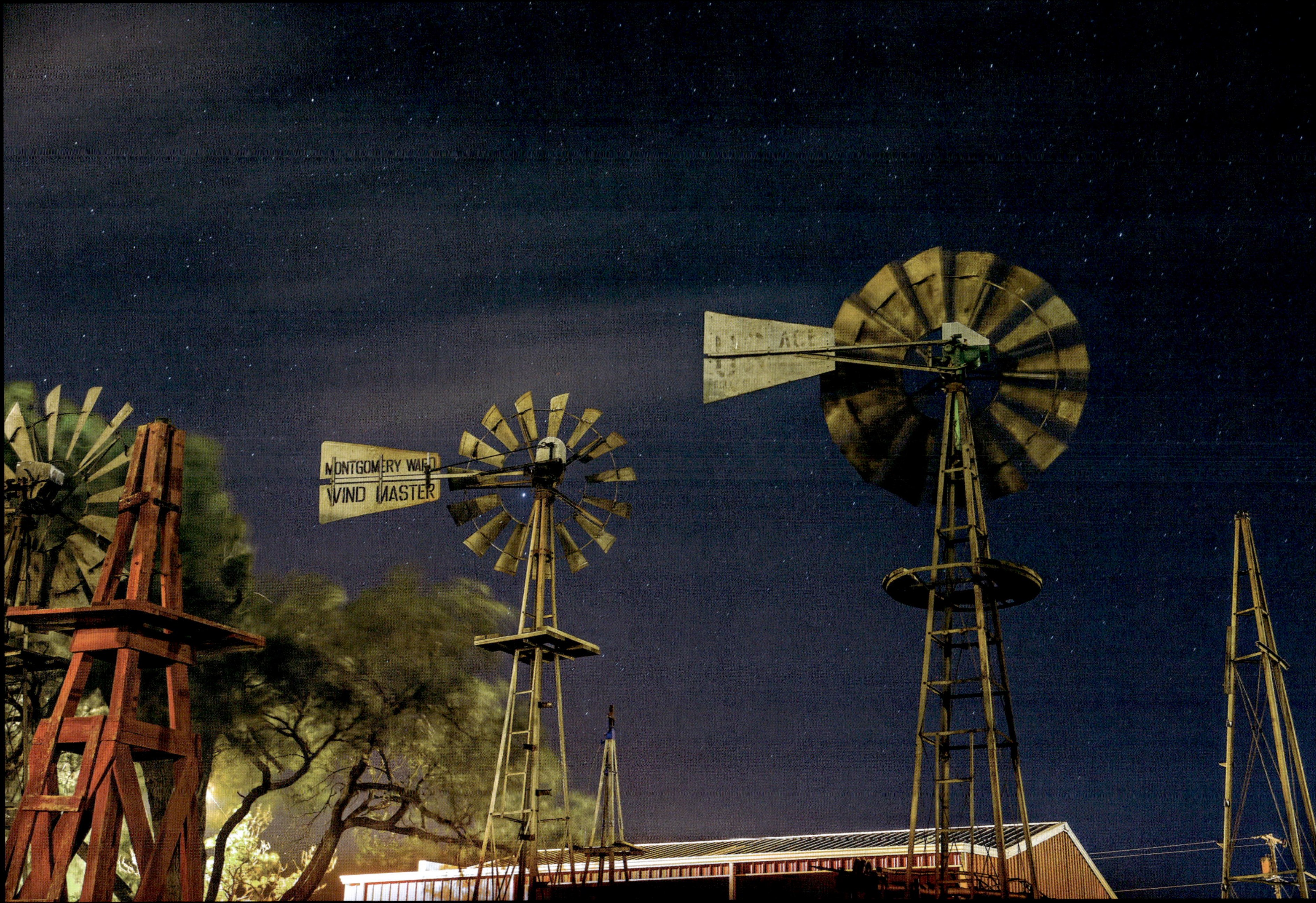

The collection of windmill water pumps at this Oklahoma museum, dating back to the mid-1800s, shows us another side of Manifest Destiny: farms, and what their views of the stars may have been once upon a time.

Shattuck Windmill Museum, Shattuck, Oklahoma

4

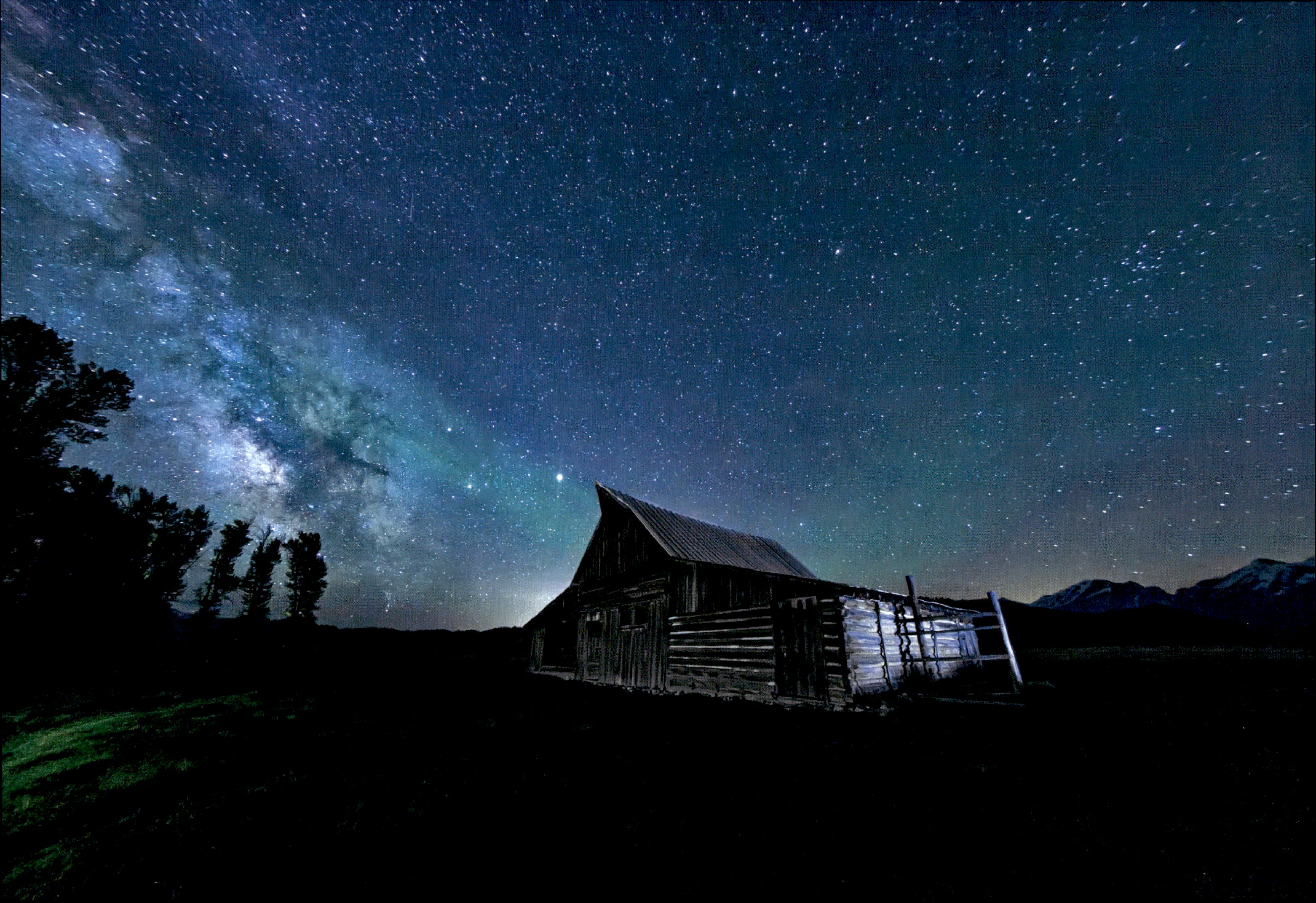

Due to the excellent quality of skies there, Mormon Row farms of Grand Teton National Park are like flashbacks to the past. Even the faint Aurora Borealis, seen here above T.A. Moulton's barn, may still be visible to the naked eye.

Mormon Row, Grand Teton National Park, Wyoming

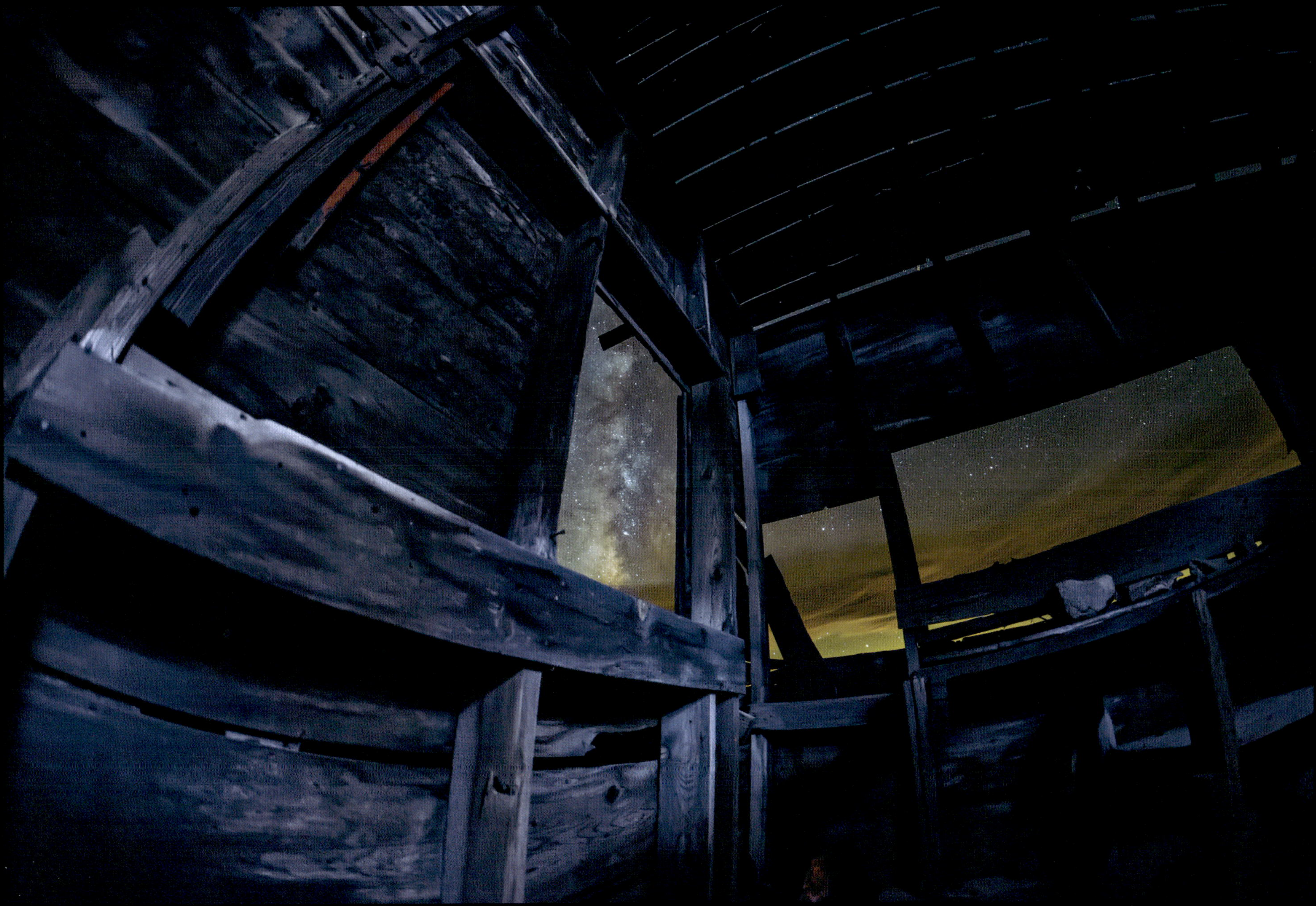

The American Frontier was soon covered with tens of thousands of farms, barns, towns, and cities, with each structure basking in dark skies like this one, depicted outside a window on the eastern side of California's Sierra Nevada Mountains.

Chemung Mine, California

1

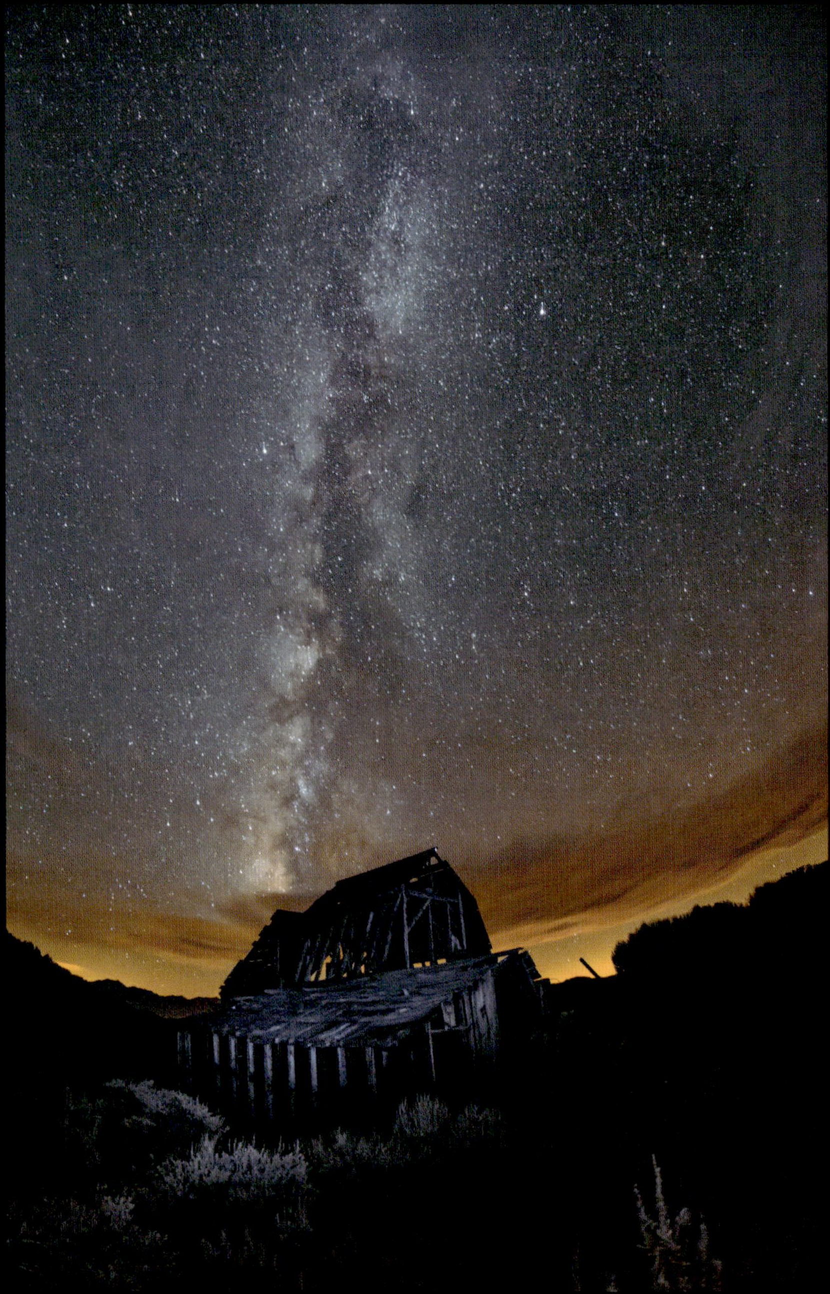

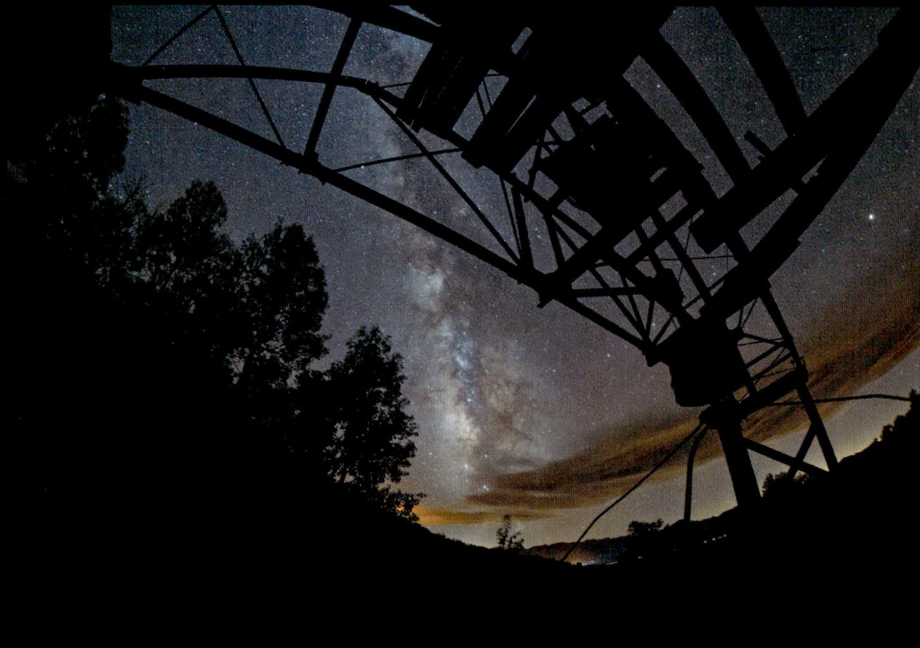

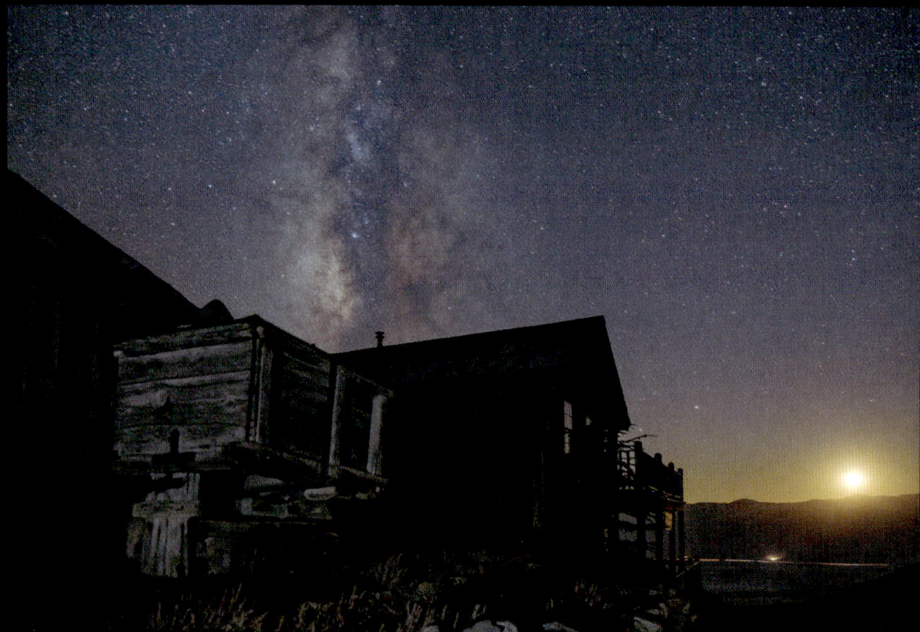

LEFT & TOP RIGHT: **Chemung Mine, California**
BOTTOM: **Cerro Gordo Ghost Town, California**

1

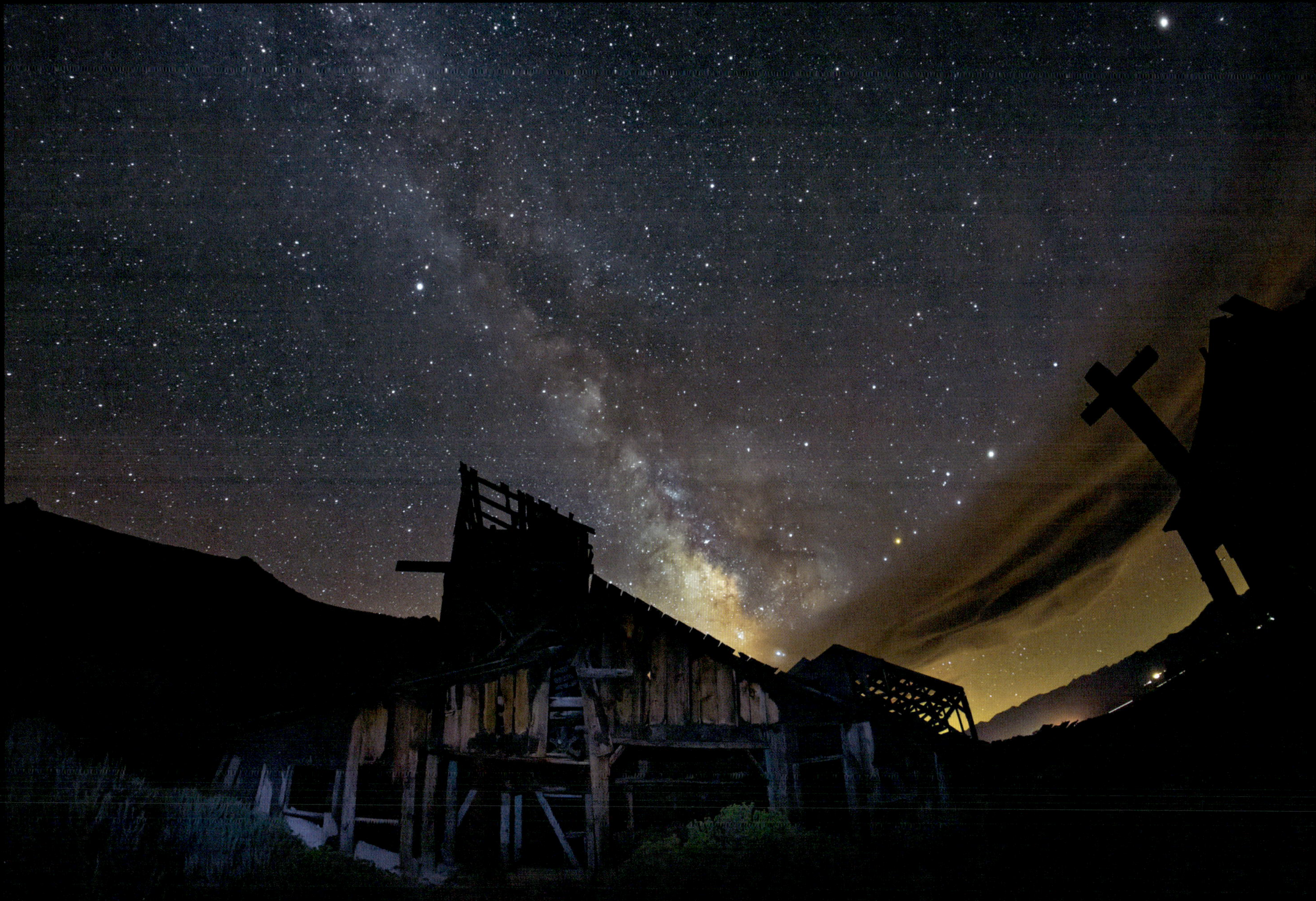

The gold and silver mines of California were the last stops on the long journey west. The Gold Rush created the permanent entrenchment of white settlers in the American West, and new technologies that made it easier to travel to and live in those harsh, mountainous climates began to arise.

Chemung Mine, California

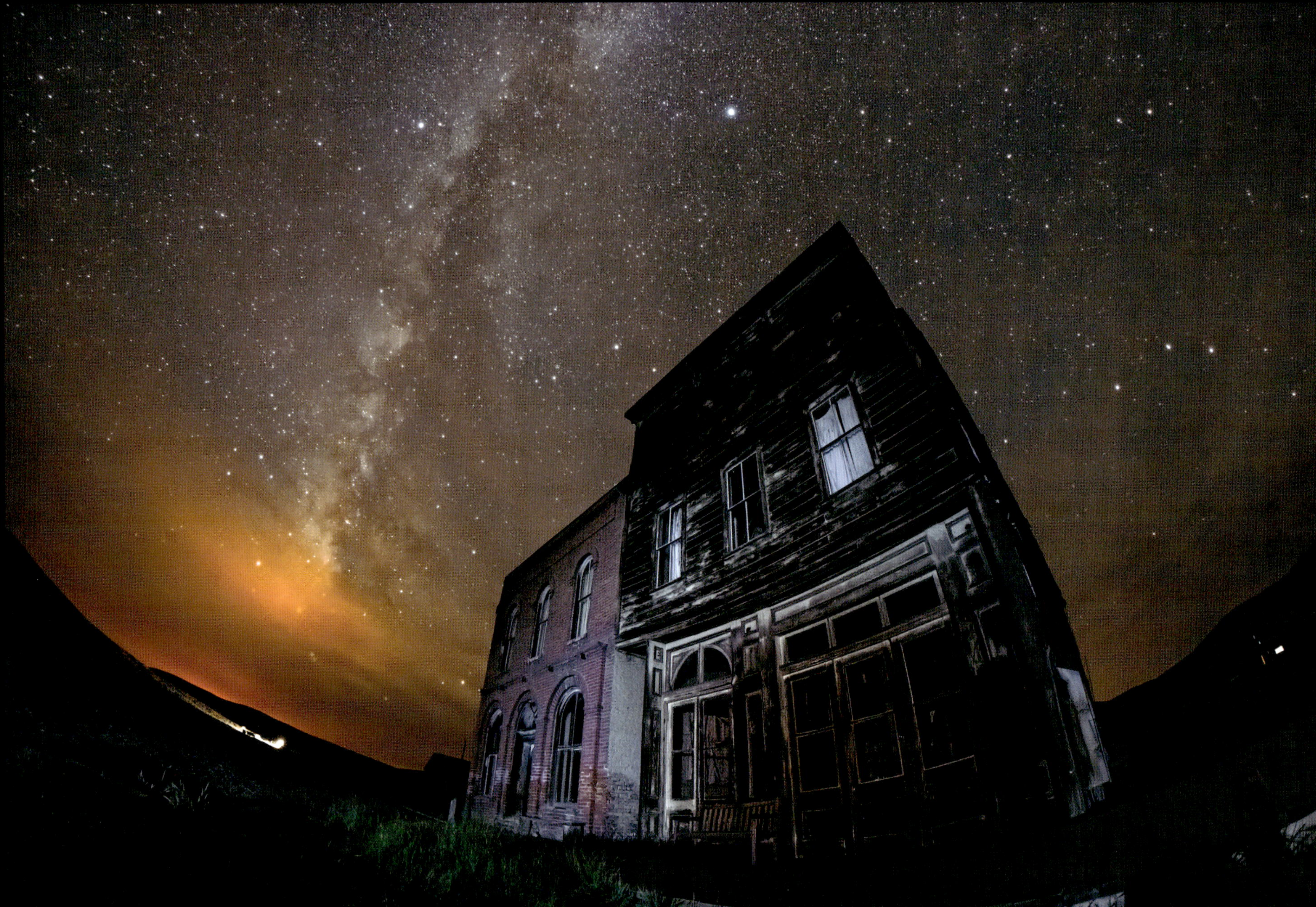

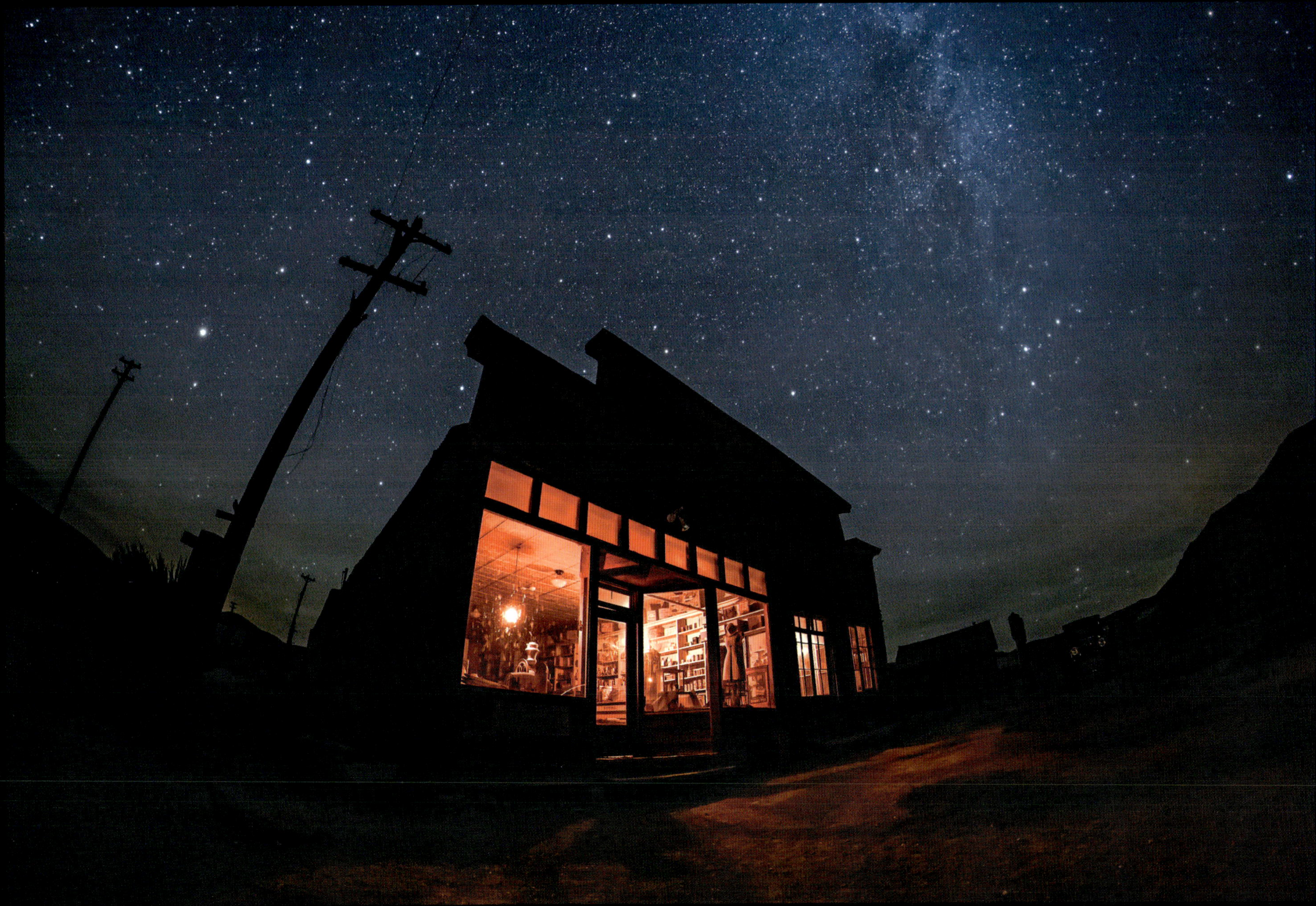

Electricity, followed by electric light, arrived in California in the late-1880s, along with steam engine train technologies, and soon thereafter the growth of the oil industry and automobiles.

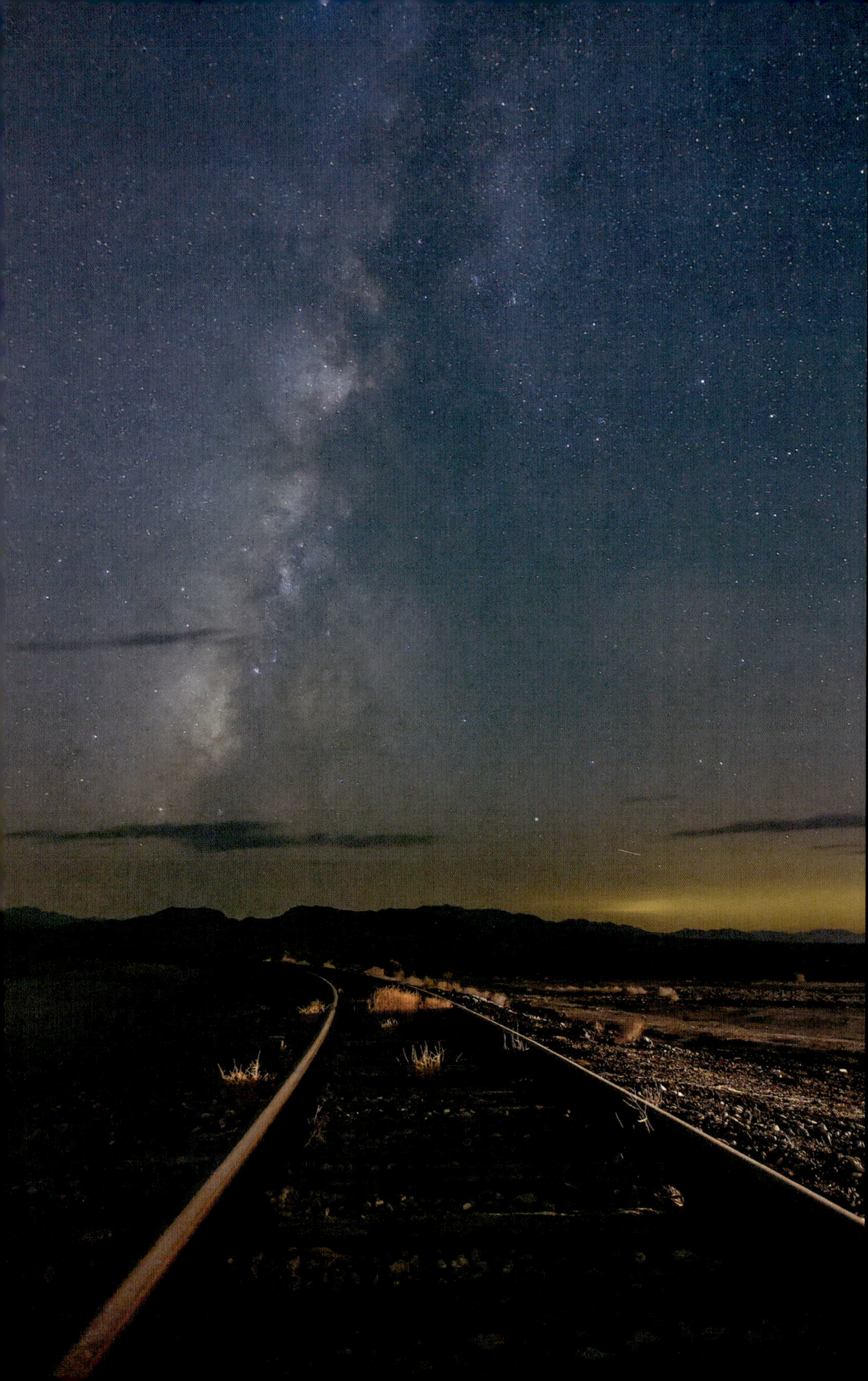

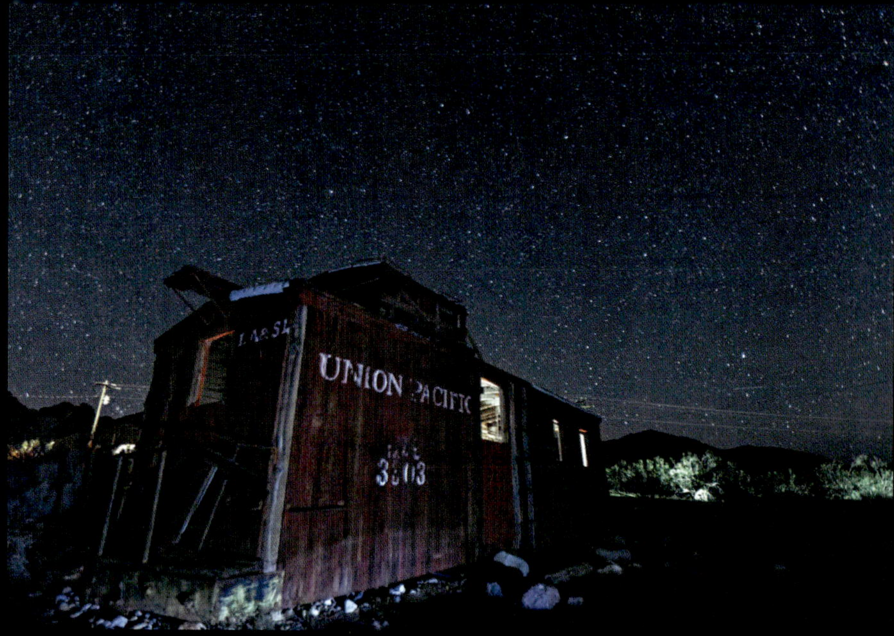

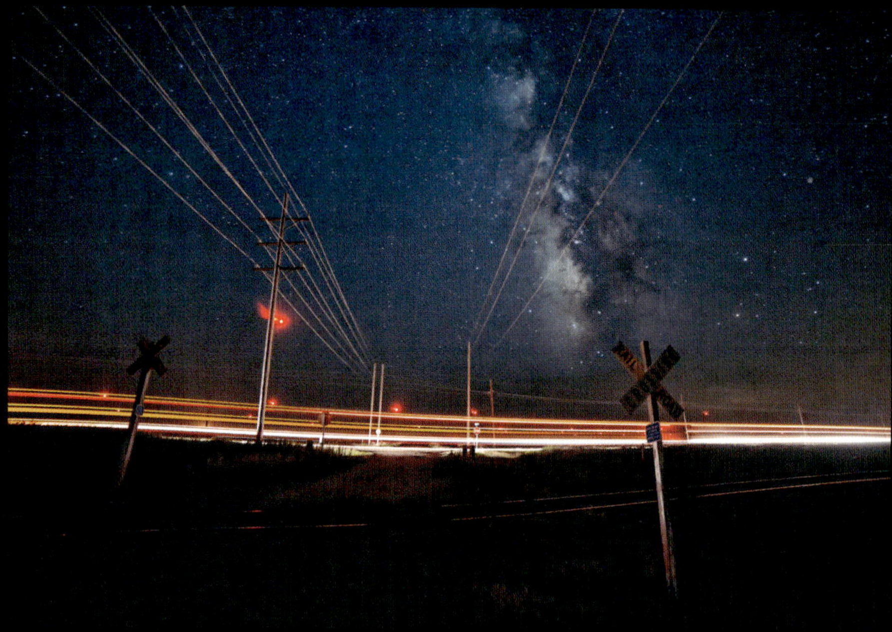

LEFT: **Railroad Tracks, Rice Ghost Town, California**

TOP: **Late 1800s Train Caboose, Rhyolite Ghost Town, Nevada**

BOTTOM: **Power Lines over a Railroad Crossing, Montezuma, Kansas**

1

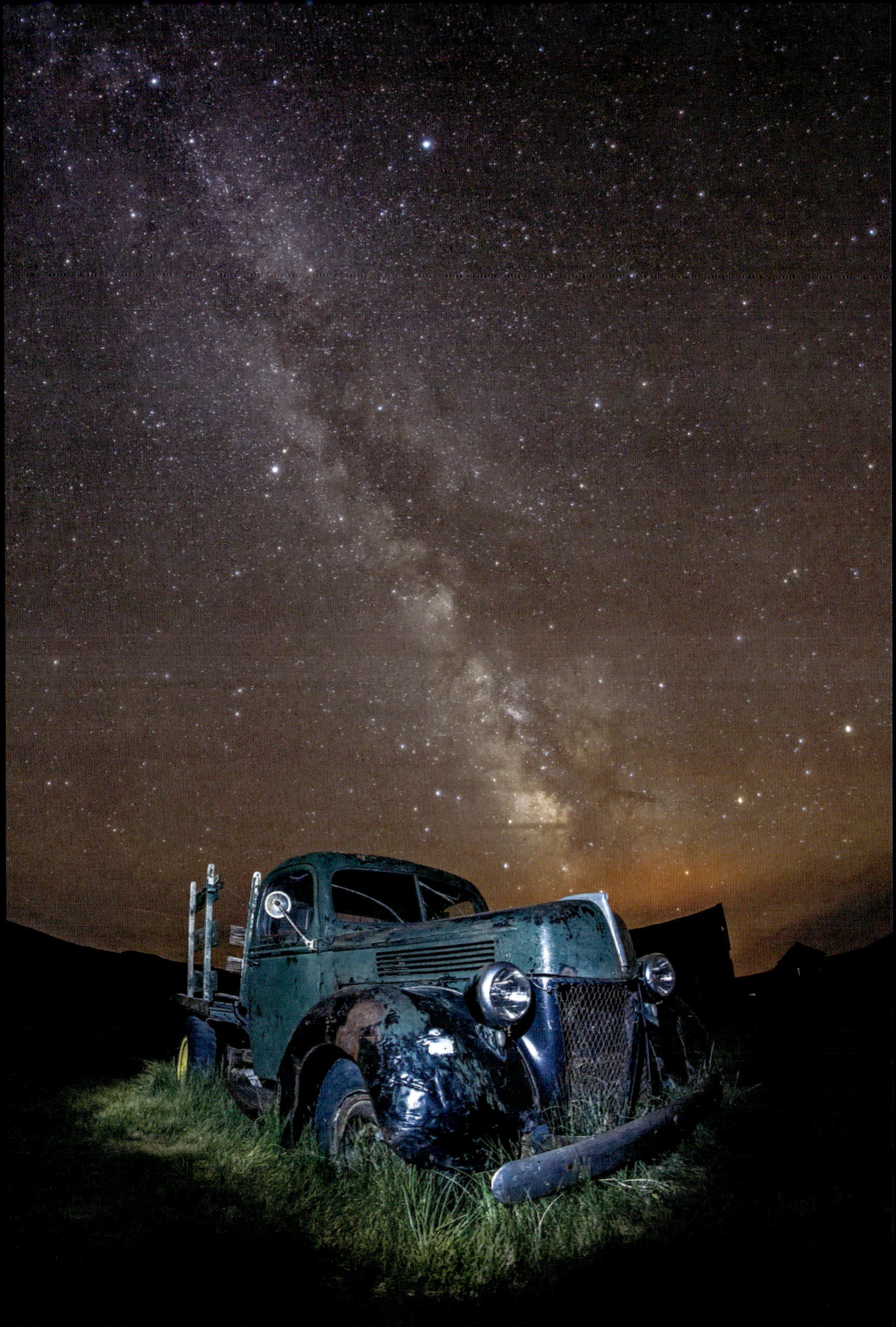

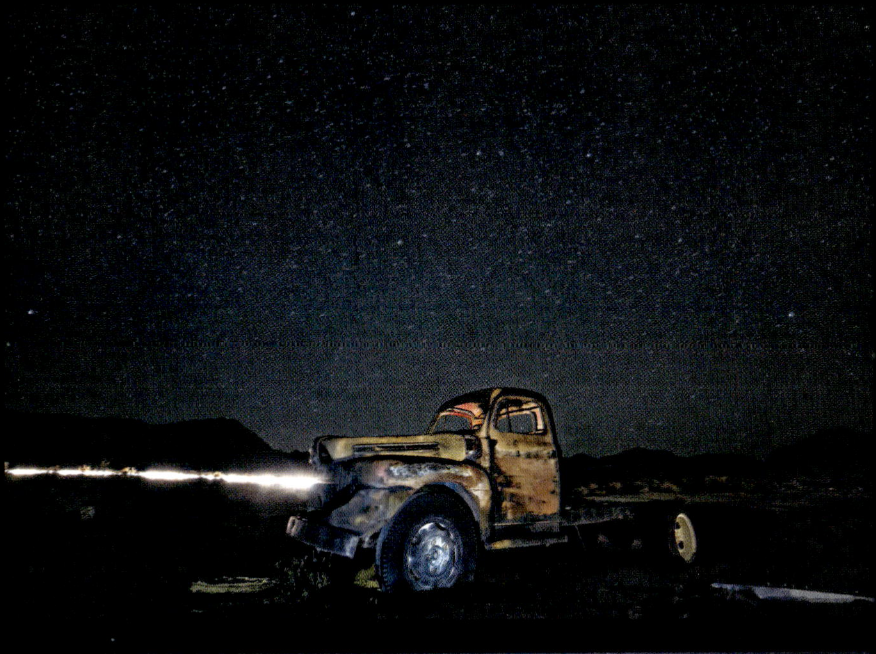

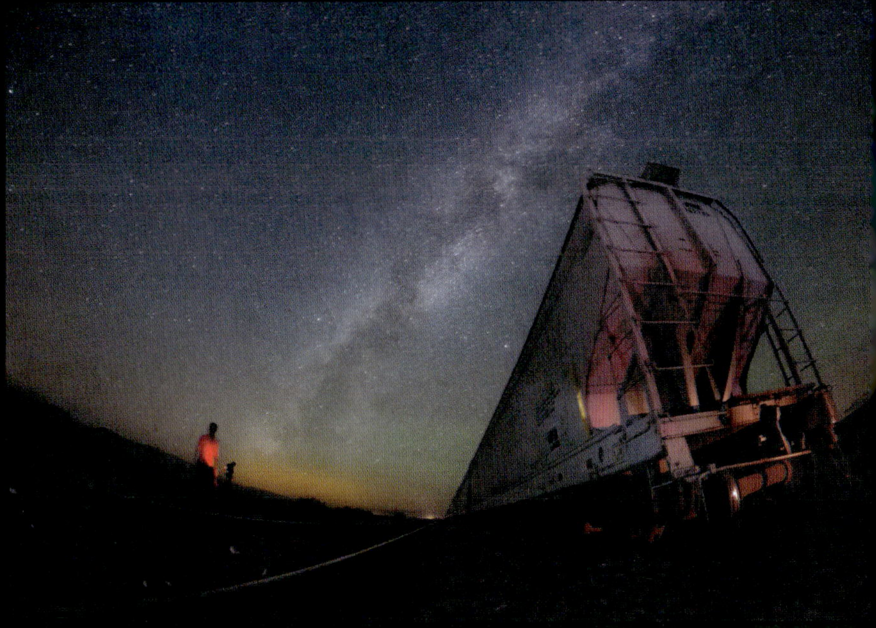

LEFT: **Bodie State Historic Park, California**
TOP: **Rhyolite Ghost Town, Nevada**
BOTTOM: **Rice Ghost Town, California**

1

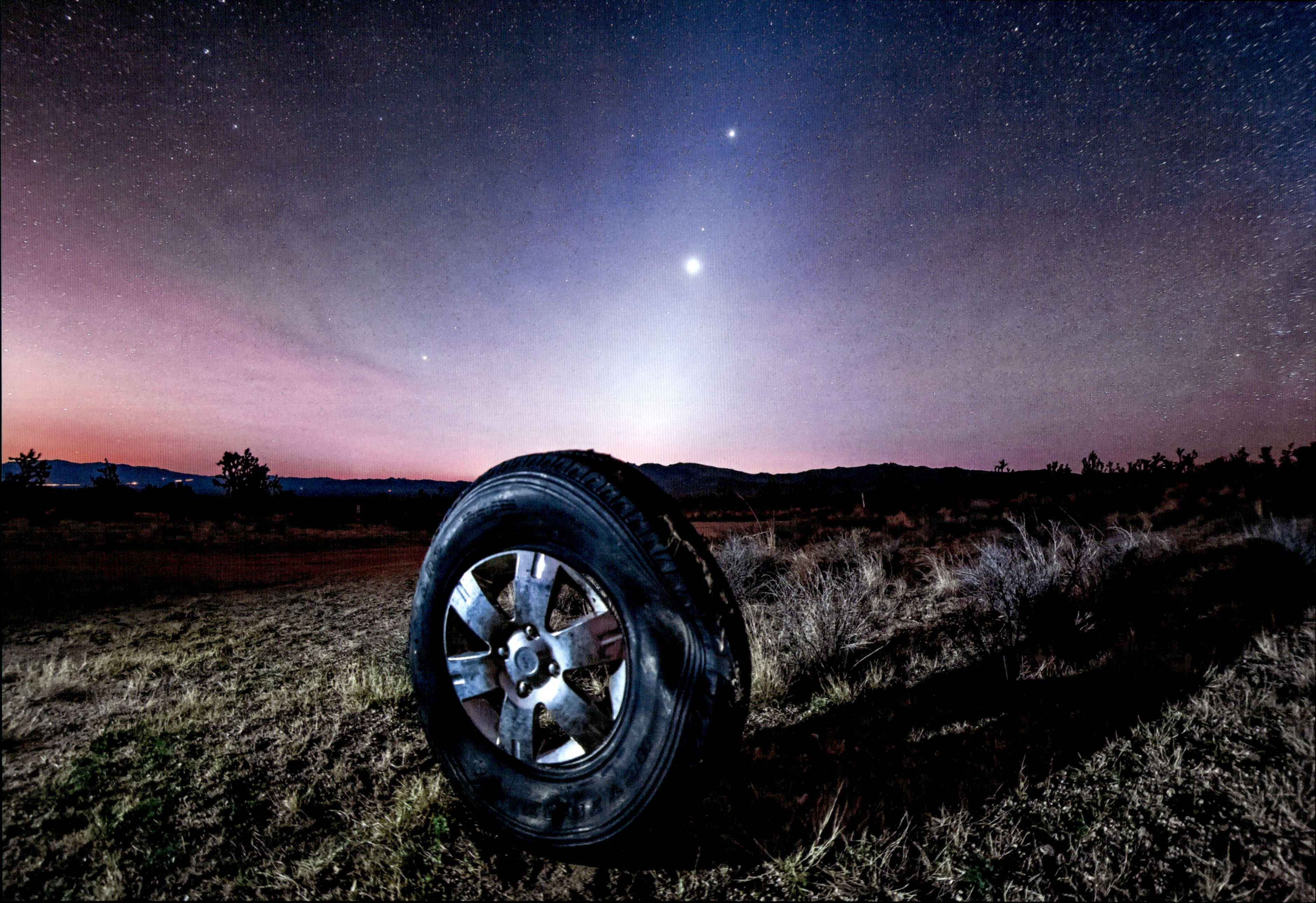

Mojave National Preserve, California

2

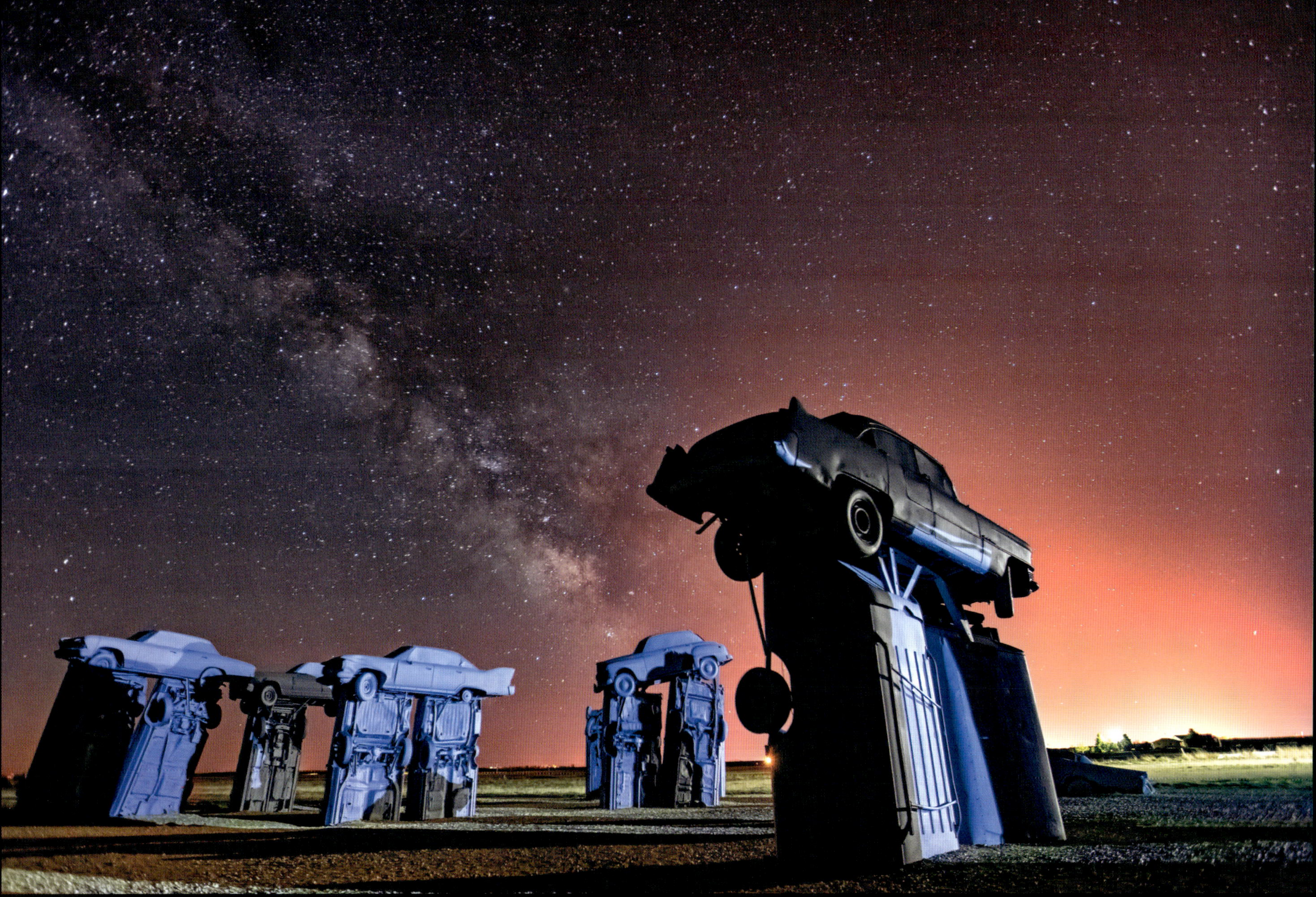

Carhenge, Alliance, Nebraska

4,5

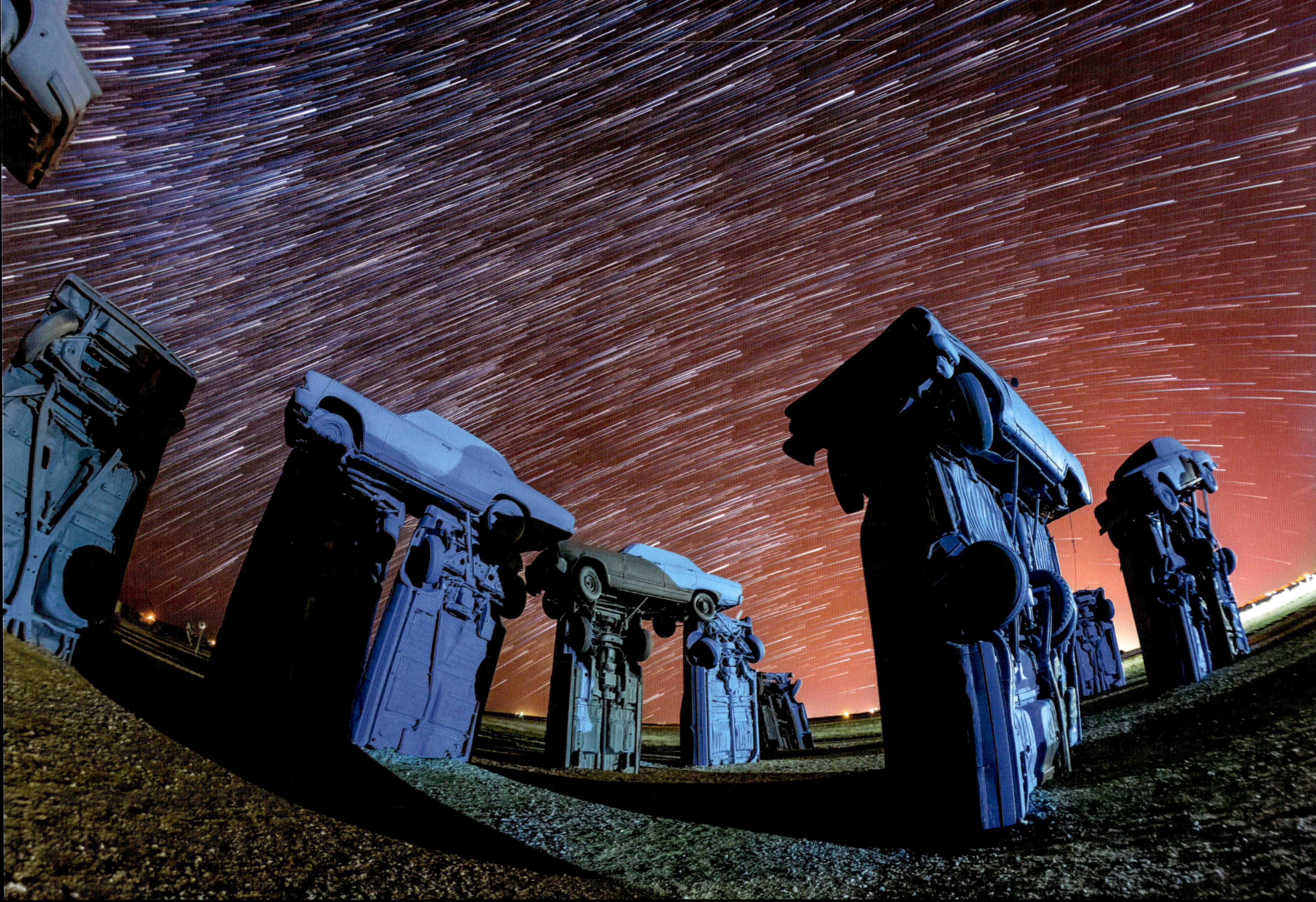

Just a few hundred years after Native American archaeoastronomy sites covered the land, Manifest Destiny had led to the near-destruction of many Native American cultures. All that remains now are Americana throwbacks to ancient European archaeoastronomy--sites such as Carhenge in Nebraska.

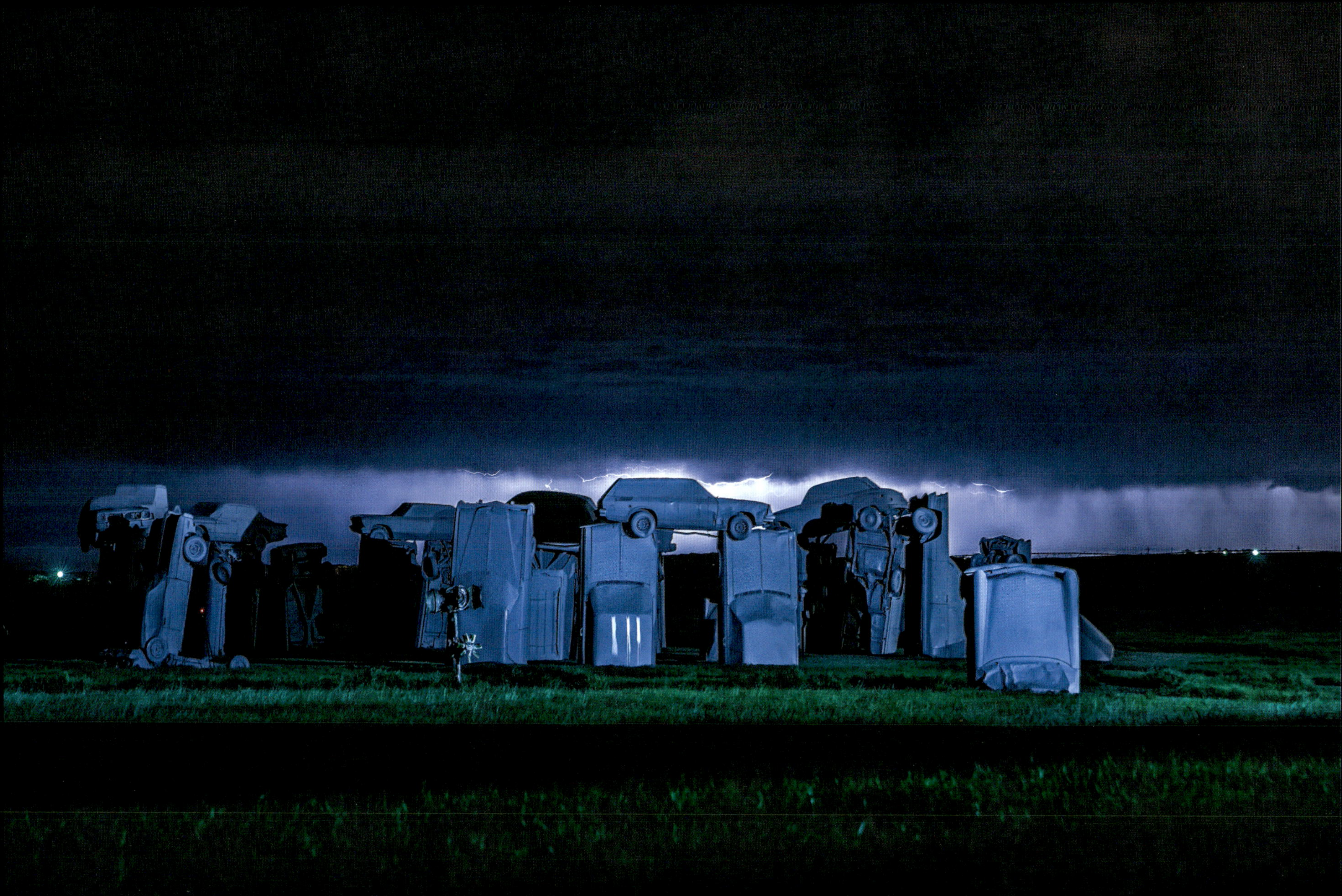

The age of dark skies, with us from the very beginning of humanity, has come to an abrupt end.

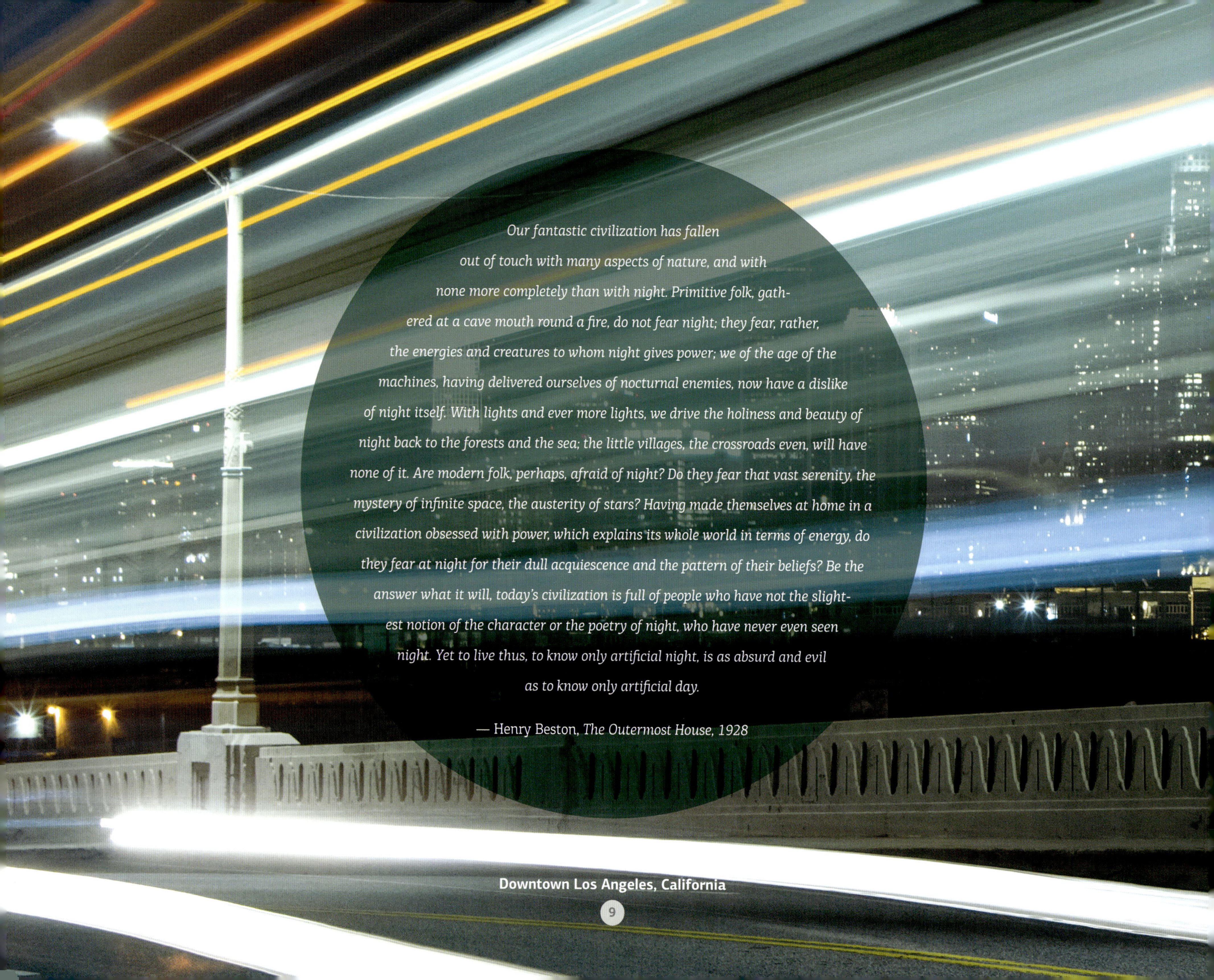

Our fantastic civilization has fallen out of touch with many aspects of nature, and with none more completely than with night. Primitive folk, gathered at a cave mouth round a fire, do not fear night; they fear, rather, the energies and creatures to whom night gives power; we of the age of the machines, having delivered ourselves of nocturnal enemies, now have a dislike of night itself. With lights and ever more lights, we drive the holiness and beauty of night back to the forests and the sea; the little villages, the crossroads even, will have none of it. Are modern folk, perhaps, afraid of night? Do they fear that vast serenity, the mystery of infinite space, the austerity of stars? Having made themselves at home in a civilization obsessed with power, which explains its whole world in terms of energy, do they fear at night for their dull acquiescence and the pattern of their beliefs? Be the answer what it will, today's civilization is full of people who have not the slightest notion of the character or the poetry of night, who have never even seen night. Yet to live thus, to know only artificial night, is as absurd and evil as to know only artificial day.

— Henry Beston, *The Outermost House*, 1928

Downtown Los Angeles, California

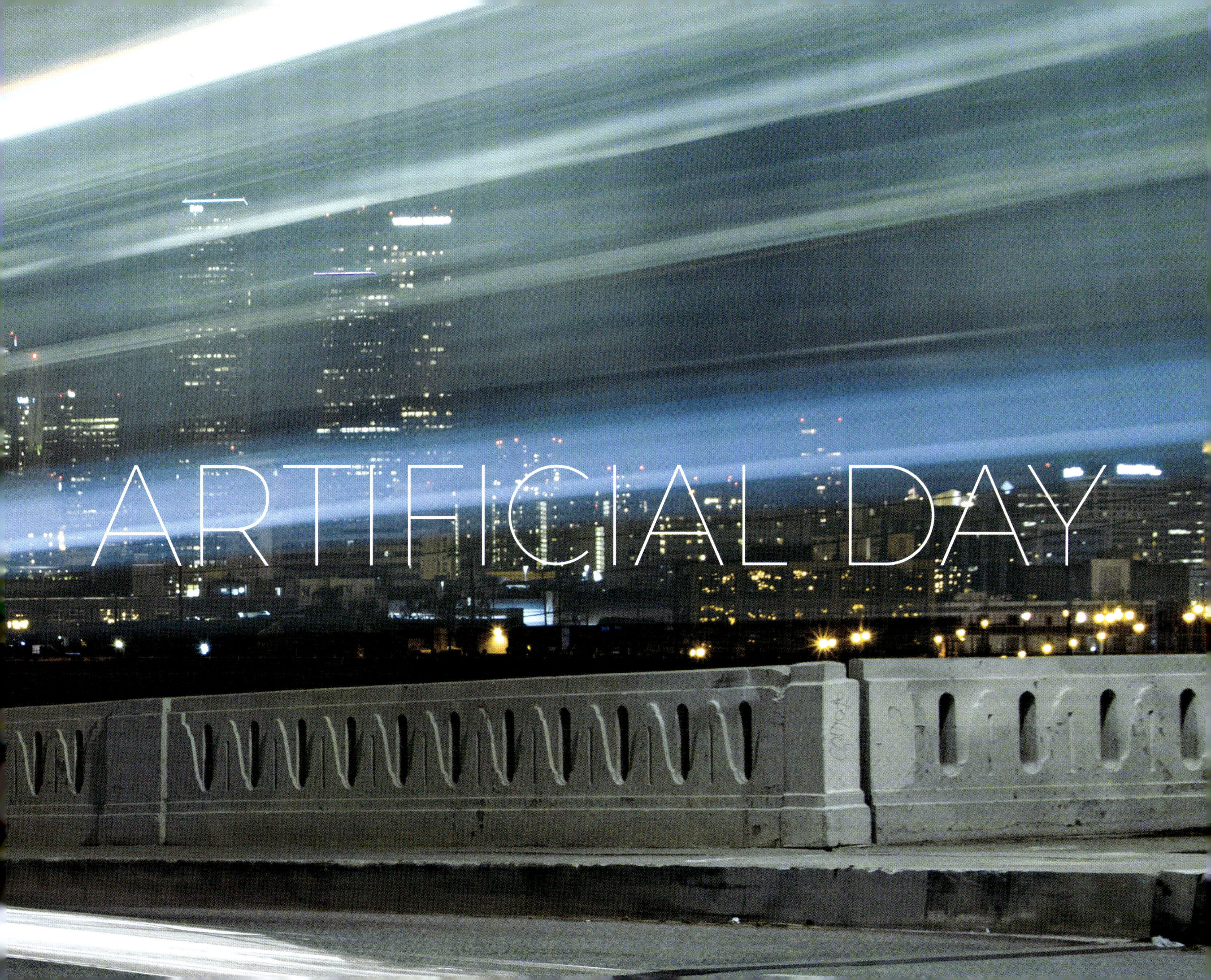
ARTIFICIAL DAY

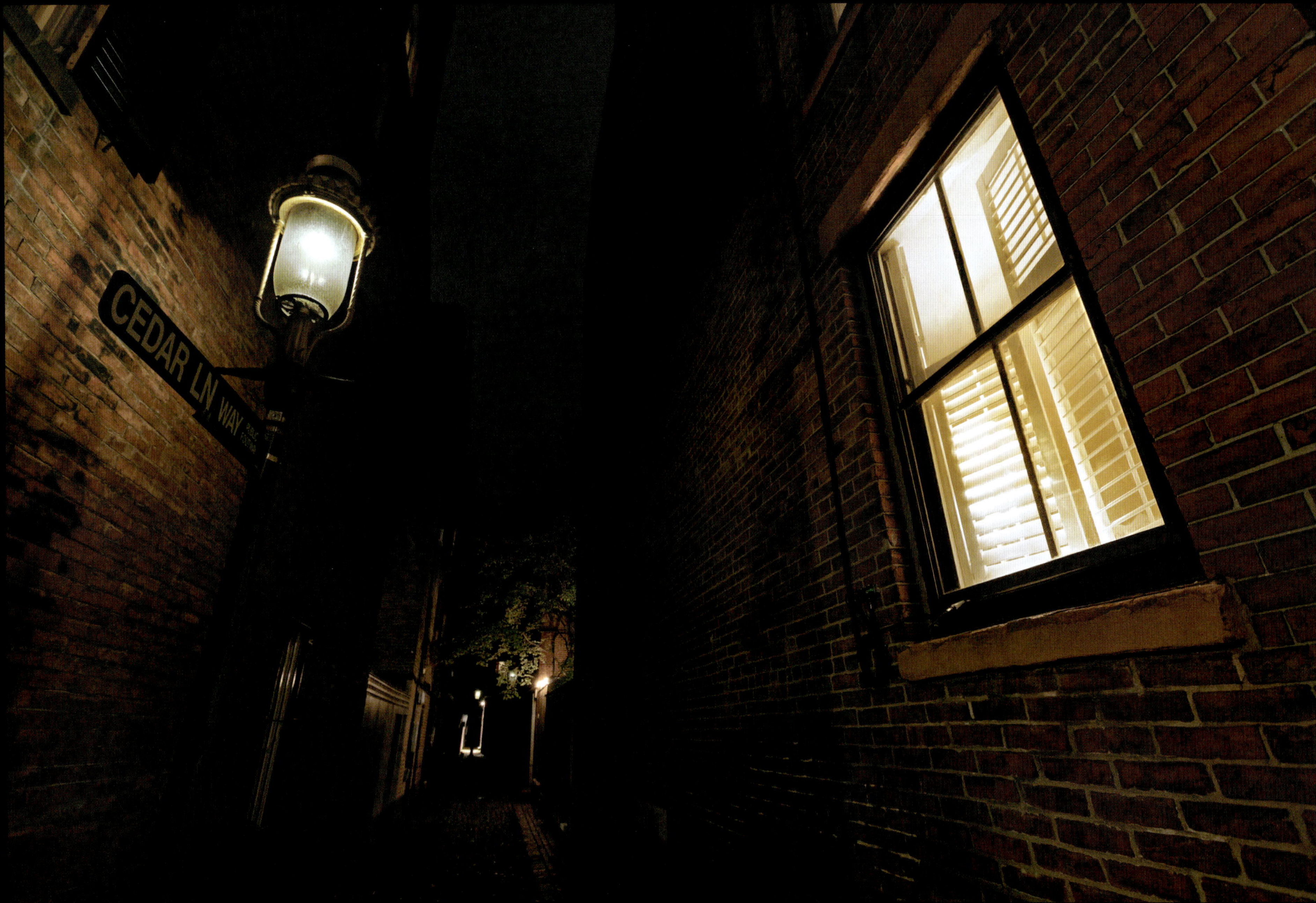

Historic Beacon Hill, Boston, Massachusetts

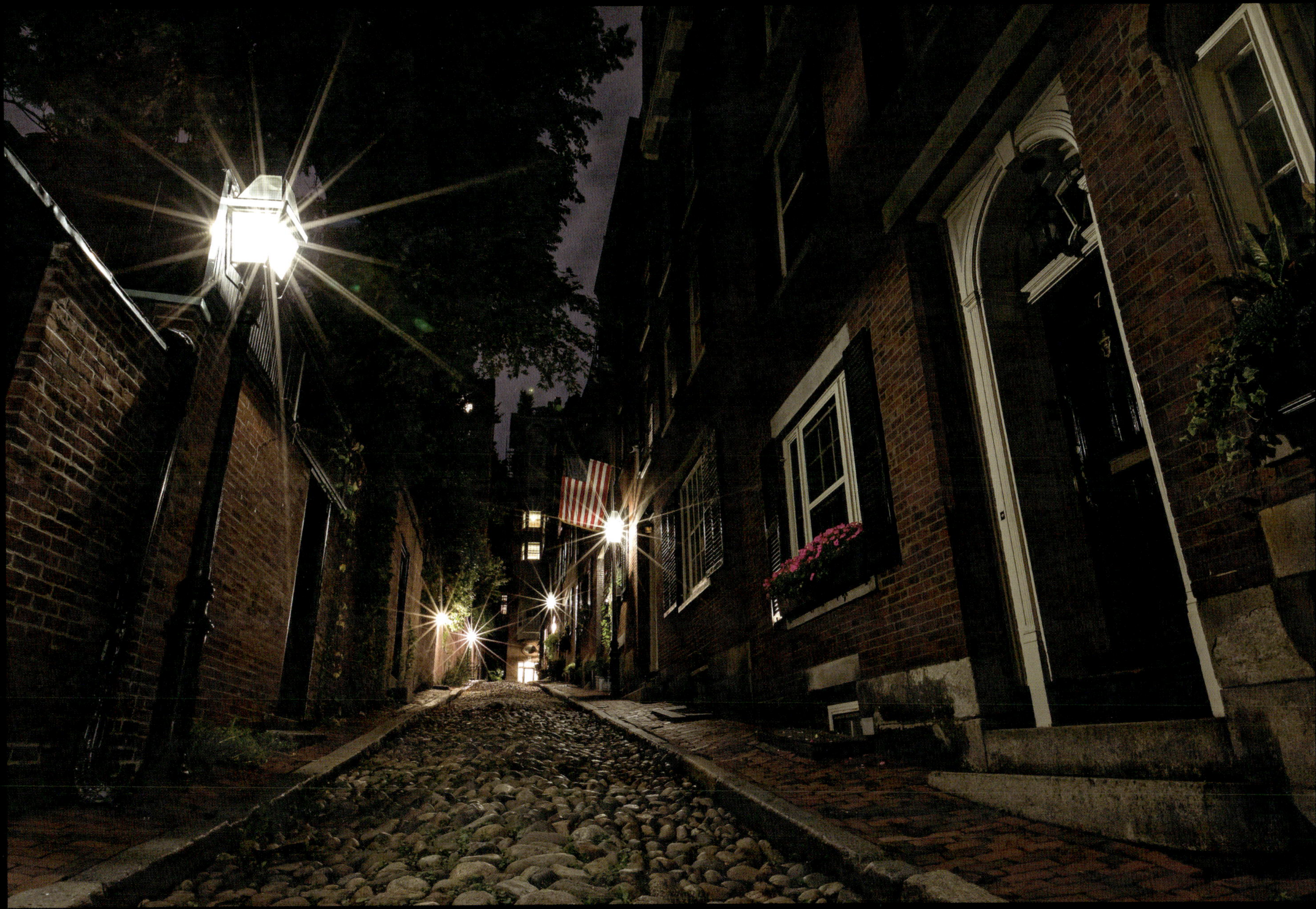

Offering a rare glimpse into what it would be like to take a night walk through a city prior to electrical light, Boston's Beacon Hill features gas lamps, which started in one form as early as the 1400s and became popular during the 1800s as technology advanced, replacing torch-lit or completely unlit city streets with widespread street lighting.

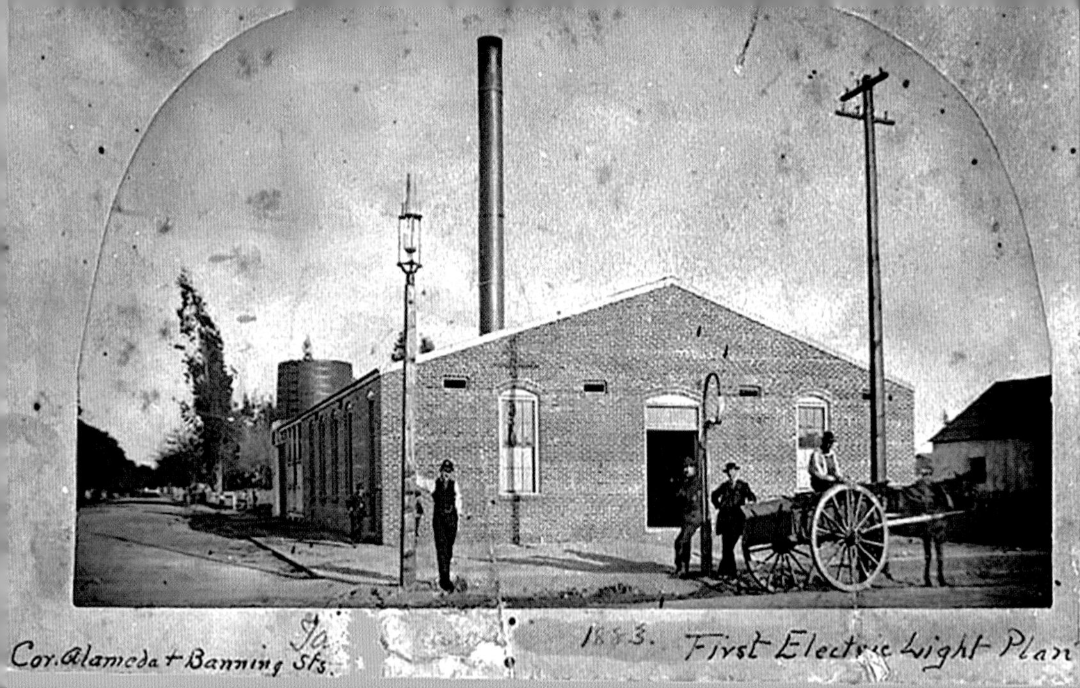

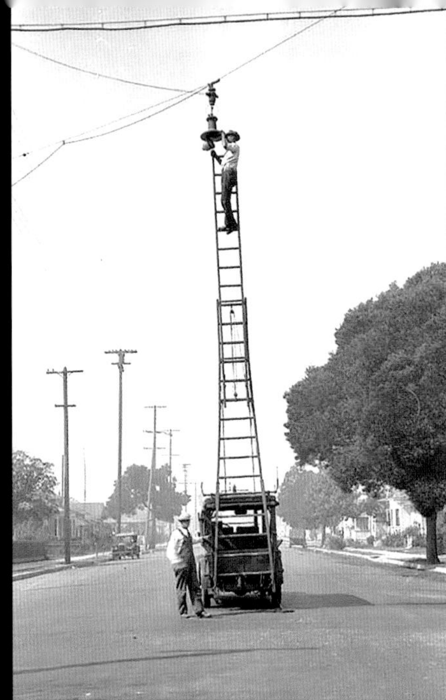

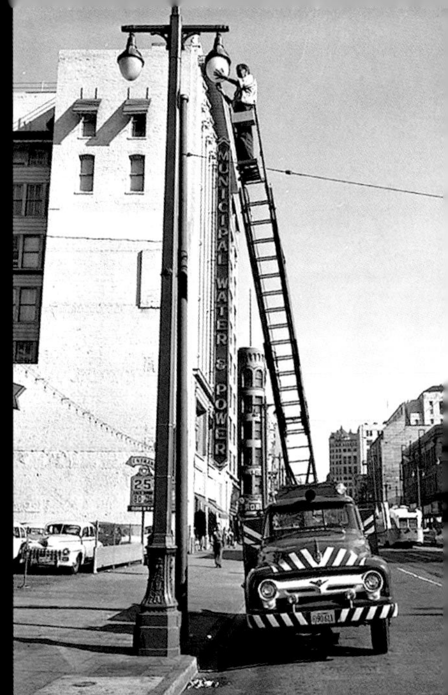

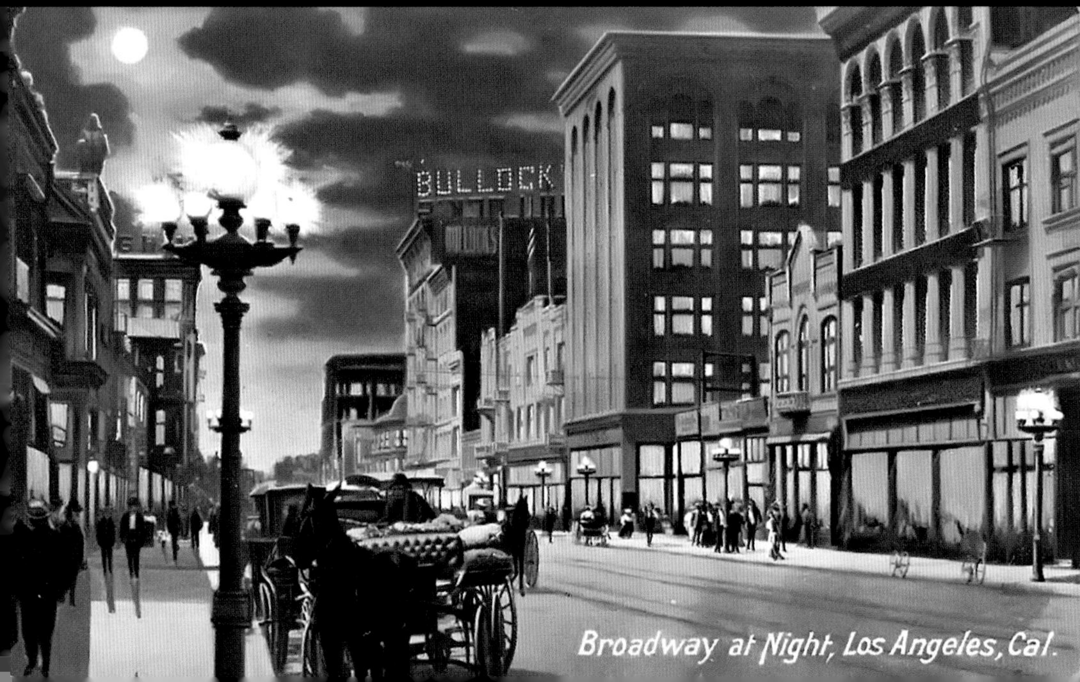

Archival Photos of early street lights in Los Angeles, California

Courtesy of Department of Water & Power Photo Archive

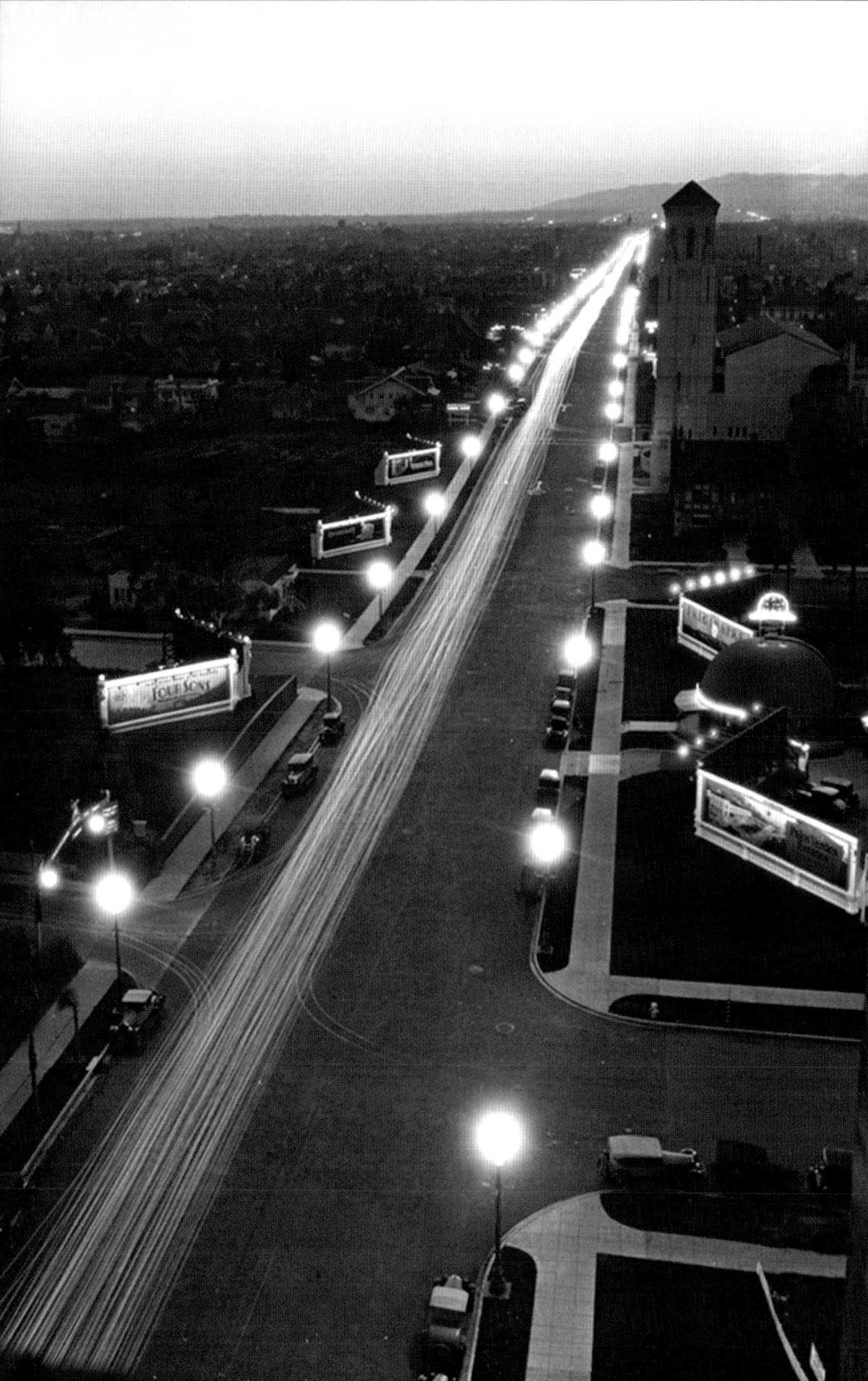

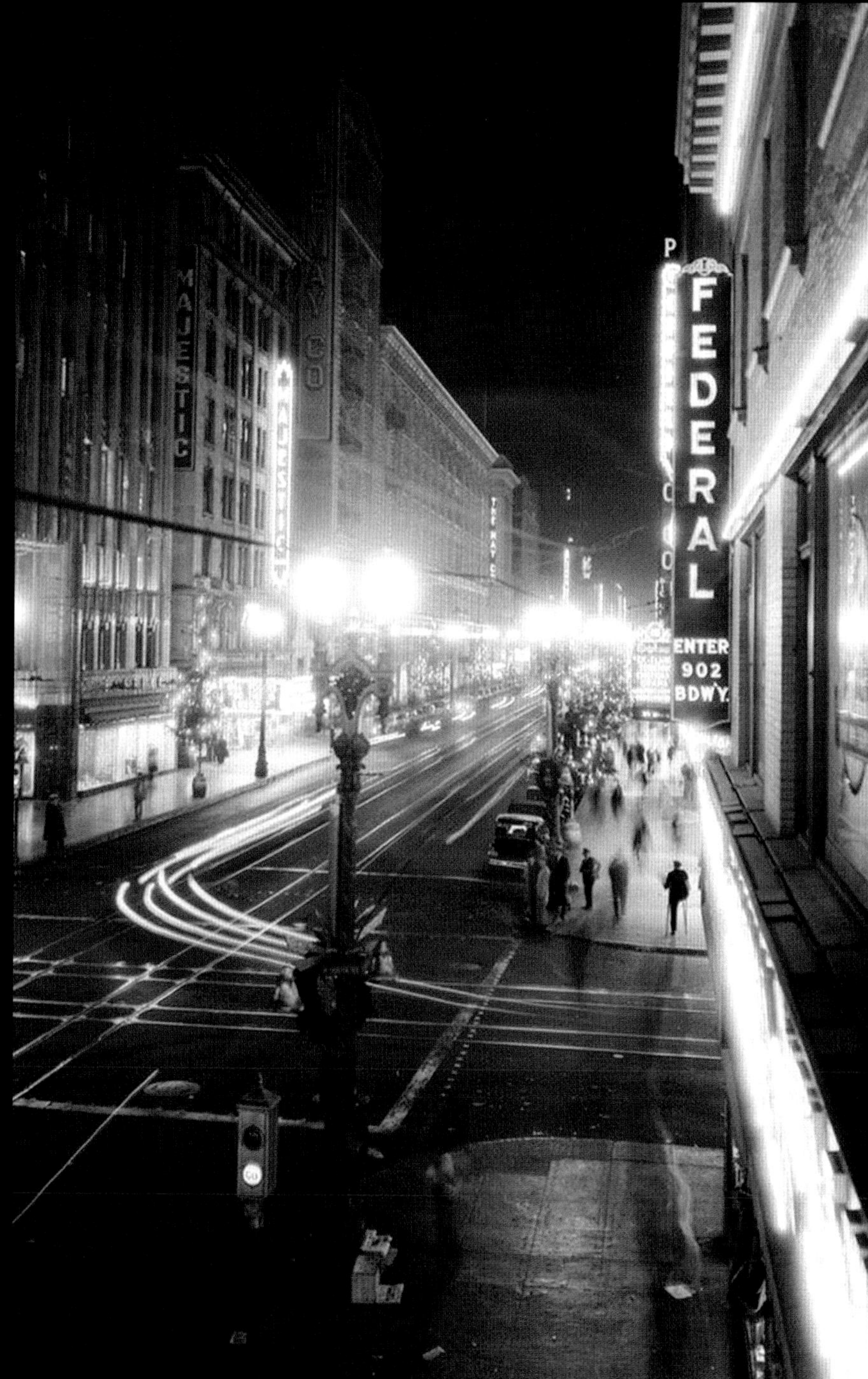

After a short-lived period of gas lamps, electricity soon swept across every major city in the United States, with each generation of street lamps brighter than the last.

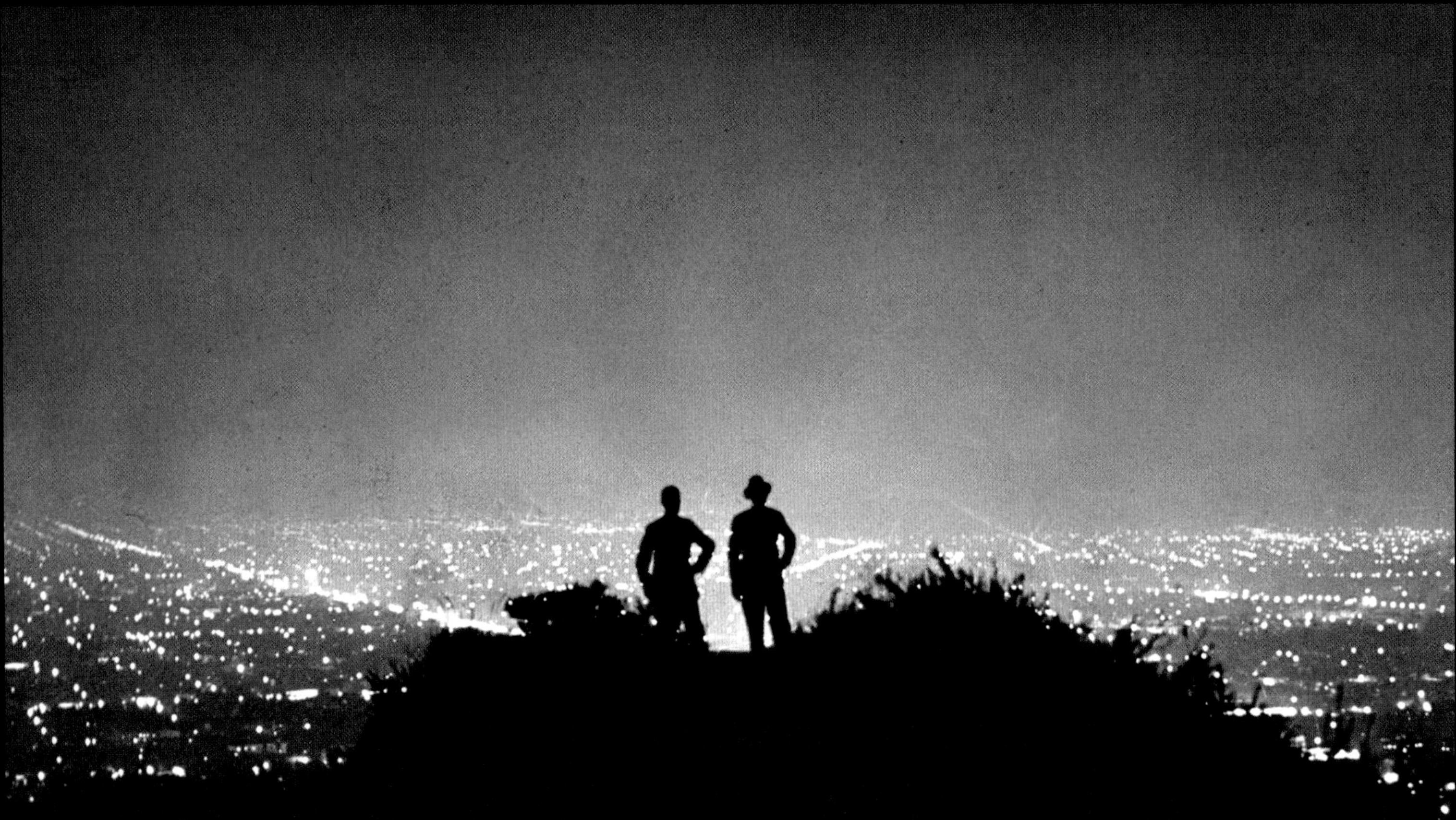

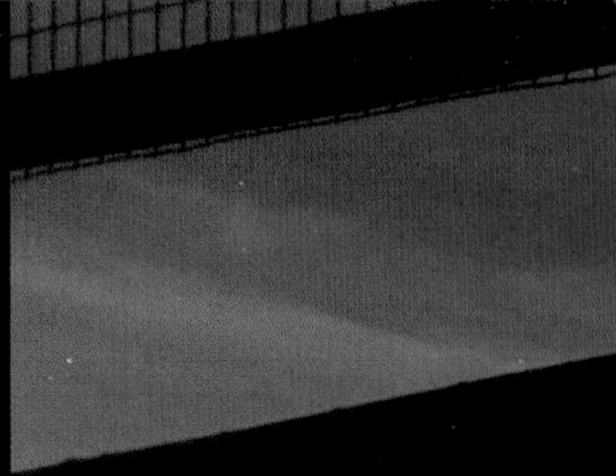

LEFT:
Alfred Eisenstaedt's photo of men overlooking Los Angeles skyglow from Griffith Park in 1936

RIGHT:
Harun Mehmedinovic and Gavin Heffernan overlooking Los Angeles from Mount Wilson in 2015

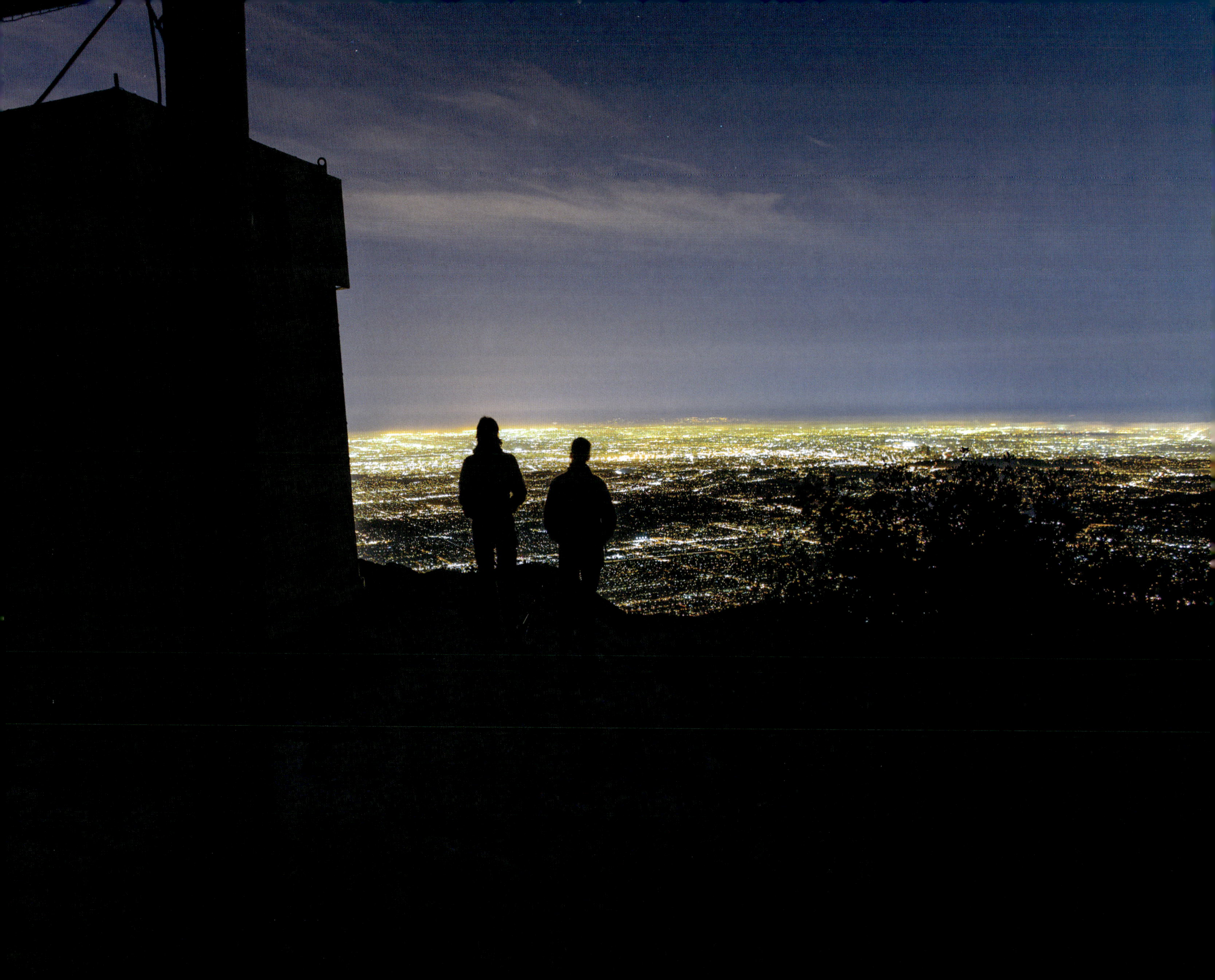

After donating a piece of land in the Hollywood Hills in 1896 with a dream that common people would have access to astronomy, Colonel Griffith J. Griffith would never live to see his Observatory built forty years later. What he did have a chance to see, shortly before his death, was the first Los Angeles municipal power pole on the corner of Pasadena Avenue and Piedmont Street, installed by the Bureau of Power and Light. Shortly thereafter, expansion of the highways brought thousands of electric street lamps, and Los Angeles soon became overrun with light. Much of this light, due to the design of unshielded outdoor fixtures installed in thousands of new homes, apartment buildings, and businesses, was pointed directly into the atmosphere.

By the time the Griffith Observatory was completed in 1935, the Colonel's dream ended just as soon as it was realized. Because of light pollution in his city, no serious scientific observation would be possible from the facility, and the public he so desperately wanted to educate about astronomy would never get to look into the heavens and observe the faint stars of the Milky Way. Instead,

Griffith Park overlook of Los Angeles, California

8

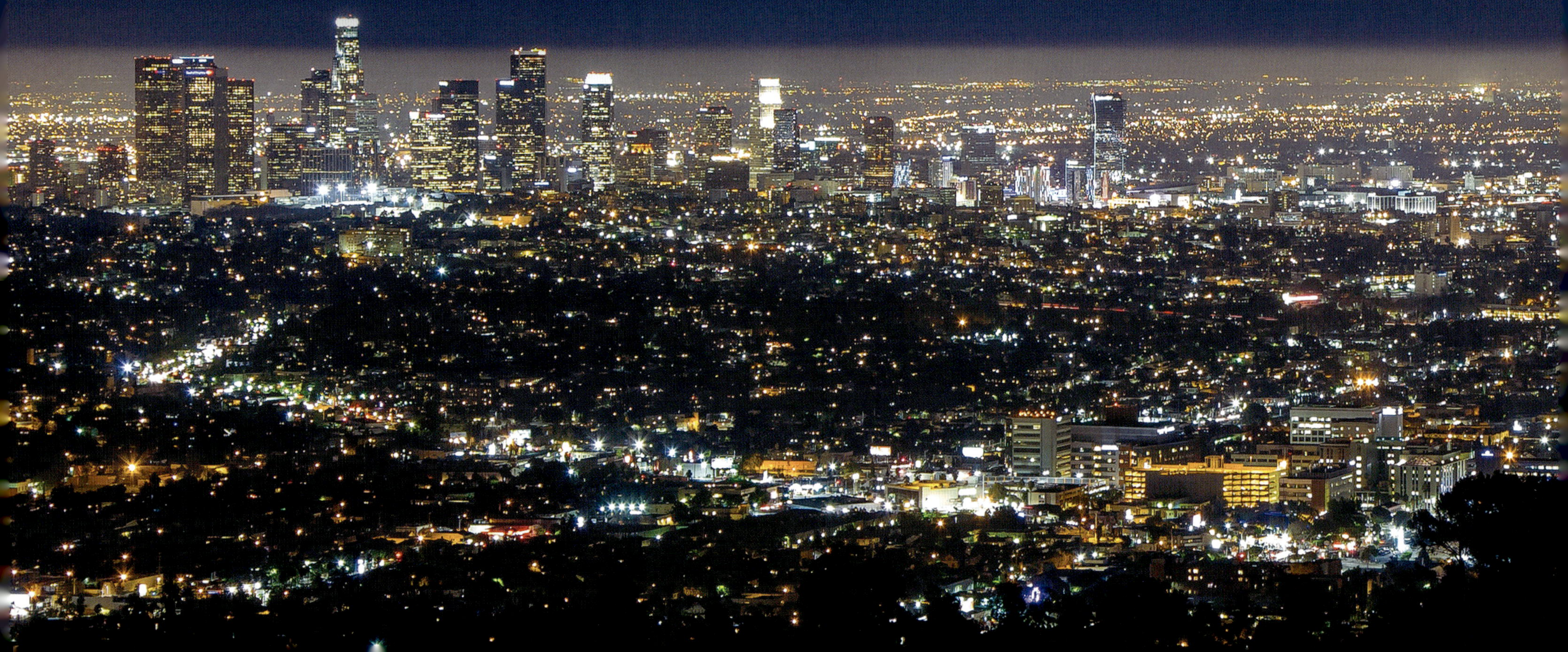

when they looked up, citizens would see bright glow emanating from the sky. This glow, caused by the electrical light bouncing around the smog, dust, oxygen particles, and water vapor in the atmosphere, meant that humans have created what is in essence artificial daylight. This phenomenon, which shortly spread around the developed world, was dubbed "skyglow."

Over the next few decades, the Los Angeles area would grow to become the most widely light-polluted region in the world, drowning out the stars from the shores of Santa Monica all the way to the suburbs of Palm Desert, and ensuring that no child would ever see the galaxy from front porches anywhere between Tijuana and San Francisco. What remained of the dark skies that inspired the imaginations of our ancestors and gave birth to our civilization became confined to the desert areas of eastern Joshua Tree and Death Valley. As the Los Angeles area expanded and Las Vegas shortly followed as a spectacle of light, those very few dark-sky areas continued to shrink each year.

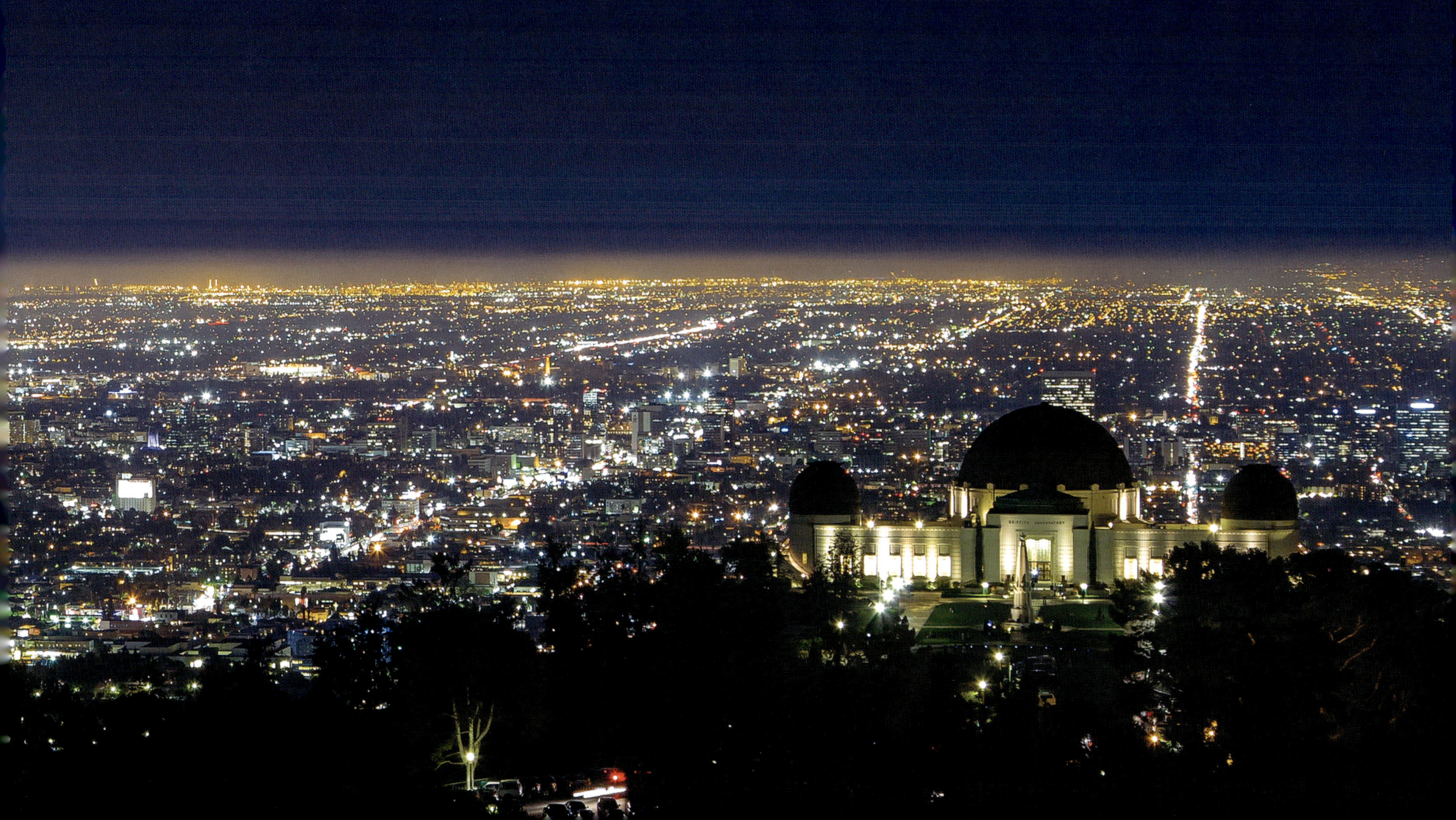

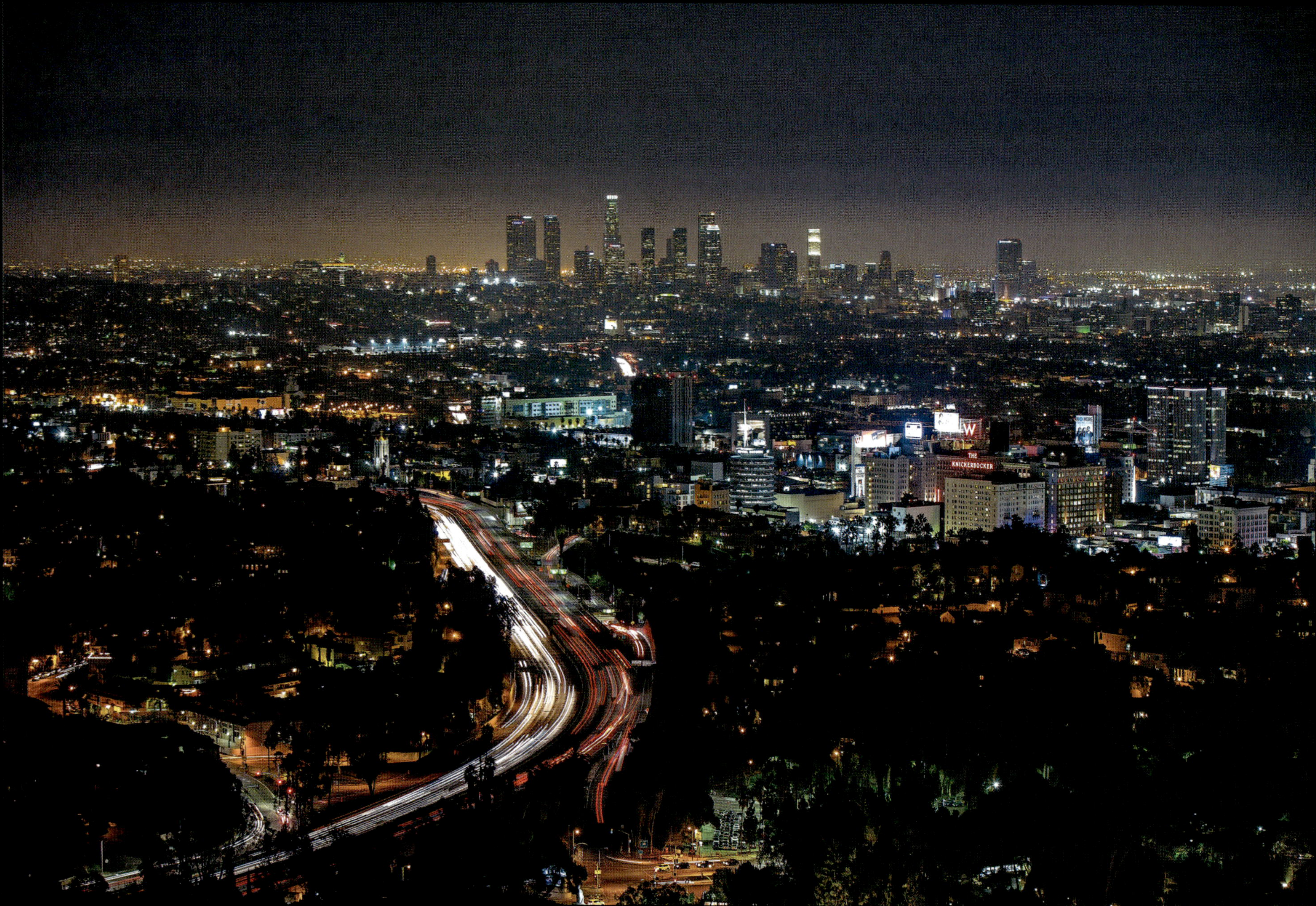

Hollywood Bowl overlook of Los Angeles, California

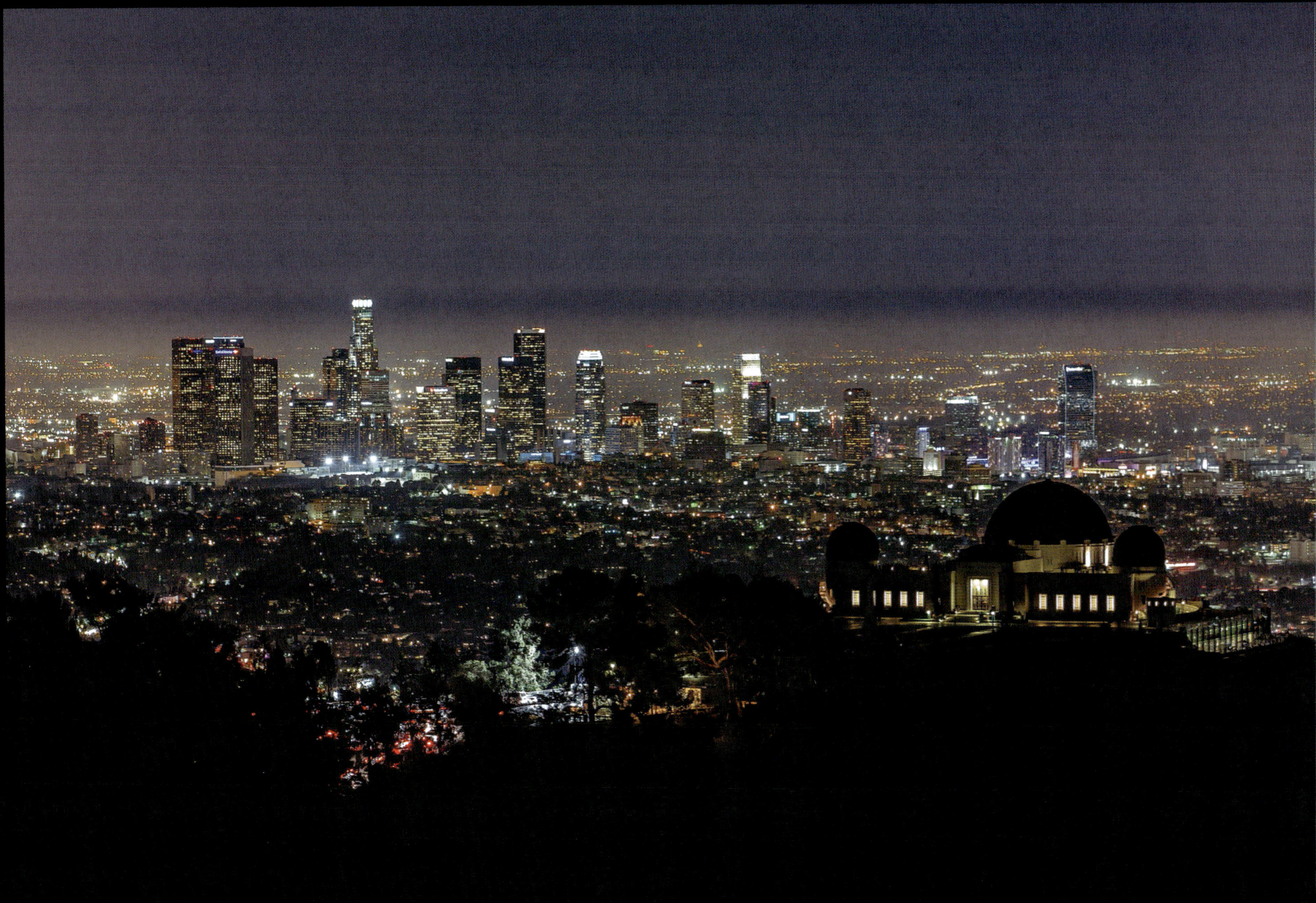

Today, Griffith Observatory towers over the increasingly bright Los Angeles panorama.

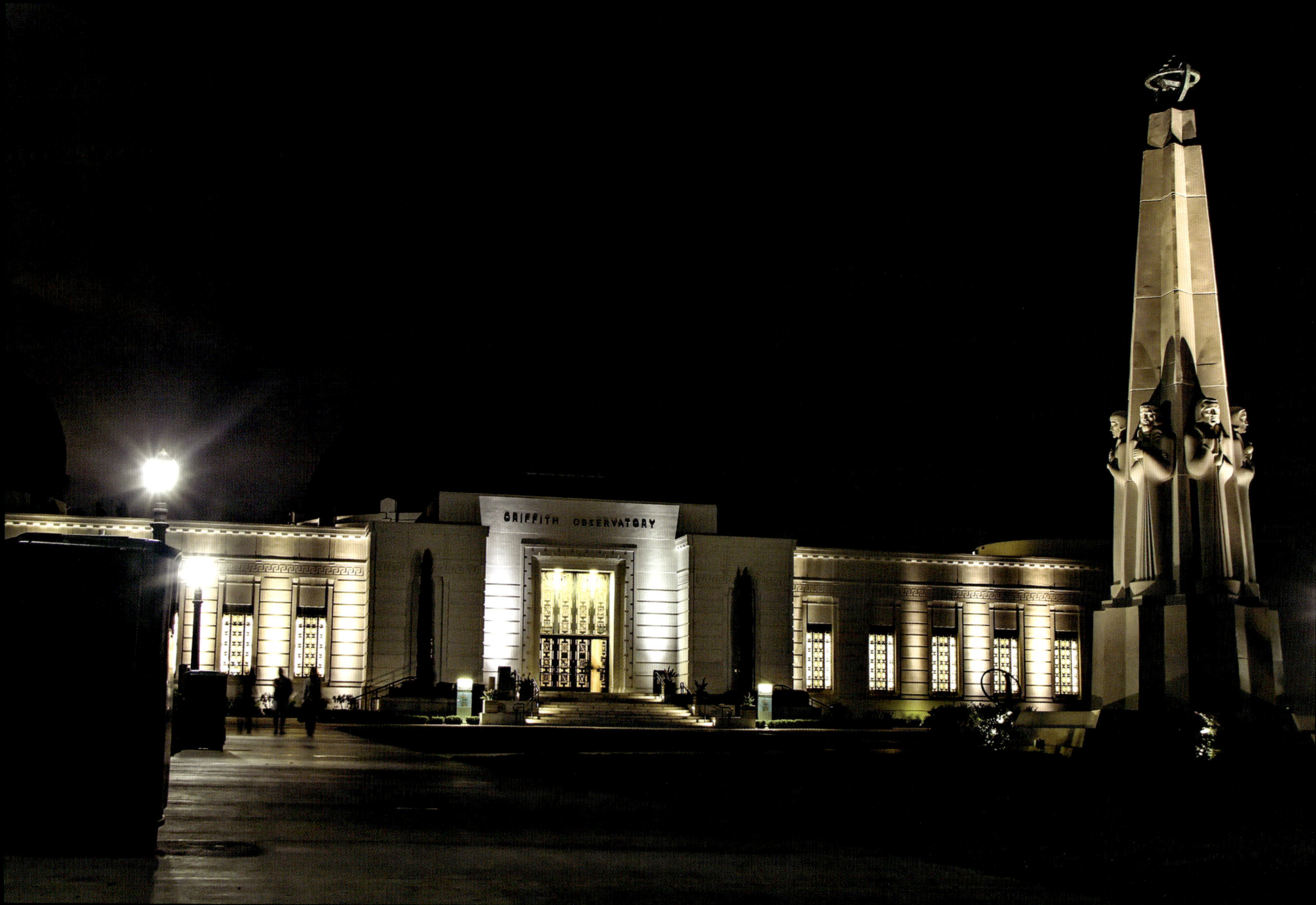

The extensive lighting of the Griffith Observatory building itself immediately signals to visitors the doomed state of observational astronomy there.

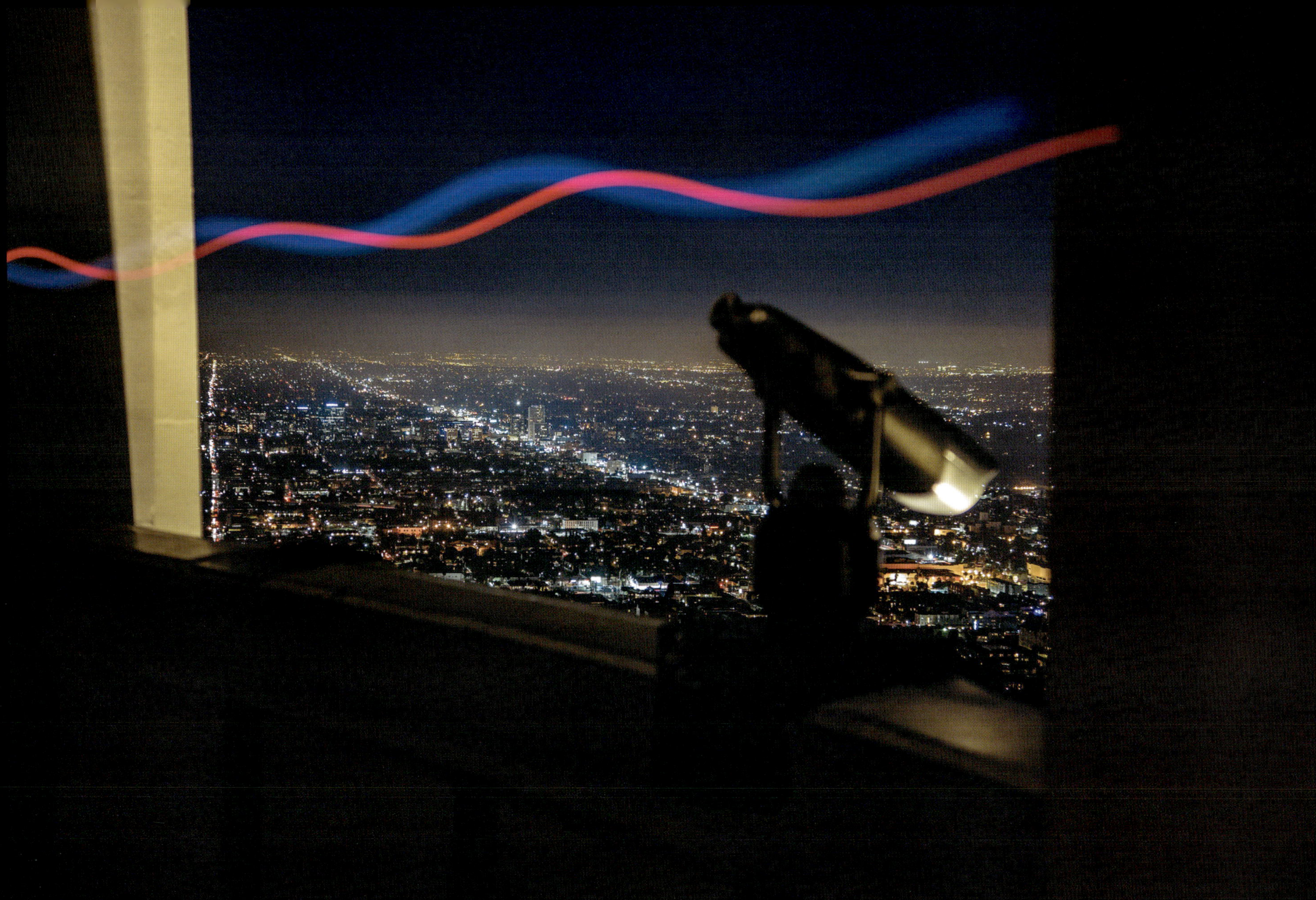

With no stars in sight, all that's left to see are billions of shining bulbs in the city below.

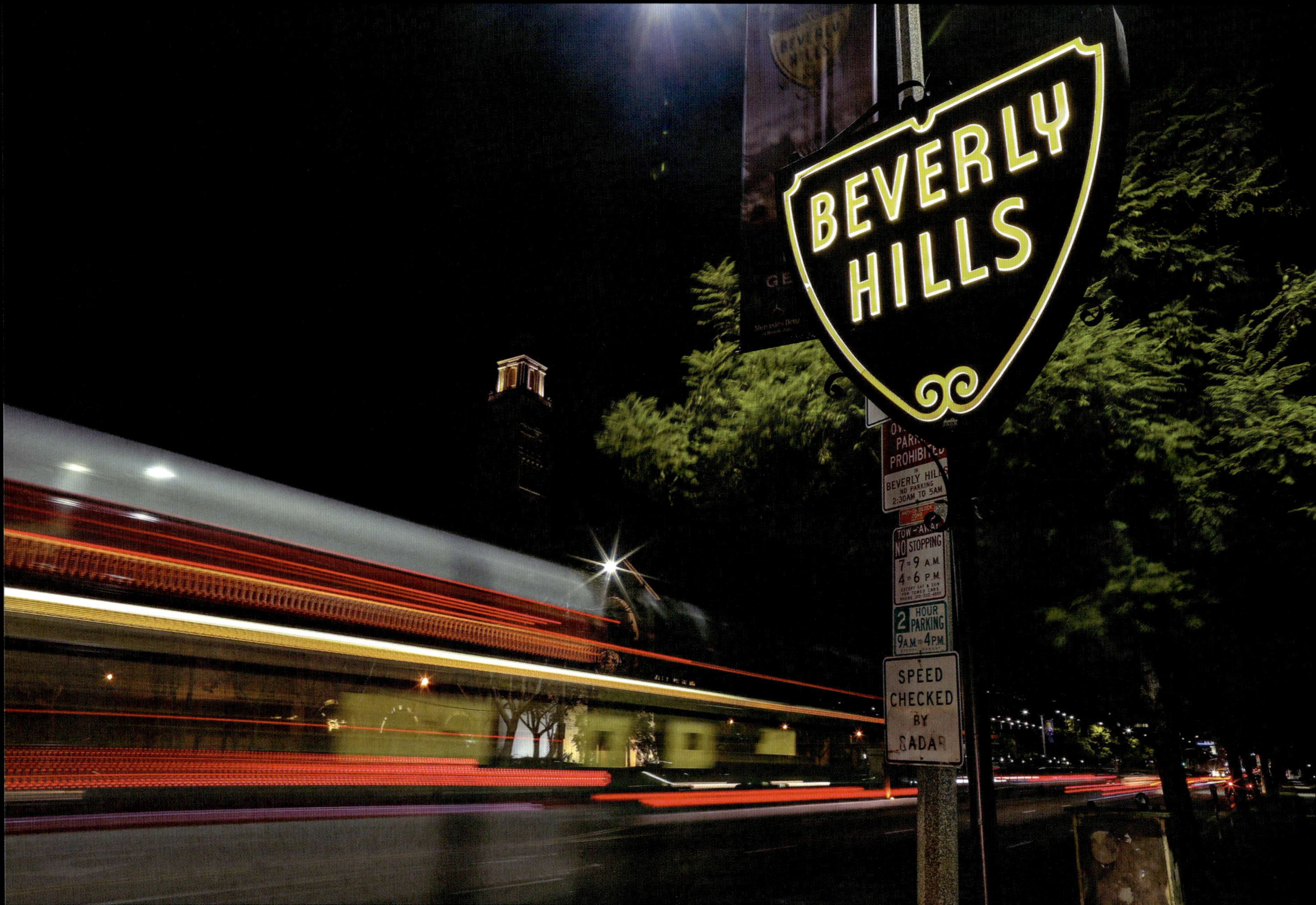

Beverly Hills, California

9

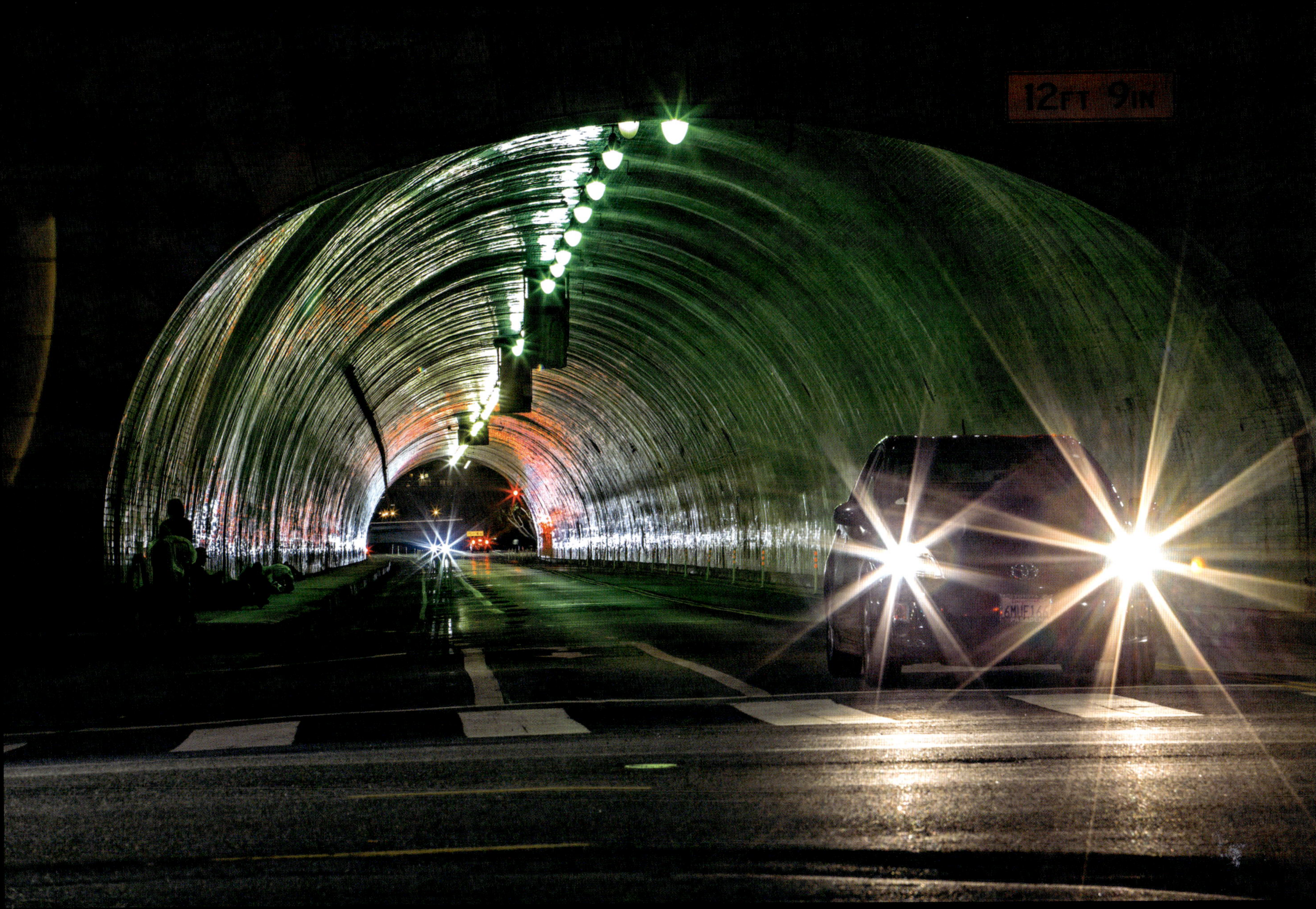

2nd Street Tunnel, Downtown Los Angeles, California

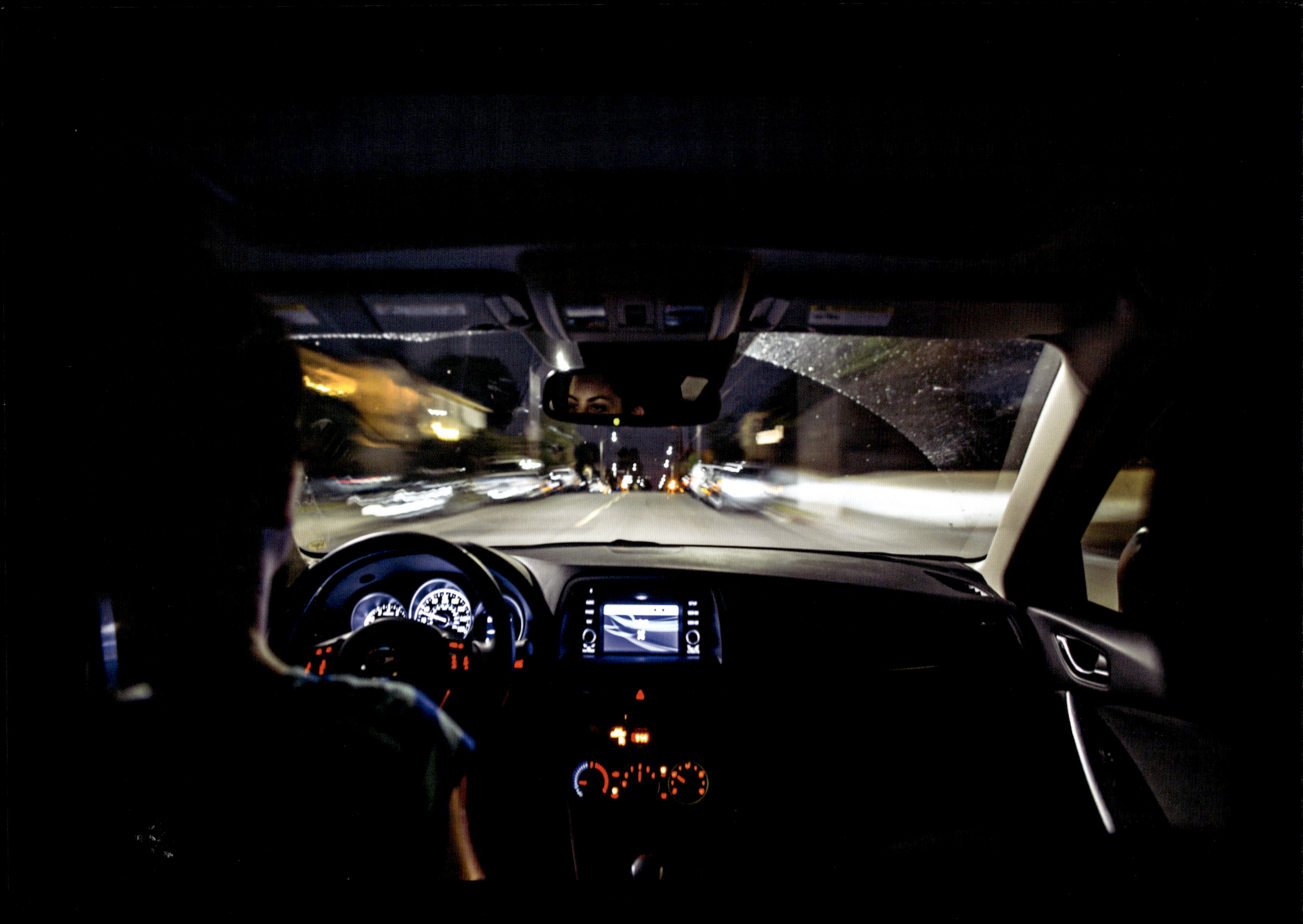

Downtown Los Angeles, California

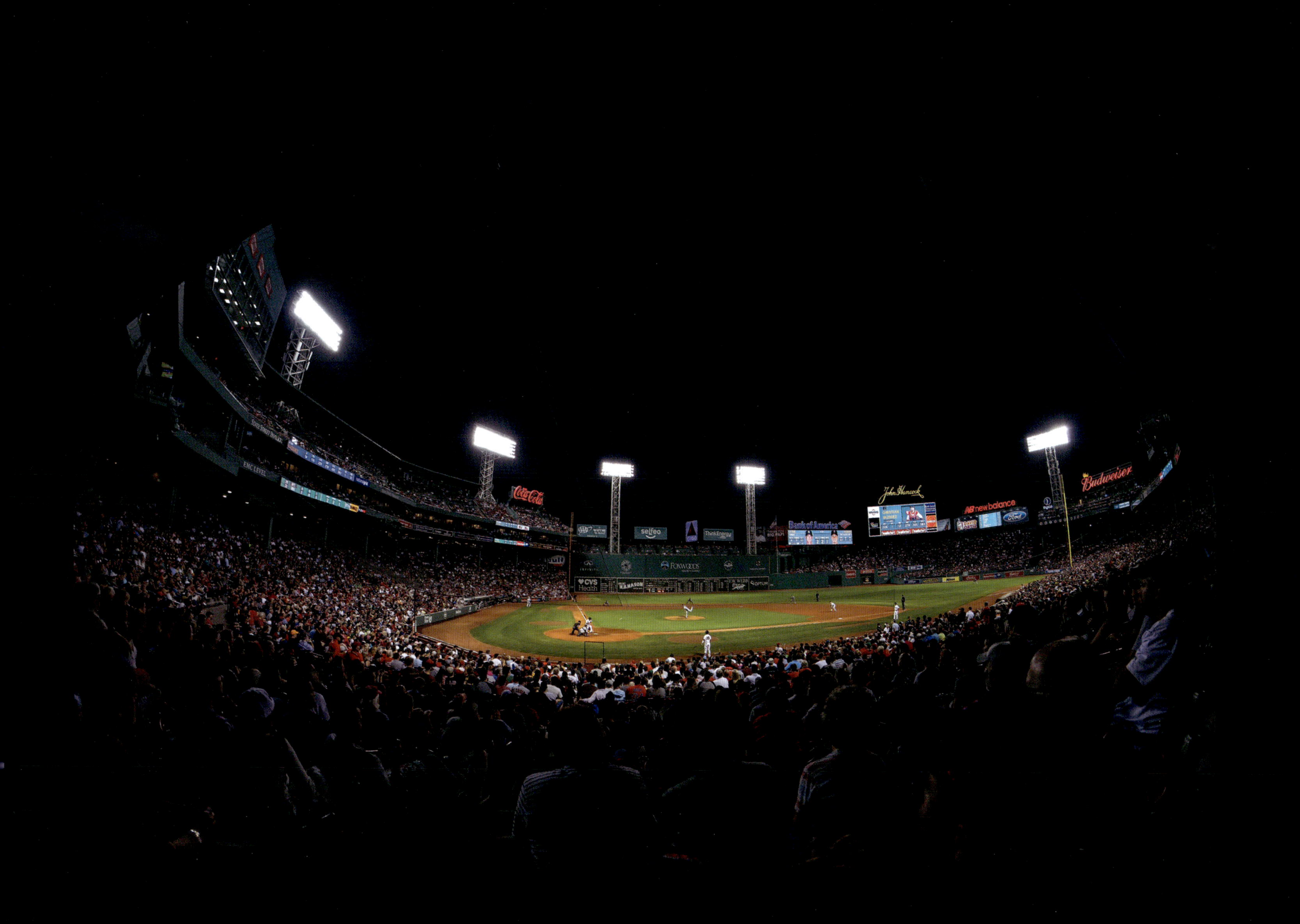

Fenway Park, Boston, Massachusetts

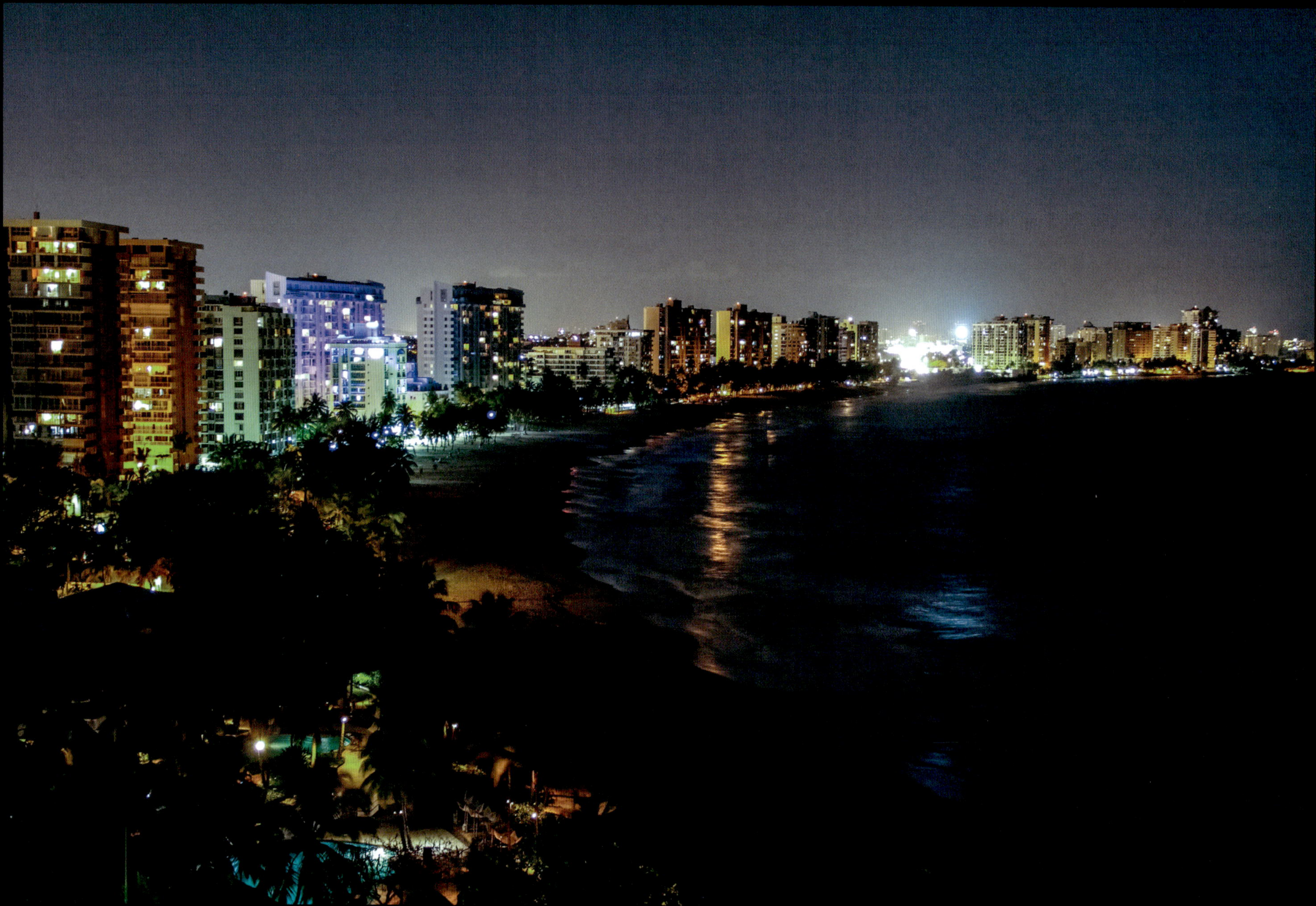

Even escaping to the beaches often isn't enough to outrun the artificial light that erases thousands of faintly shining stars above.

Playa de Isla Verde, San Juan, Puerto Rico

6

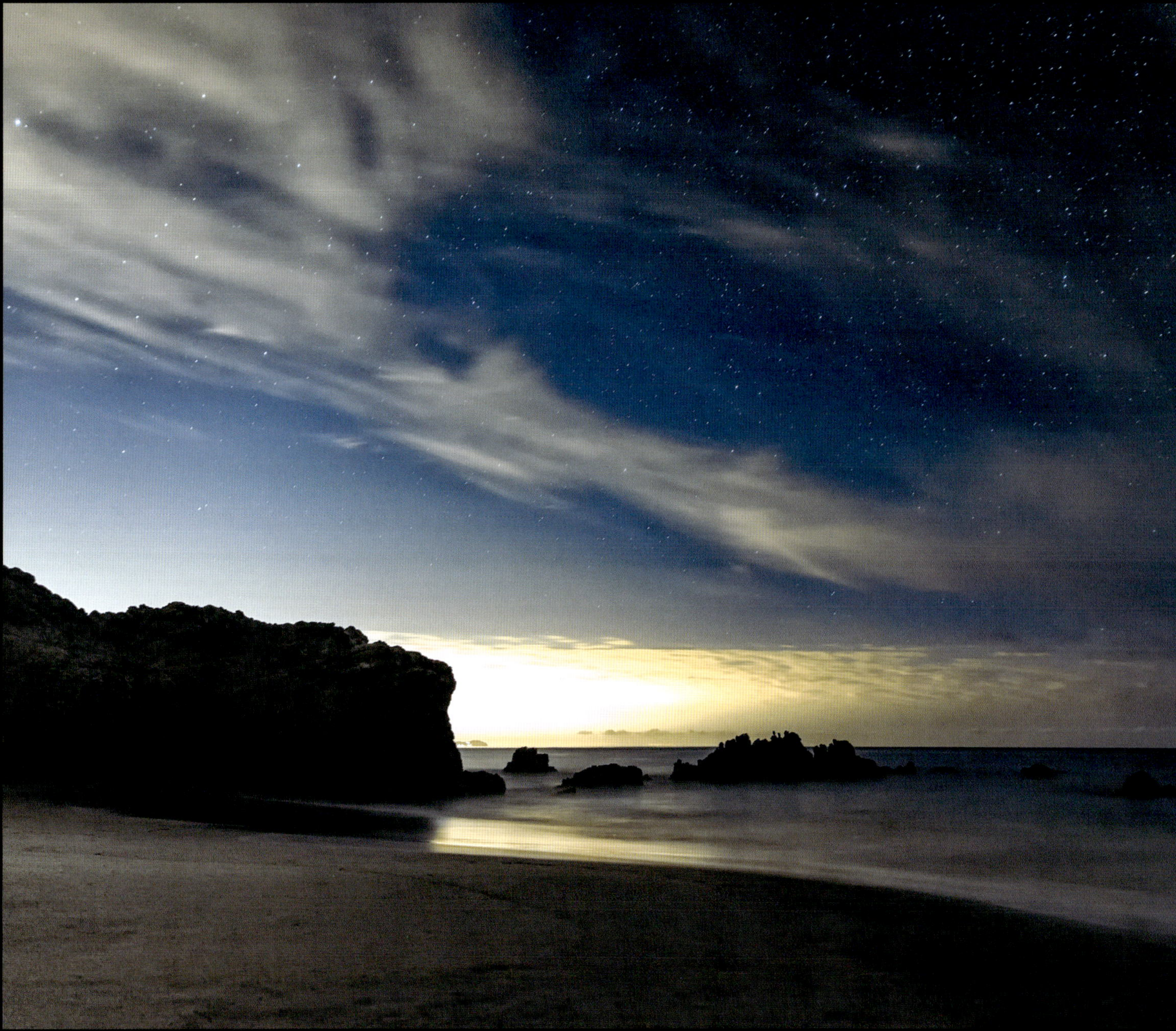

Offering an almost real-life demonstration of the Bortle Scale, Malibu views towards Los Angeles and the Pacific Ocean
gradation from perfectly dark skies over the distant ocean to what appears to be a nuclear mushroom cloud o

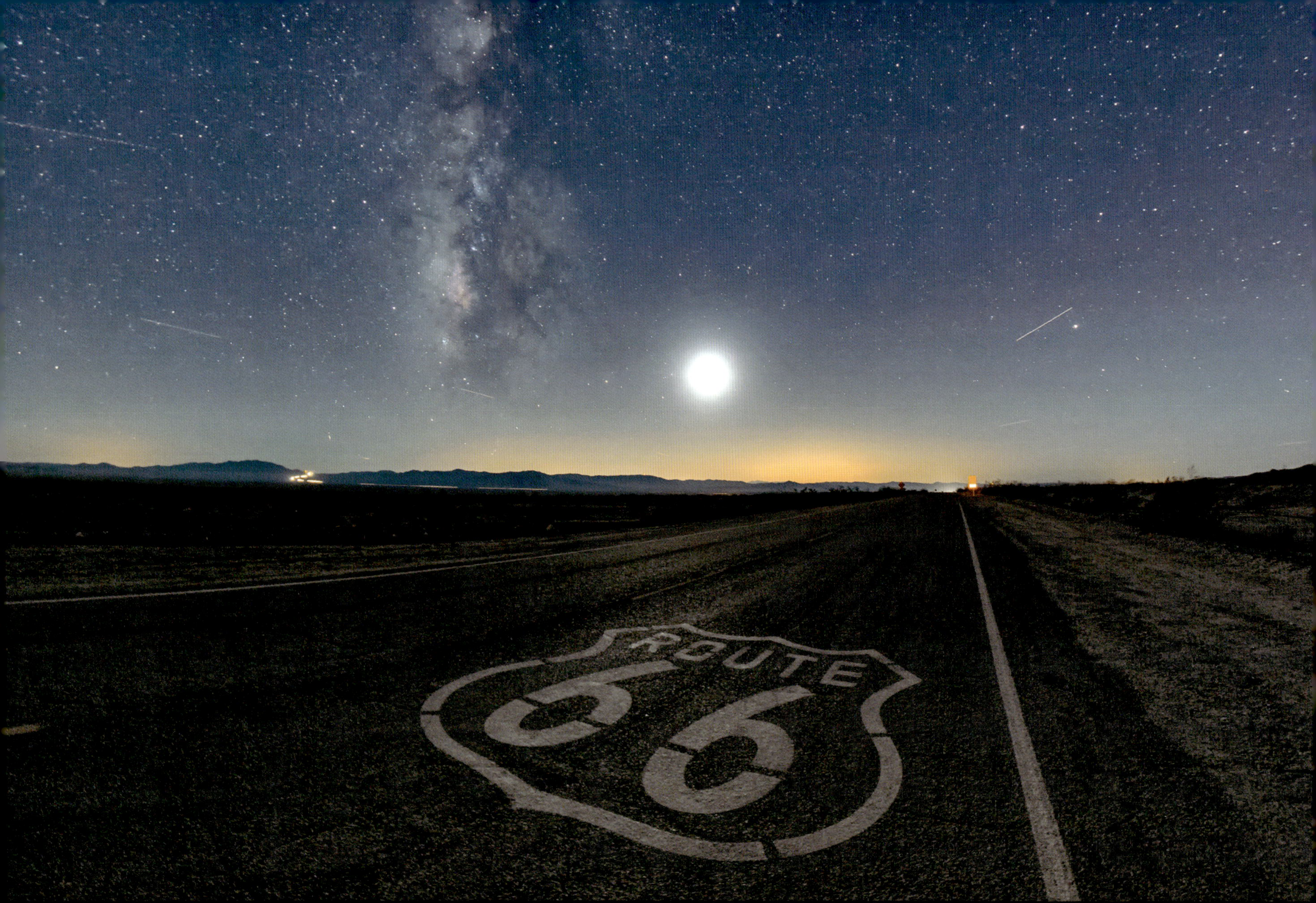

Distant Los Angeles Area Light Pollution from Route 66, Amboy, California

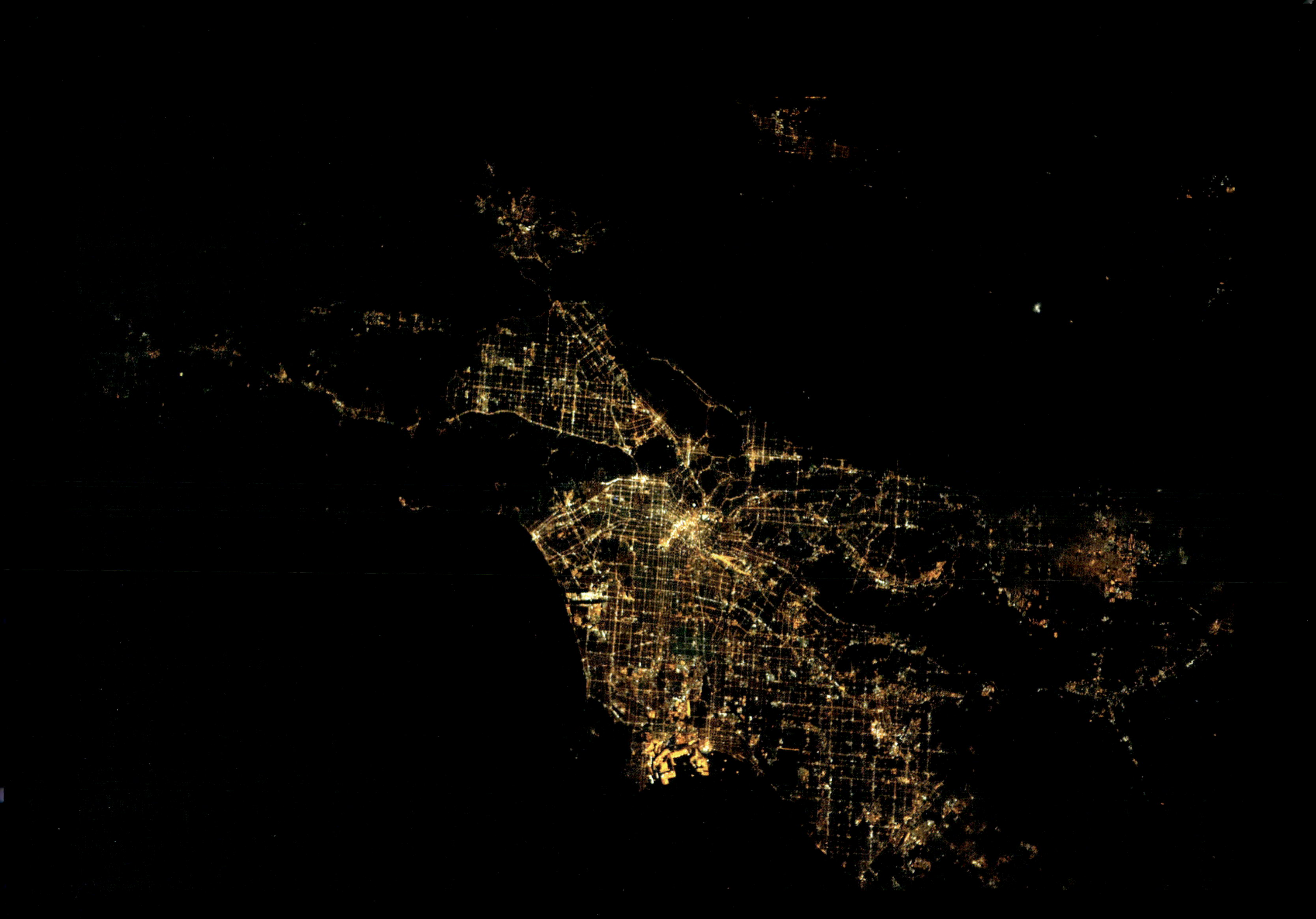

International Space Station view of Los Angeles
Image courtesy of NASA

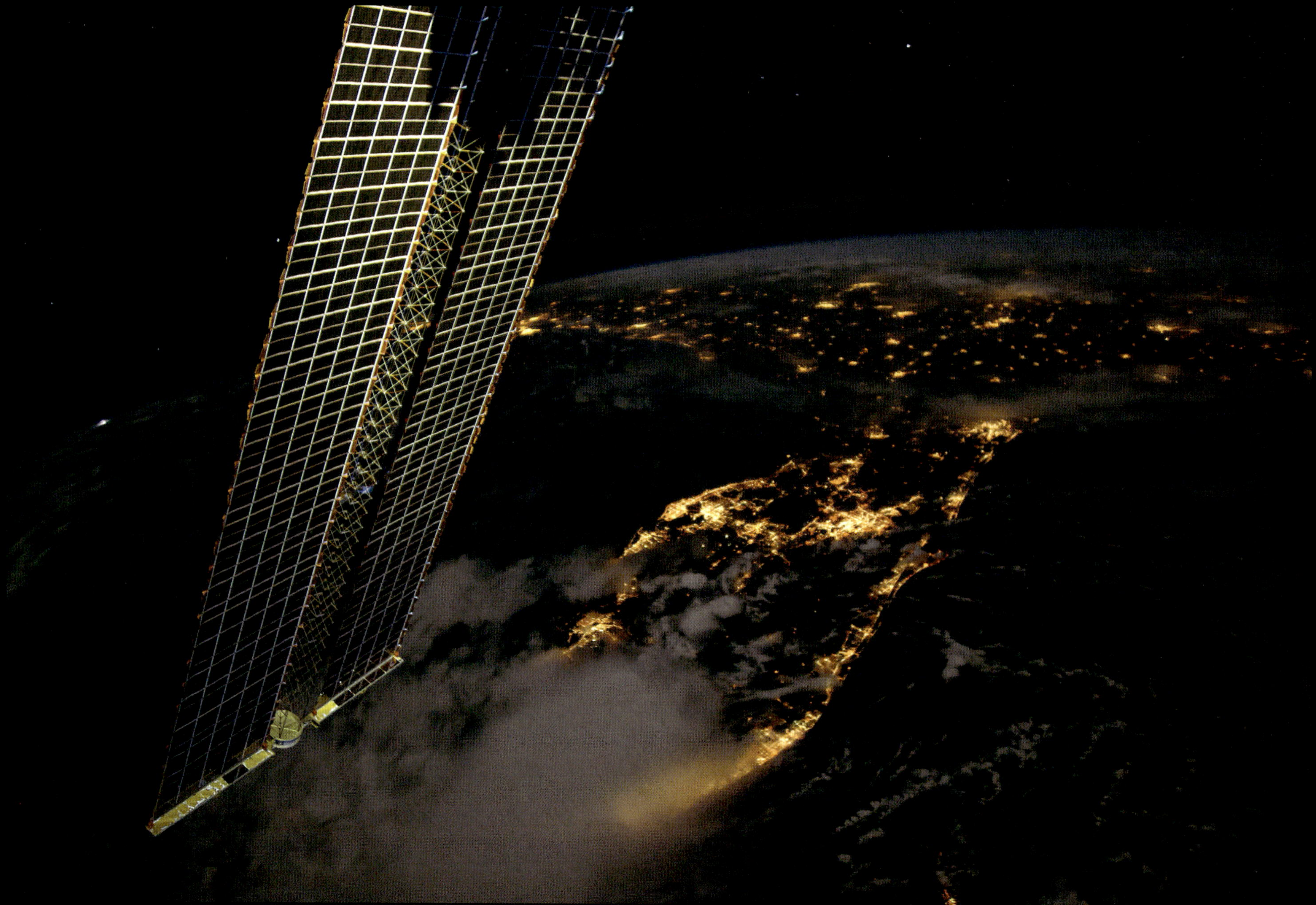

International Space Station view of Southern United States
Image courtesy of NASA

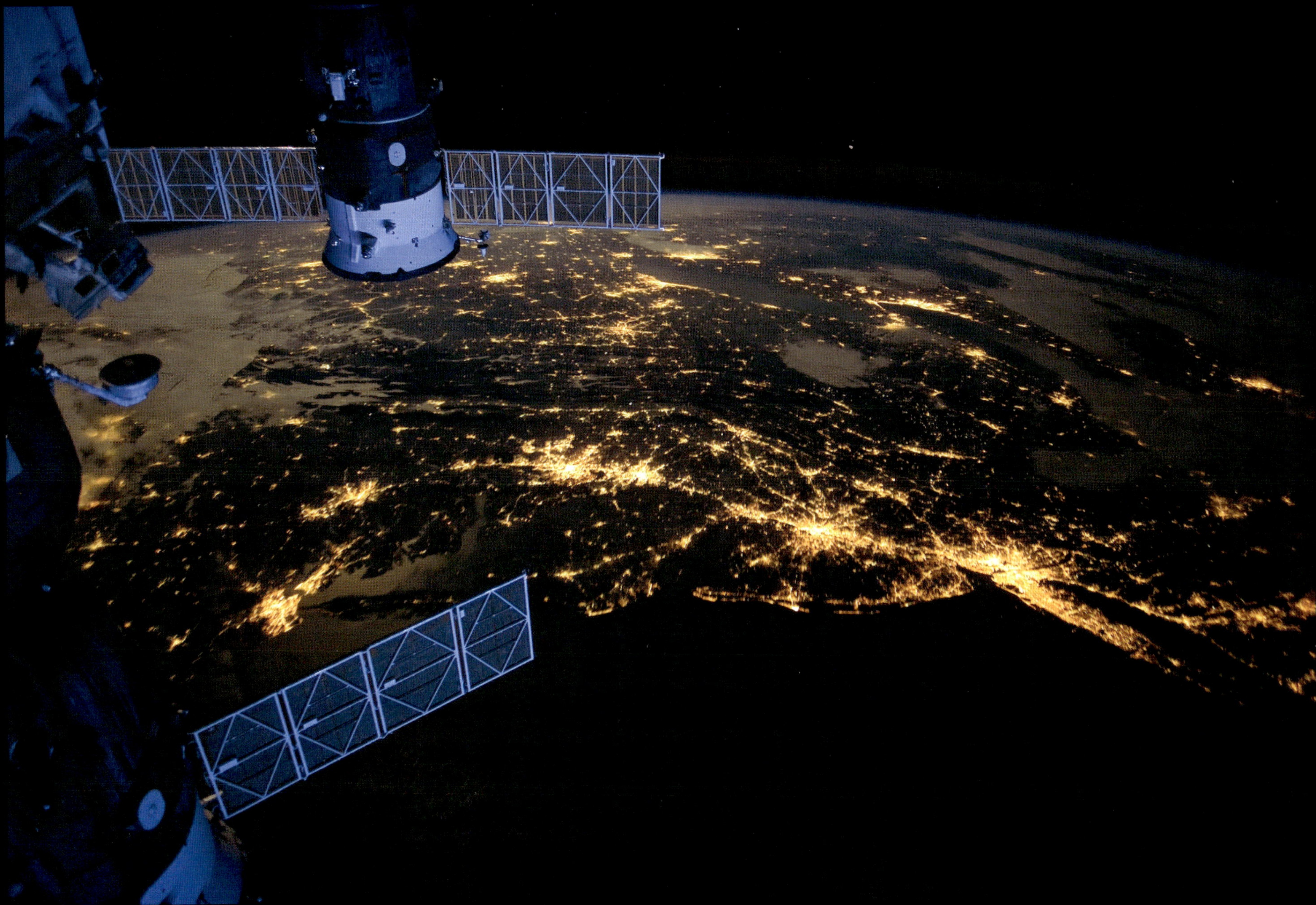

International Space Station view of United States Eastern Seaboard
Image courtesy of NASA

The White-Light Night

By Eric Betz

Gazing down from the International Space Station, thousands of tiny, shimmering suns make shapes on the surface of the Earth. They cluster along coastlines and follow rivers, forming grids across the largest metropolises. Streetlights have given the sky a warm glow since before astronauts could marvel at it from above. But recently, the familiar golden hue from big cities has shifted toward a brighter shade of white. It's dawn for a new technology.

In places like New York and Los Angeles, hundreds of thousands of LED streetlights have visibly brightened cityscapes even from orbit. These lights are bright, cheap and energy-efficient. Experts predict the technology will dominate the lighting industry by the end of the decade, making light pollution worse.

Every creature evolved with a 24-hour cycle of day and night. And much of that natural cadence has disappeared. Artificial lights are all around us — in our homes as well as out our bedroom windows. Humans now bathe in the glow of myriad electronics at all hours. Our office computers greet us in the morning; smartphones and televisions rock us to sleep. Street and porch lights burn along empty roadways on a nightly vigil. The luminescence leaks from our homes, businesses and highways, altering the night for the creatures we share the planet with.

Light pollution now expands its reach by some 6 percent each year. And beyond the loss of the night sky, there's a growing body of scientific evidence that suggests conquering the night is dangerous for both humans and the environment. Like many species, humans evolved for millions of years with the regular cycles of dark and light caused by Earth's rotation. And biologists have now shown that exposure to artificial light alters the body's circadian clock.

Sleep disorders are an obvious consequence. Historians note that before electricity, people commonly slept in two 4-hour shifts — a "first sleep" and "second sleep." In between was an hour or more of quiet wakefulness. Humans exposed to a more natural day-night cycle will still revert back to this two-shift slumber.

But this cycle also controls as much as 15 percent of our genes, which means the health repercussions of keeping the lights on can spread far beyond feeling groggy at work. Even a small amount of light at night can suppress how much melatonin the body releases. This vital hormone helps regulate circadian rhythms.

Scientists are now exploring links between light pollution and everything from cancer and diabetes to heart disease and obesity. Exposure to light at night can even accelerate breast cancer tumor growth, according to a few studies. The American Medical Association now recommends avoiding screentime at night, as well as limiting shift work and cutting down light pollution in communities.

LEDs compound the problem. Most city-dwellers are familiar with the traditional yellow hues of low- and high-pressure sodium-vapor streetlamps, which shine in a narrow range. But broad-spectrum LEDs are the future. The brighter and more efficient streetlights can save cities millions of dollars on their power bills, so municipalities big and small are starting to convert.

The most common bulbs are blue with a special coating that makes them appear white to human eyes. This also causes these LEDs to peak in the short wavelengths astronomers commonly use to observe

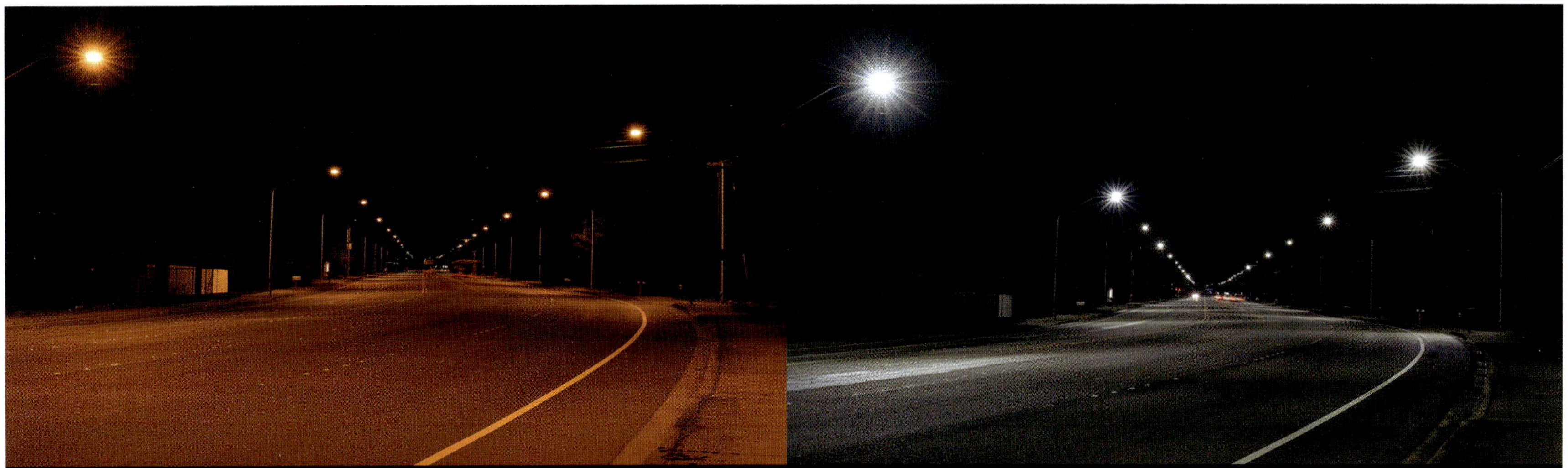

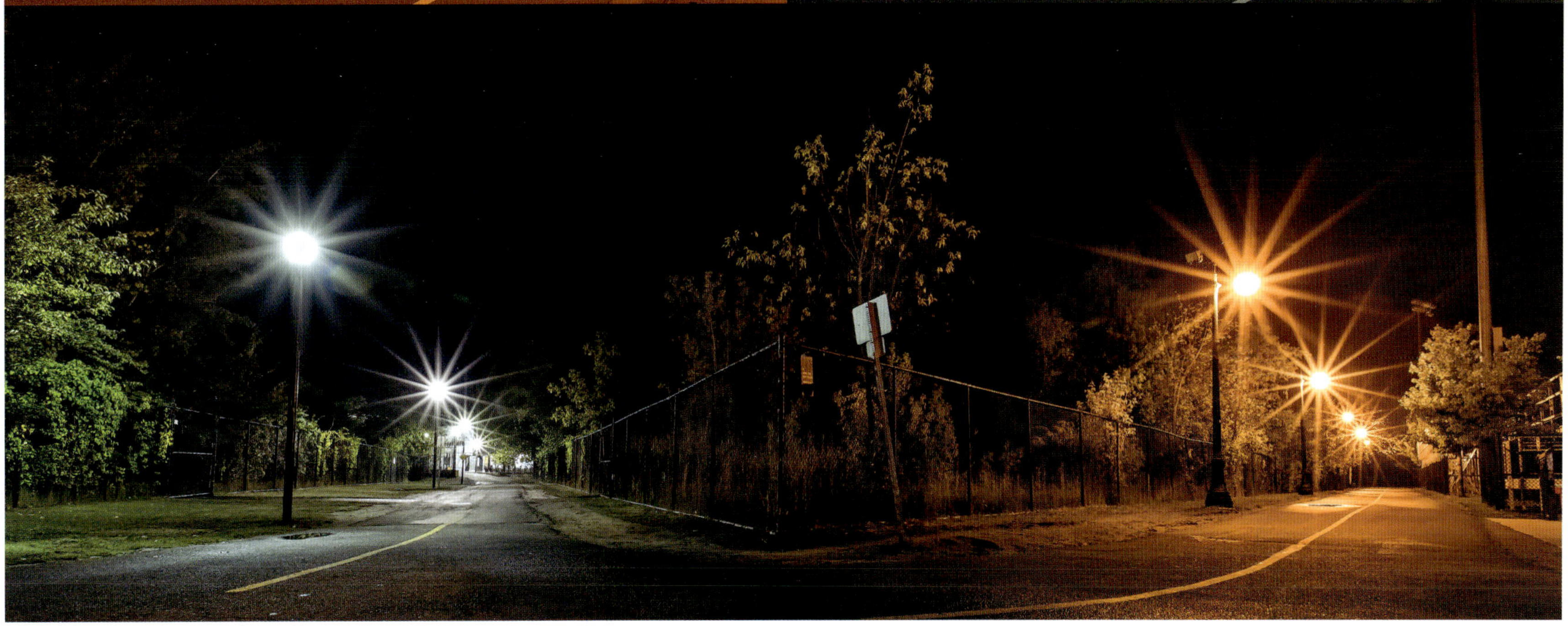

Before (left) and after (right) images of Bellingham, Washington, showing the conversion from fainter, yellow-colored temperature sodium-vapor lights to brighter white/blue LEDs.

Alewife Linear Park, Cambridge, Massachusetts, lit half by LED (left) and half by sodium-vapor lights (right).

TOP: **Bellingham, WA**

BOTTOM: **Cambridge, Massachusetts**

6

8

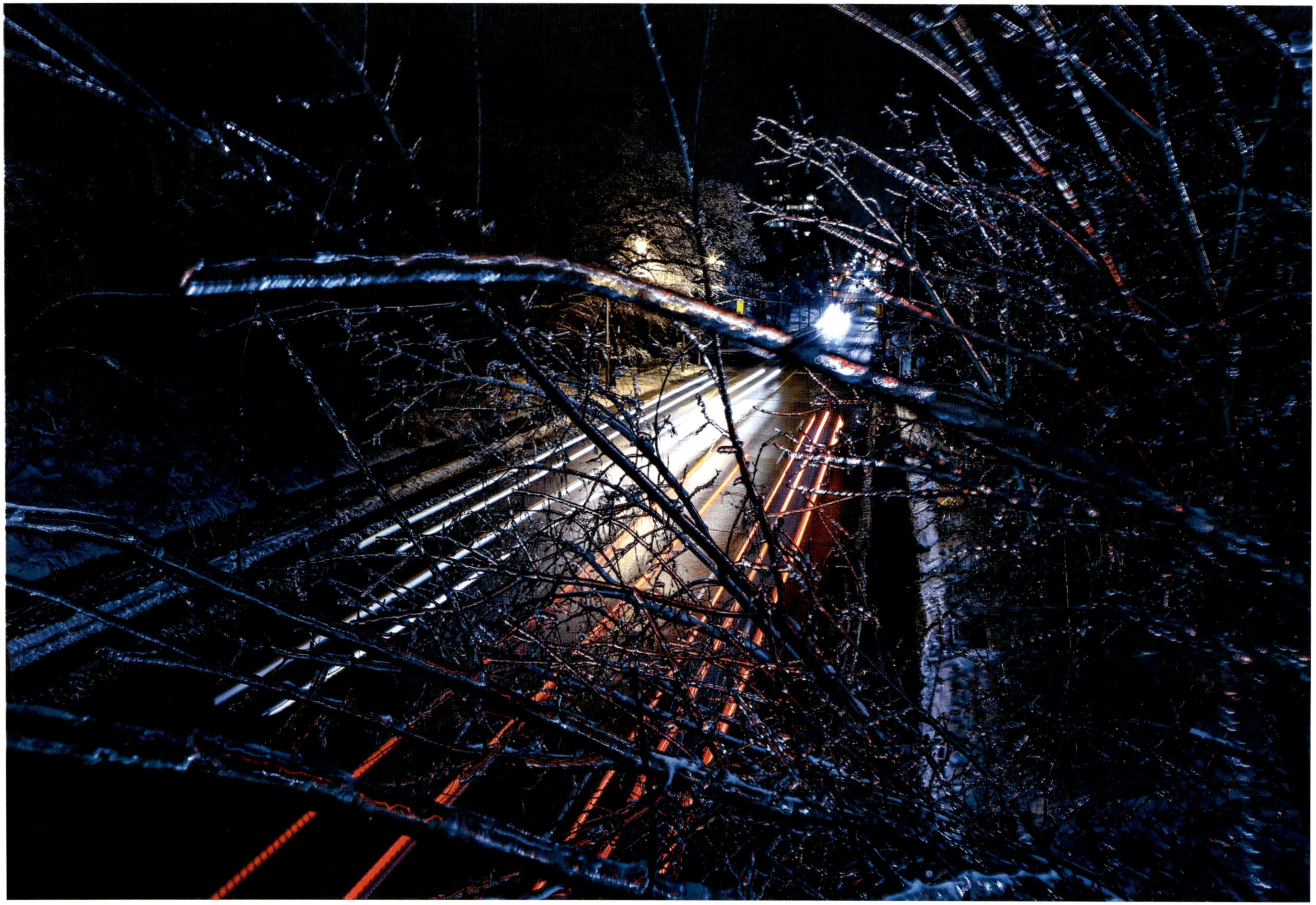

Sodium Vapor-lit Toronto street before the LED retrofitting.

Toronto, Ontario, Canada

the night sky. And these luminaries radiate their light over much of the electromagnetic spectrum and scatter much farther through the atmosphere, even when shielded. Such blue light has been shown to improve mood and increase attention and reaction times during the day. But it's disruptive at night. Studies show blue light suppresses sleep-inducing melatonin more than other lights.

This light pollution is as troubling to biologists as it is to dark-sky advocates. Around one-third of vertebrates are nocturnal. Twice as many invertebrates need the night. Endangered sea turtles can lose their way on the coast. In cities, birds crash into buildings. Fireflies struggle to mate.

The disruption from light pollution was already a concern, but the brighter LEDs can attract and kill higher numbers of certain sensitive insects. That can mean more pest insects in residential areas, like midge swarms. But more importantly, it impacts nighttime pollinators that provide vital services to ecosystems.

A recent study from the United Kingdom looked at rural moths and those that live in cities. The scientists found that artificial lights kill moths so often that it's already having evolutionary consequences. The survivors have adapted to be less attracted to city lights, but they also fly significantly less than their rural counterparts, providing less pollination.

The light pollution tricks trees too. Researchers recently examined more than a decade of records logging when trees got their first buds. They compared the data with satellite imagery of nighttime lighting, springtime temperatures and even accounted for the urban heat island effect, where cities warm up their environs. The scientists found that three out of the four tree species they examined grew buds earlier in areas with more light pollution.

One species, the European Ash, budded an average of 7.5 days early. That doesn't only impact the tree. The birth cycles of insects like caterpillars are timed to those first buds, so the authors suspect these changes likely have ecosystem-scale impacts.

Incredibly, artificial lights even have an impact underwater. LED lamps installed in marinas and harbors have been shown to discourage filter feeders like coral to settle in an area and encourage settlement by nuisance species that like to anchor themselves to boat hulls.

Researchers are now worried that whole ecosystems could be at risk from this "white-light night."

Sadly, solutions have so far been a hard sell. Cities need to install filters that can remove the LEDs' blue wavelengths, as well as shields that better aim the lights down. Those fixes are costly. But there are some potential positives to LEDs that aren't yet widely utilized. The latest bulbs are tunable, which means streetlights might someday make use of warmer, more natural amber colors. Indoor lights and electronic displays can change their hues throughout the day, keeping you more attentive when you need to be, but dialing back the blue as the sun sets.

And the high-tech lights are also easily programmable. In Los Angeles, city officials say they can now control hundreds of thousands of streetlights from a single laptop. It's on dark-sky advocates, doctors and biologists to push officials to use their newfound technology to turn off the excess lights. If they don't, more than just dark skies are at risk.

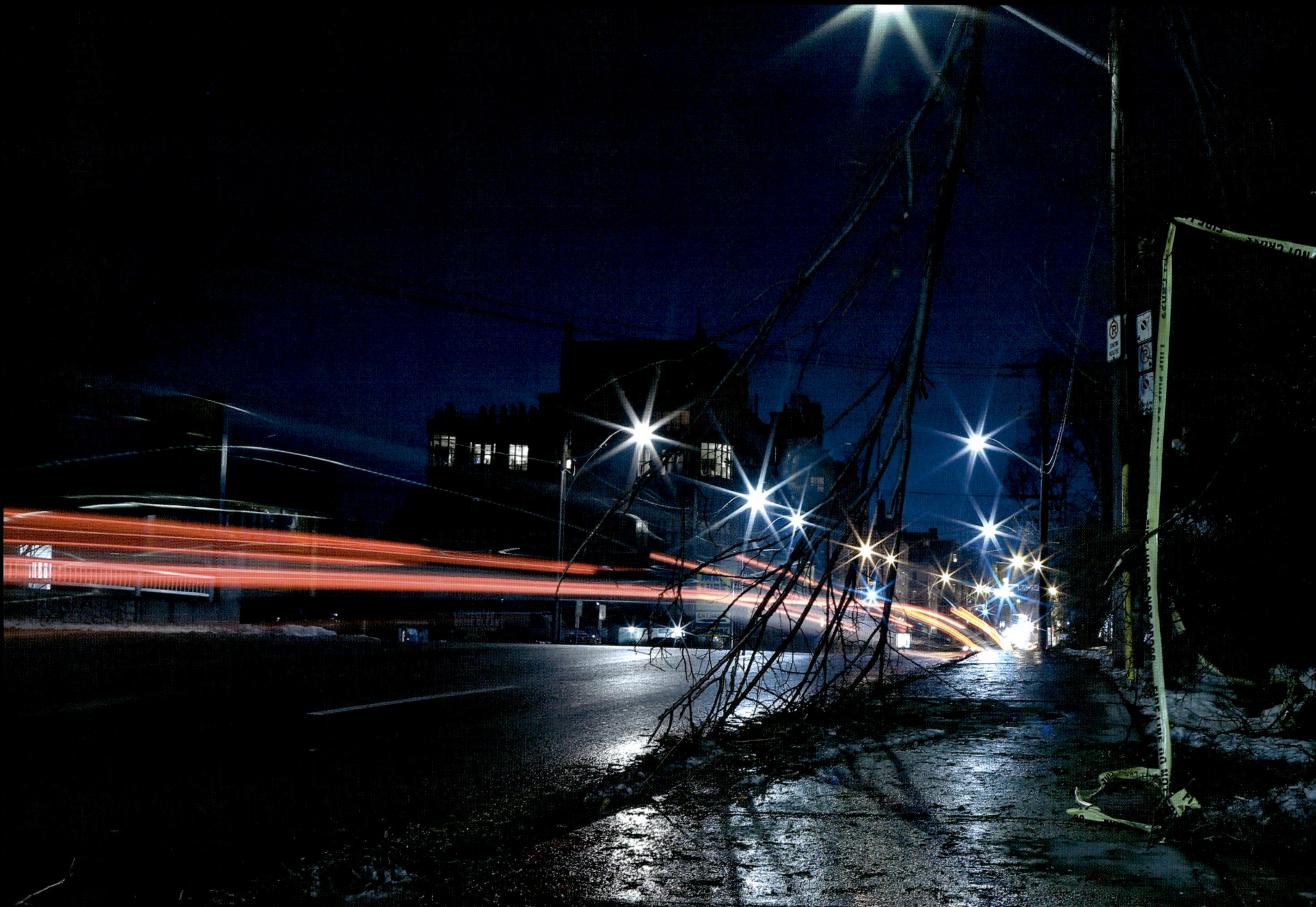

A Toronto street undergoing LED retrofitting, lit half by sodium-vapor lights and half by LED.

Toronto, Ontario, Canada

9

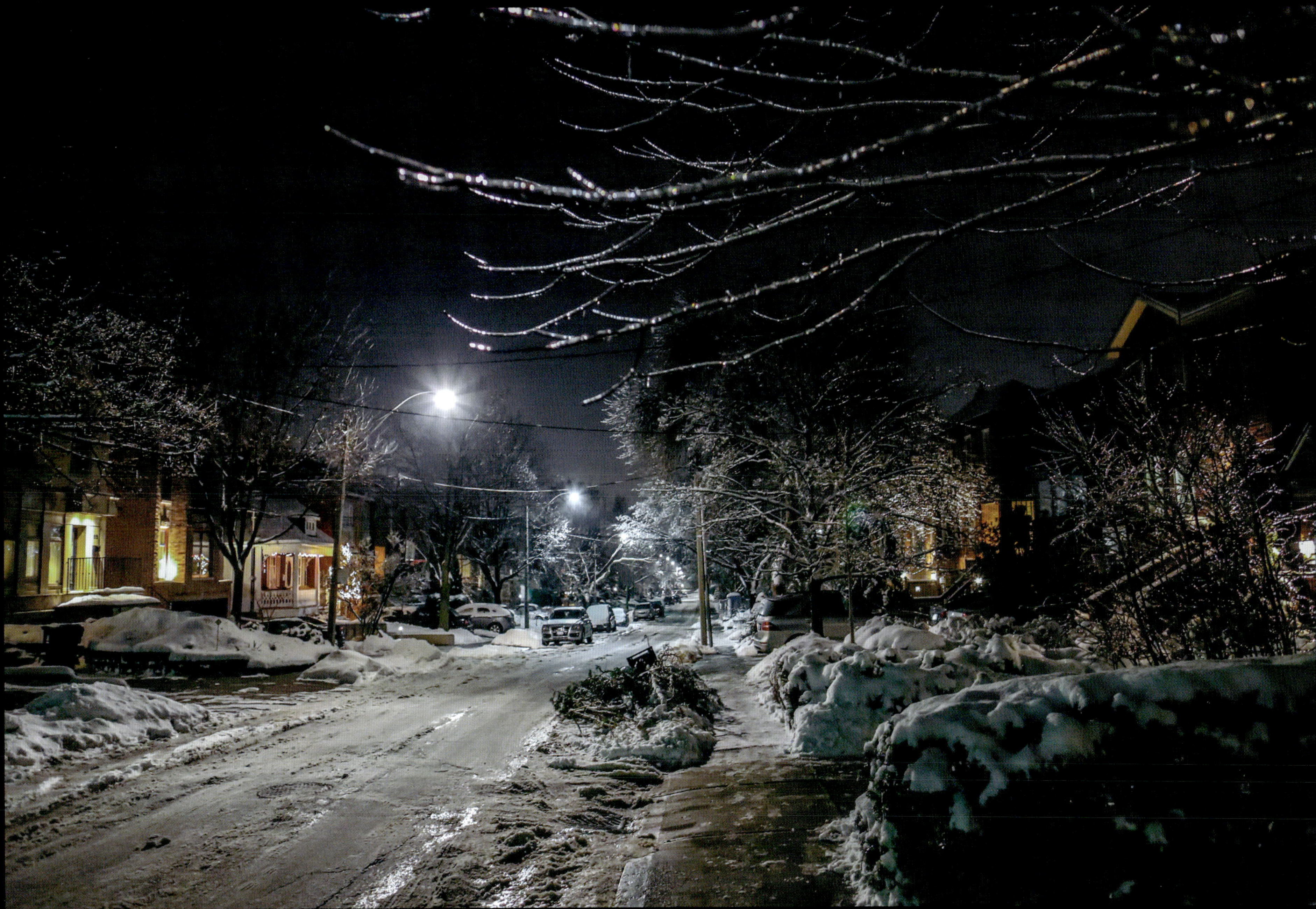

Residential Toronto street after LED retrofitting.

Toronto, Ontario, Canada

9

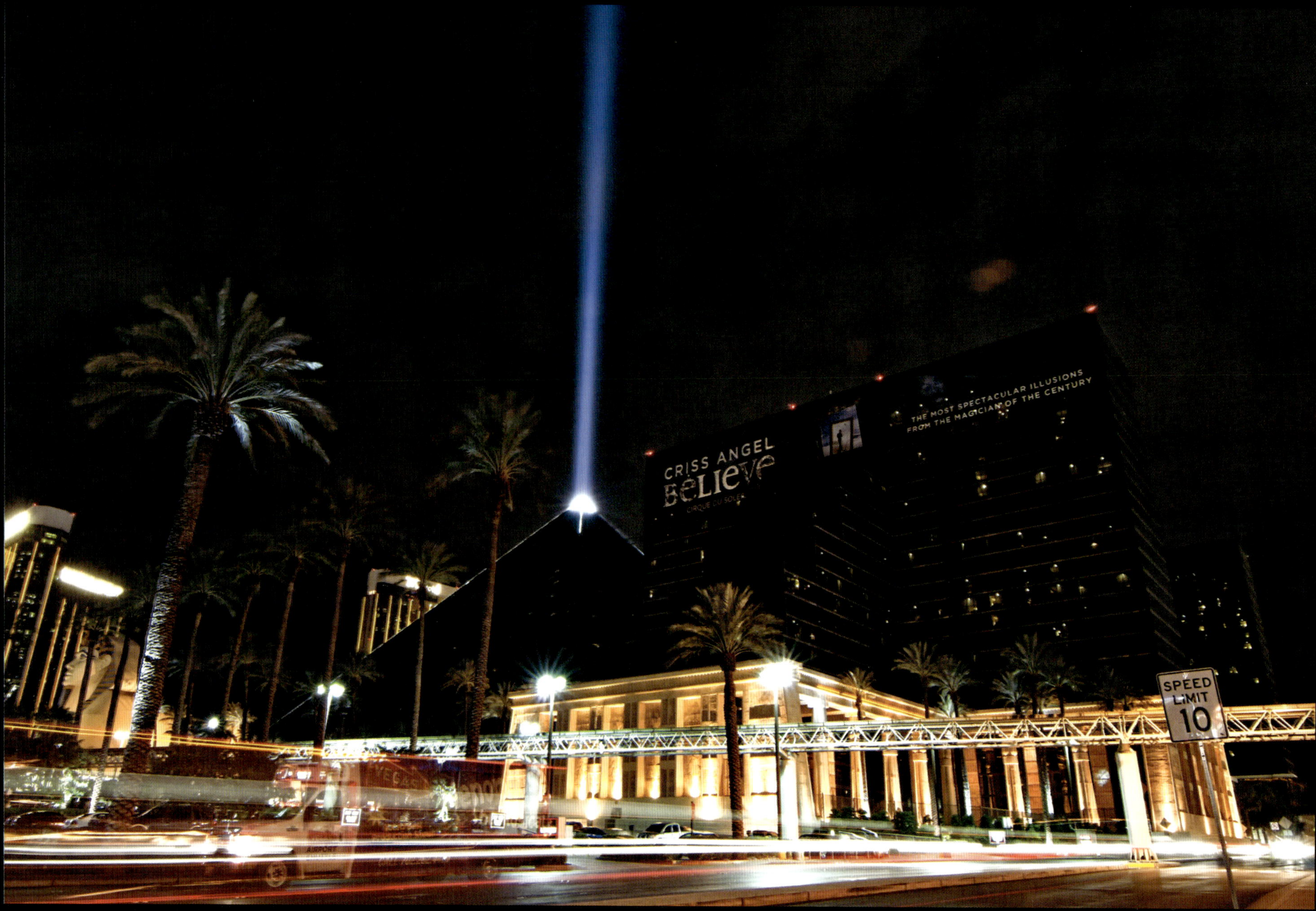

The single-brightest light pollution spot on Earth is the spotlight beam rising from Luxor Casino in Las Vegas.

Luxor Casino, Las Vegas, Courtesy of Getty Images

9

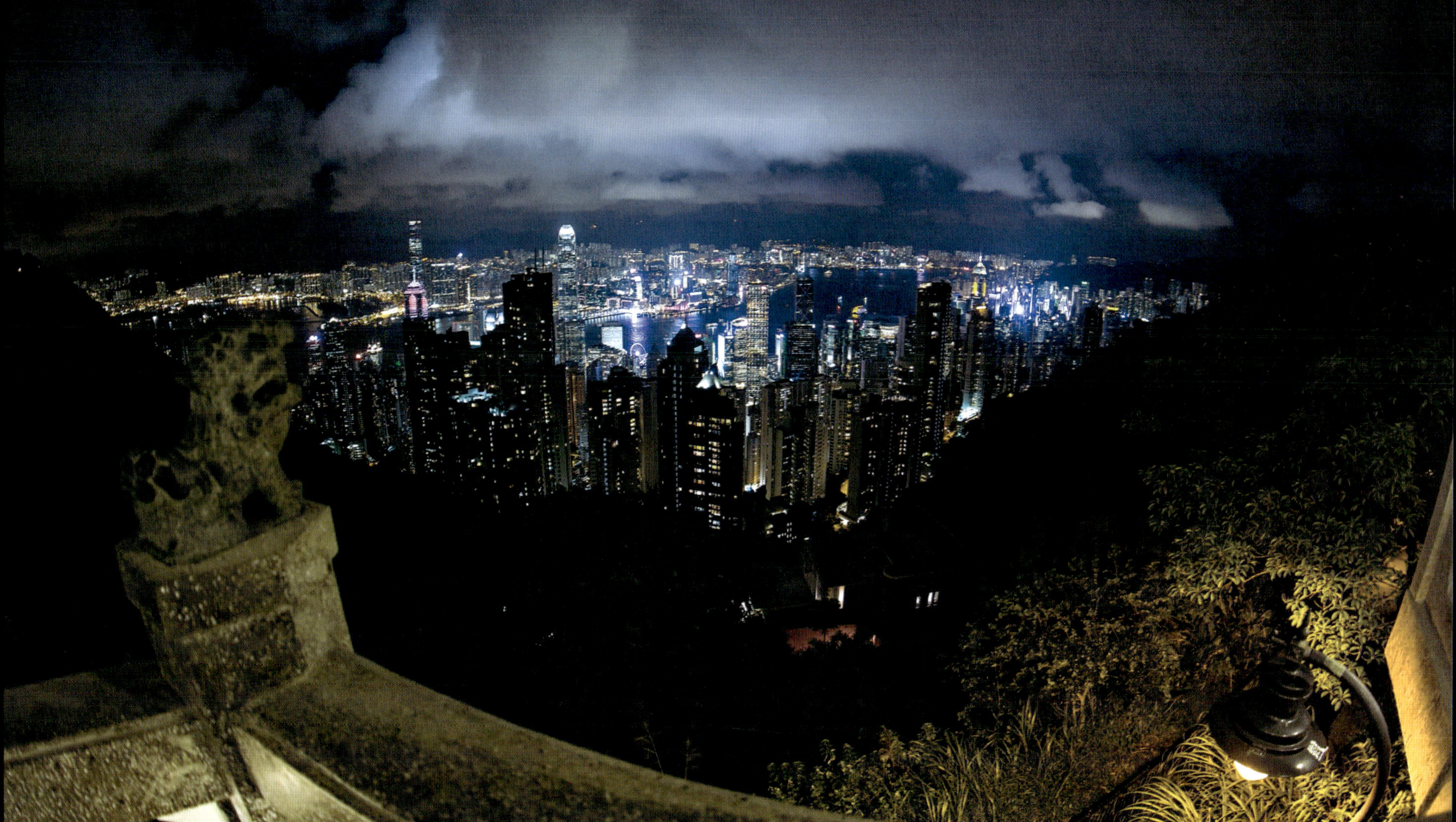

Home to over 7 million people, Hong Kong has recently taken the title of the brightest city on the planet.

Lion Rock Country Park, Hong Kong, SAR, People's Republic of China

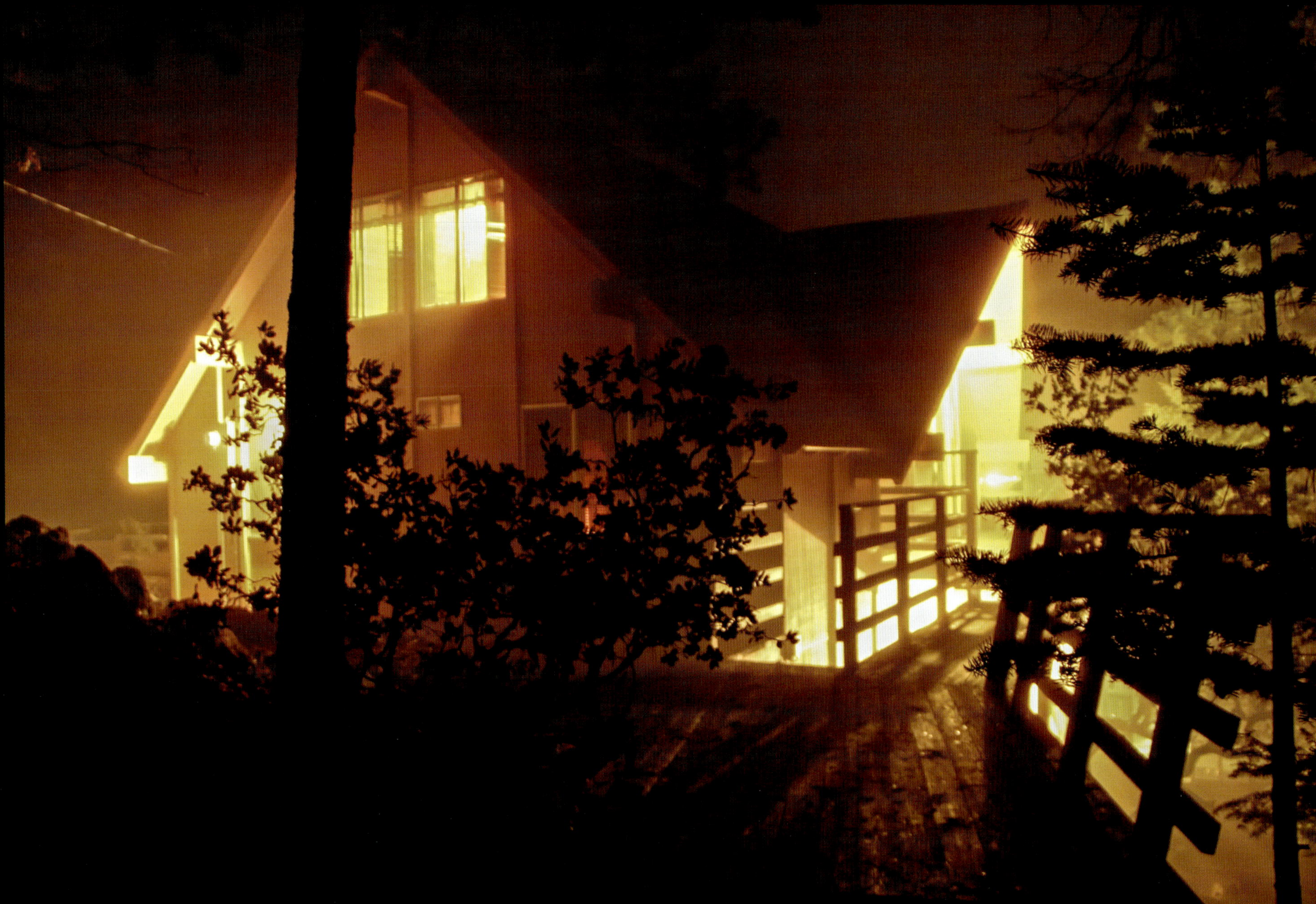

The prevalence of outdoor lighting doesn't only include streetlights, but also all lights emanating from modern human residences.

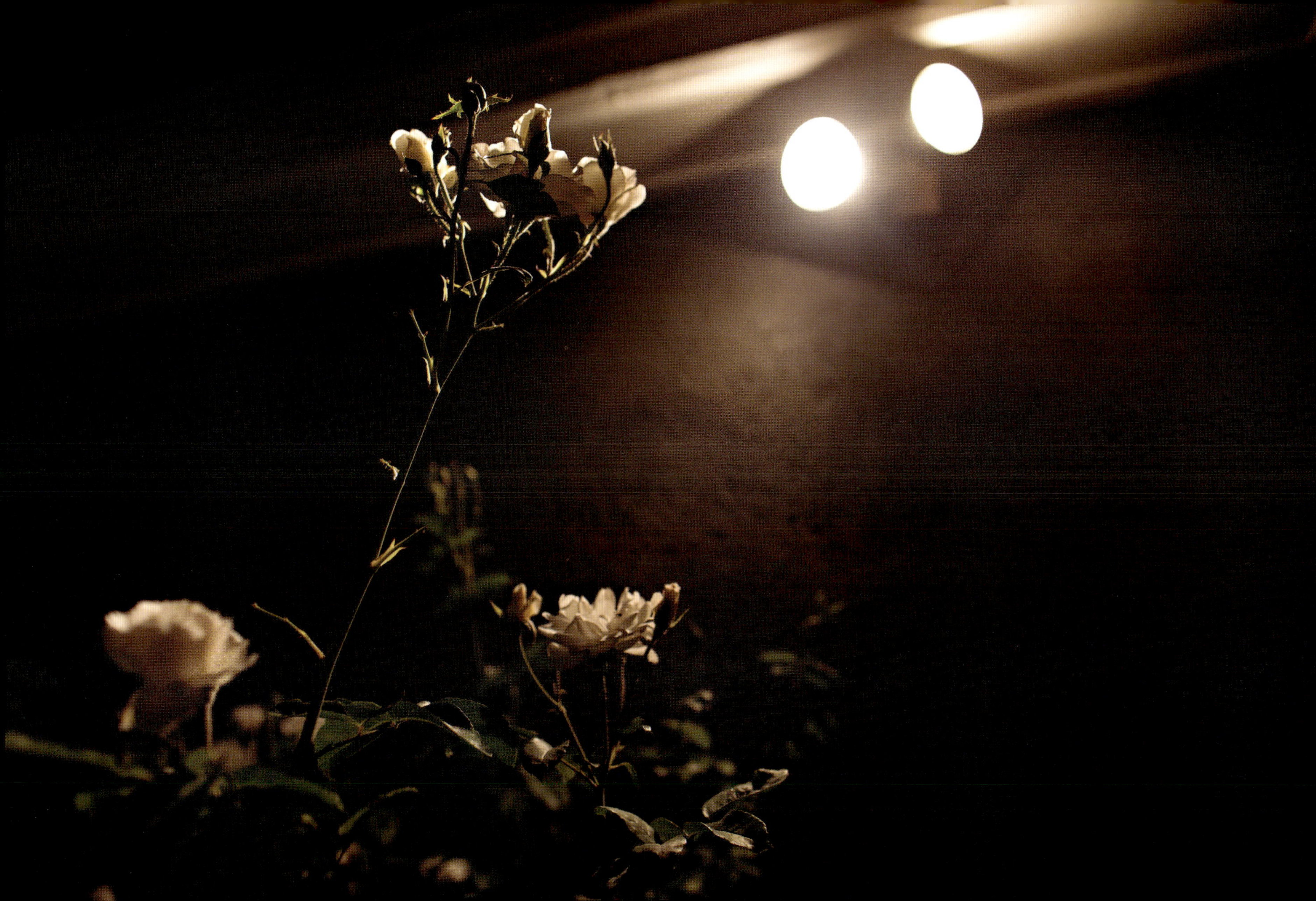

Boundless artificial night light, such as the security light in this Los Angeles house, has a profound impact on the circadian rhythm of all living things, disrupting the 24-hour cycle in the physiological processes of humans, plants, animals, fungi, and even cyanobacteria.

Los Angeles, California

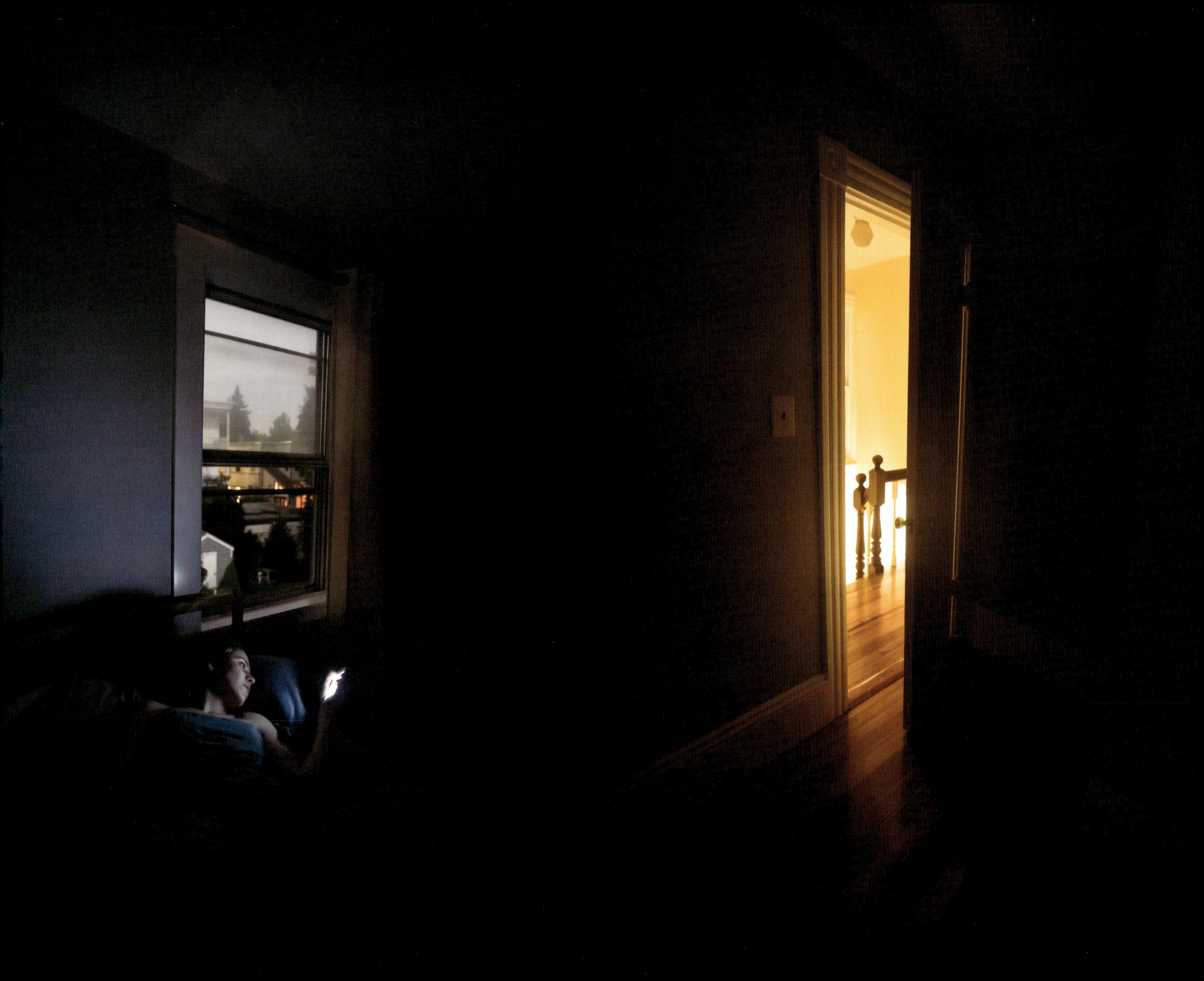

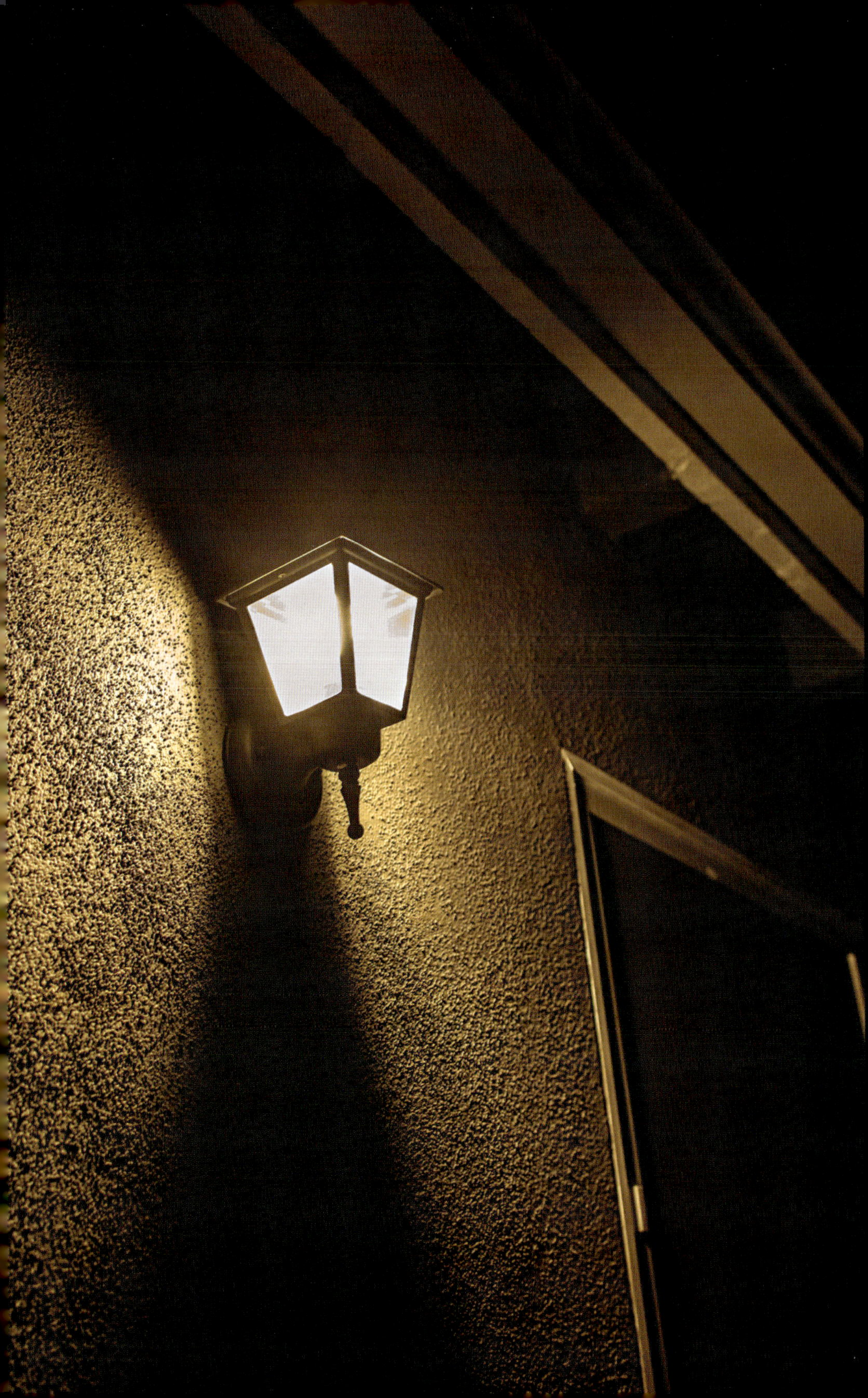

Artificial light emanating from house lights, electronics, and other sources suppresses melatonin, an antioxidant that induces sleep, boosts the immune system, lowers cholesterol, and helps the functioning of the thyroid, pancreas, adrenal glands, and various other parts of the body.

The full impact of nocturnal light on the health of human beings is not yet known as many studies are still to be conducted. Research to date suggests that artificial light's disruption of circadian rhythm is one of the causes of obesity, depression, insomnia, diabetes, and breast cancer.

LEFT PAGE:
Cambridge, Massachusetts

8

RIGHT PAGE:
Los Angeles, California

9

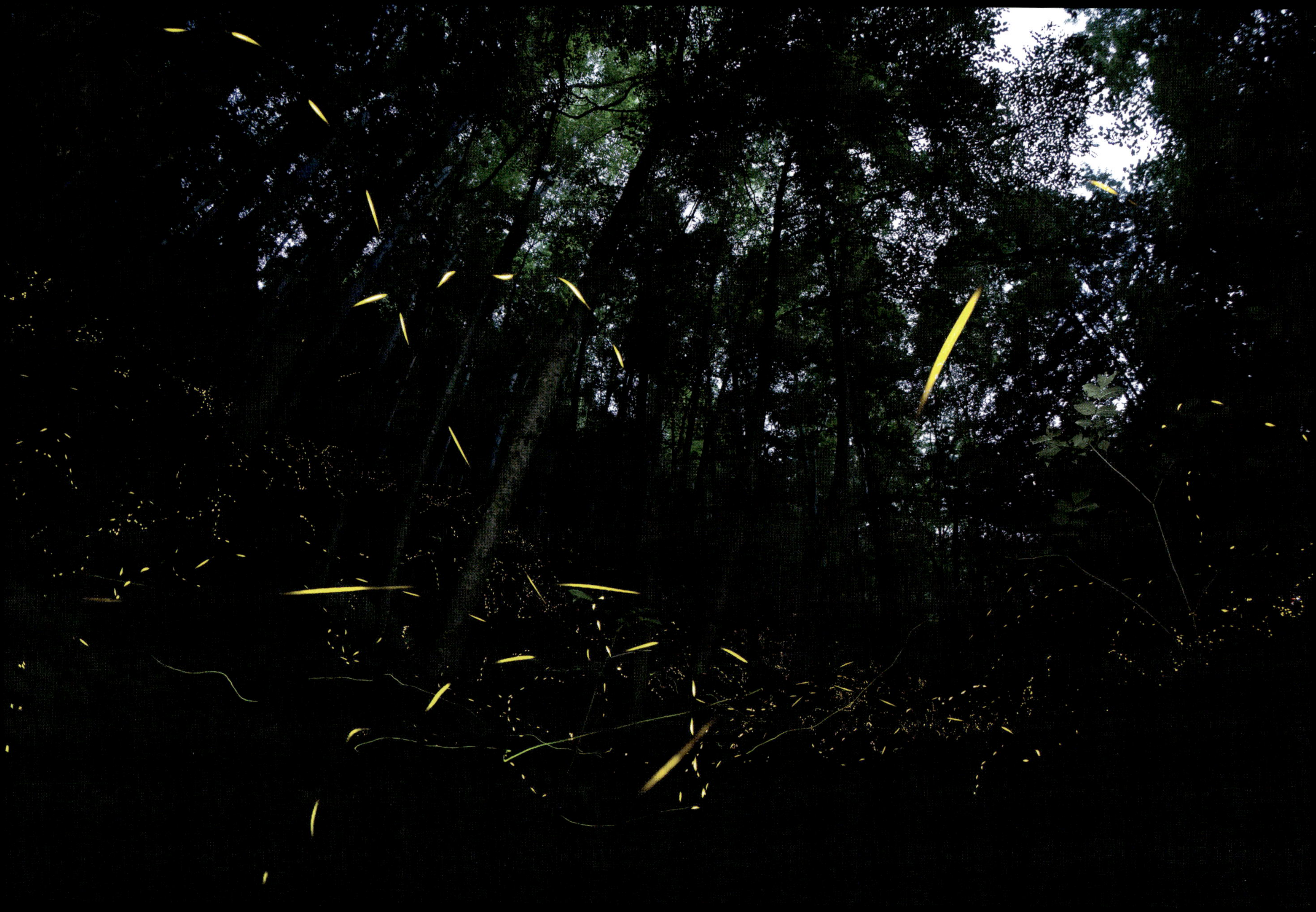

Elkmont, Great Smoky Mountains National Park, Tennessee

4.5

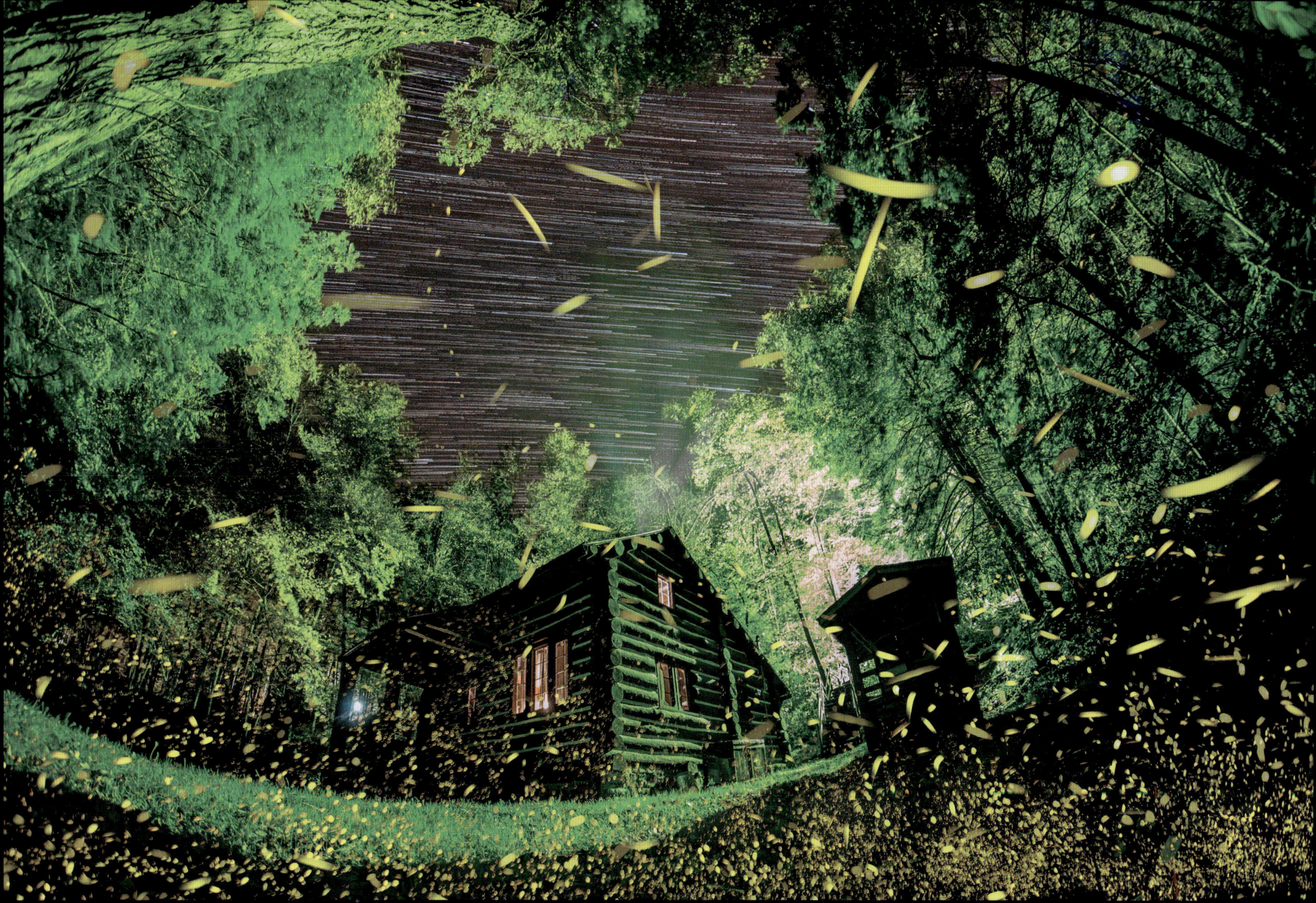

Fireflies, which used to cover most of lower North America during the months of May through July, are one of the animal species devastated by light pollution. Relying on complete darkness to perform their mating rituals, fireflies have retreated into the last remaining dark areas of their habitat, decreasing their overall population by as much as 90 percent, according to some estimates.

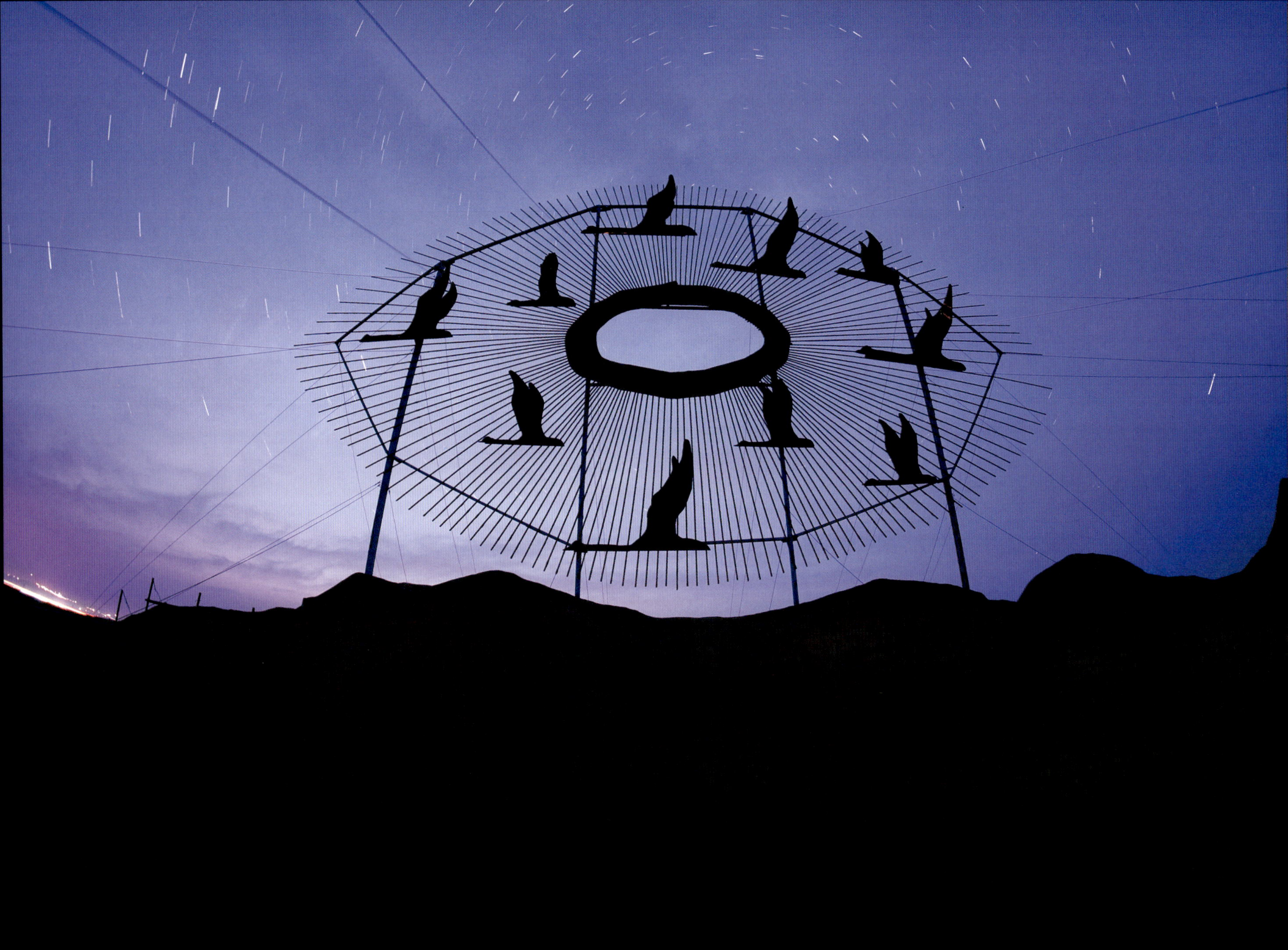

Enchanted Highway Migratory Bird Sculptures, North Dakota

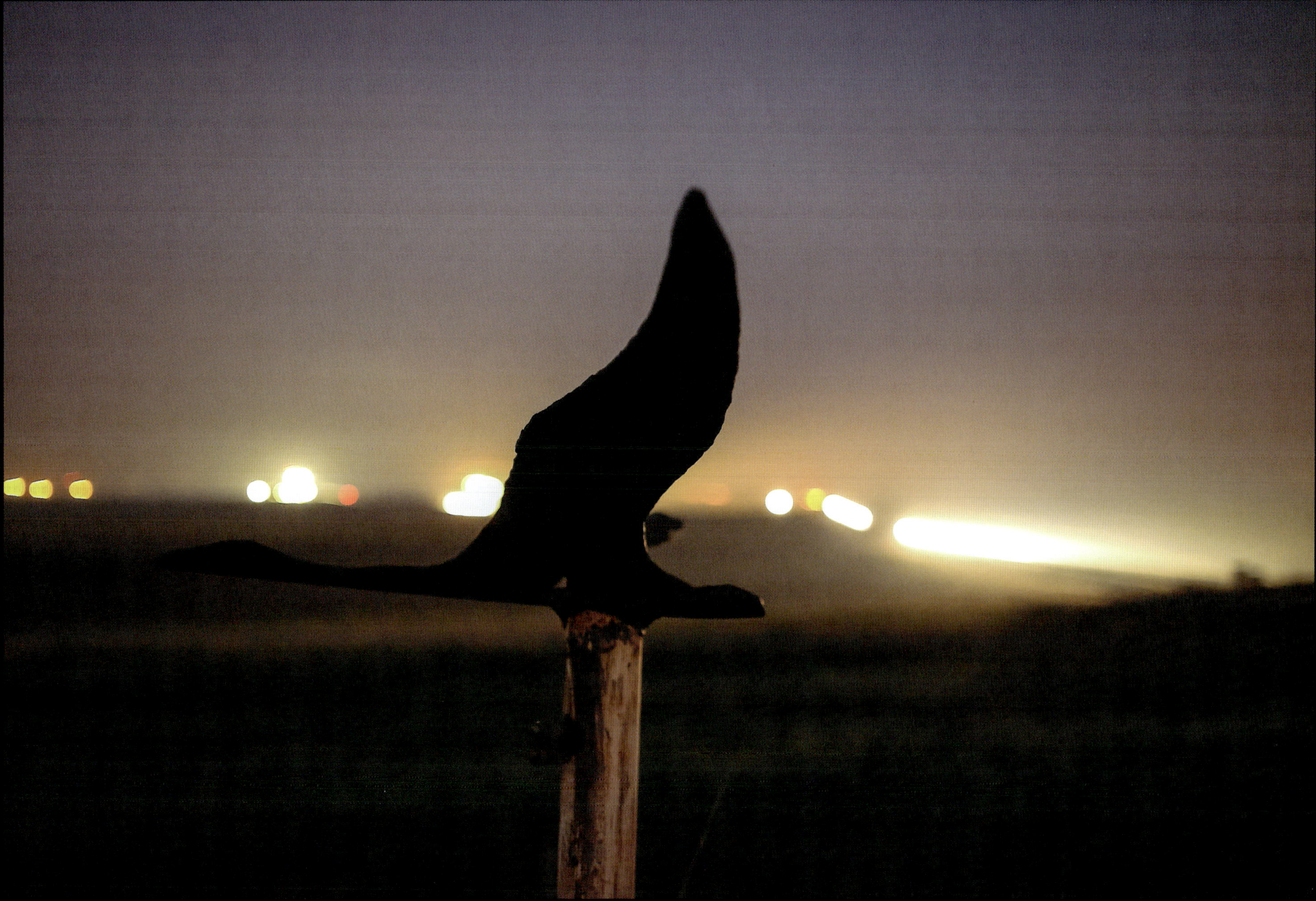

Birds that migrate or hunt at night navigate by moonlight and starlight. Artificial light can cause them to wander off course, migrate too early or too late, and miss ideal climate conditions for nesting, foraging, and other behaviors. Every year, millions of birds die colliding with needlessly illuminated buildings and towers.

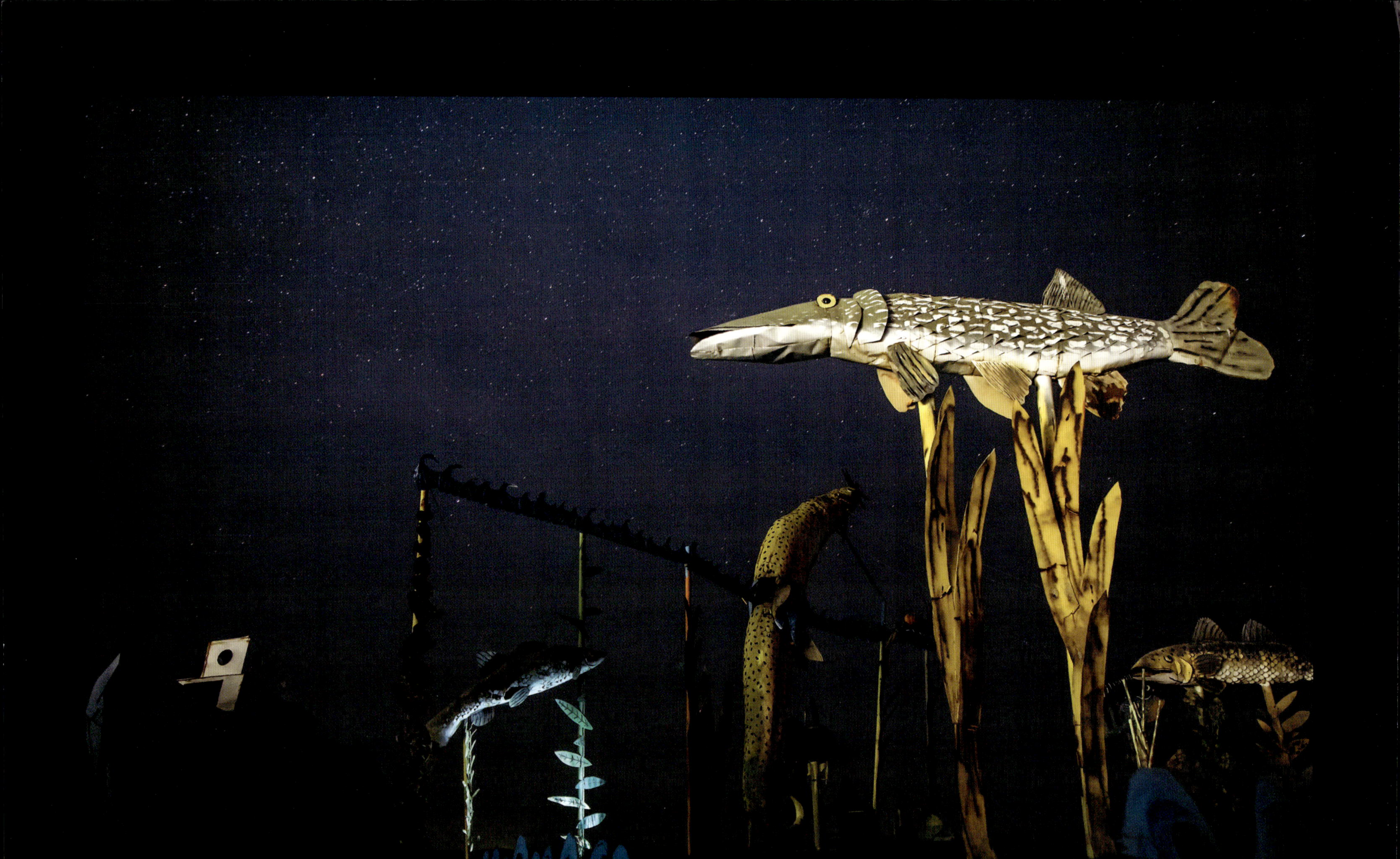

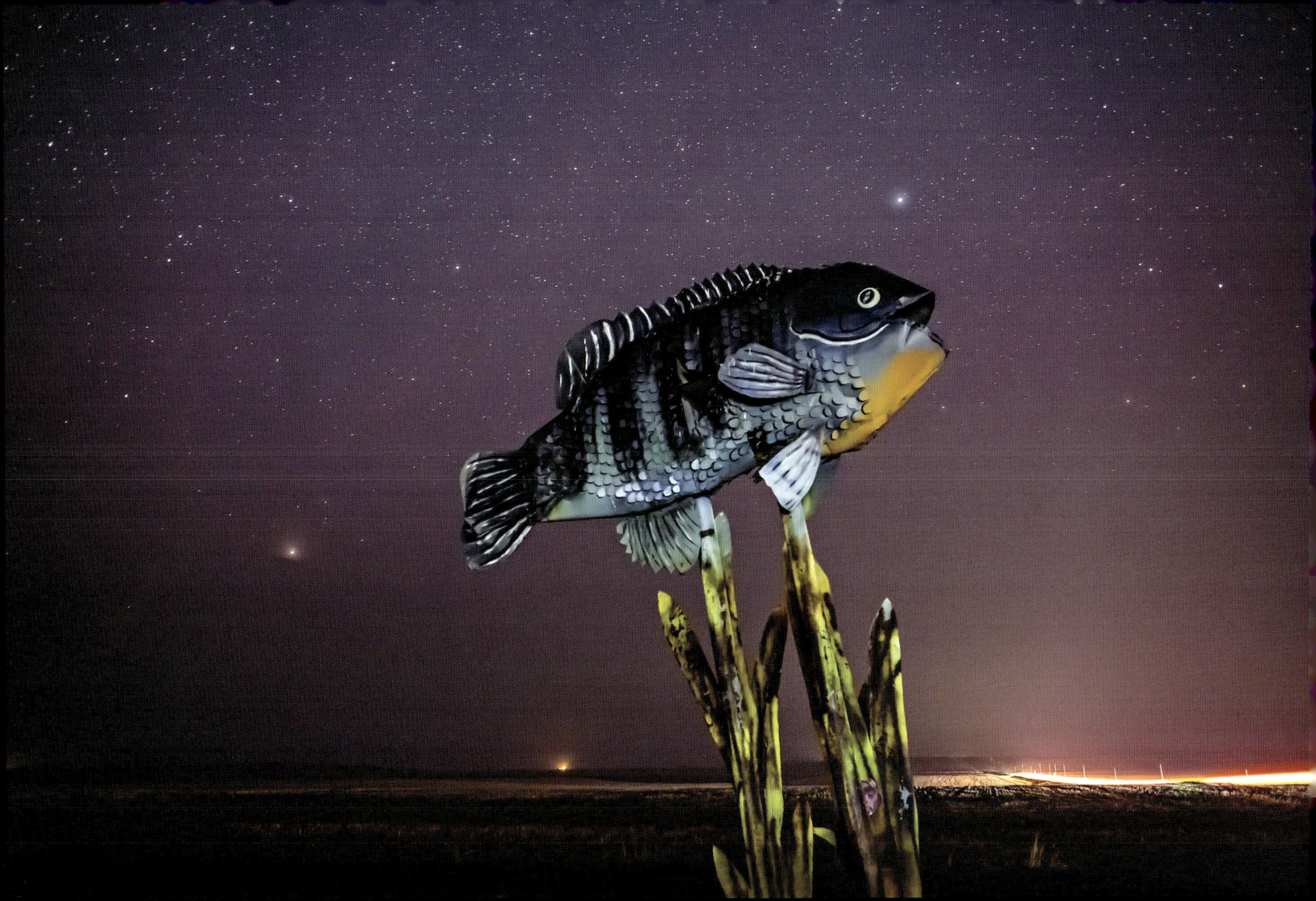

Even fish, such as salmon, are susceptible to having their mating processes disturbed by artificial light pollution. Sea turtles often mistake city lights for moonlight shimmering in the ocean.

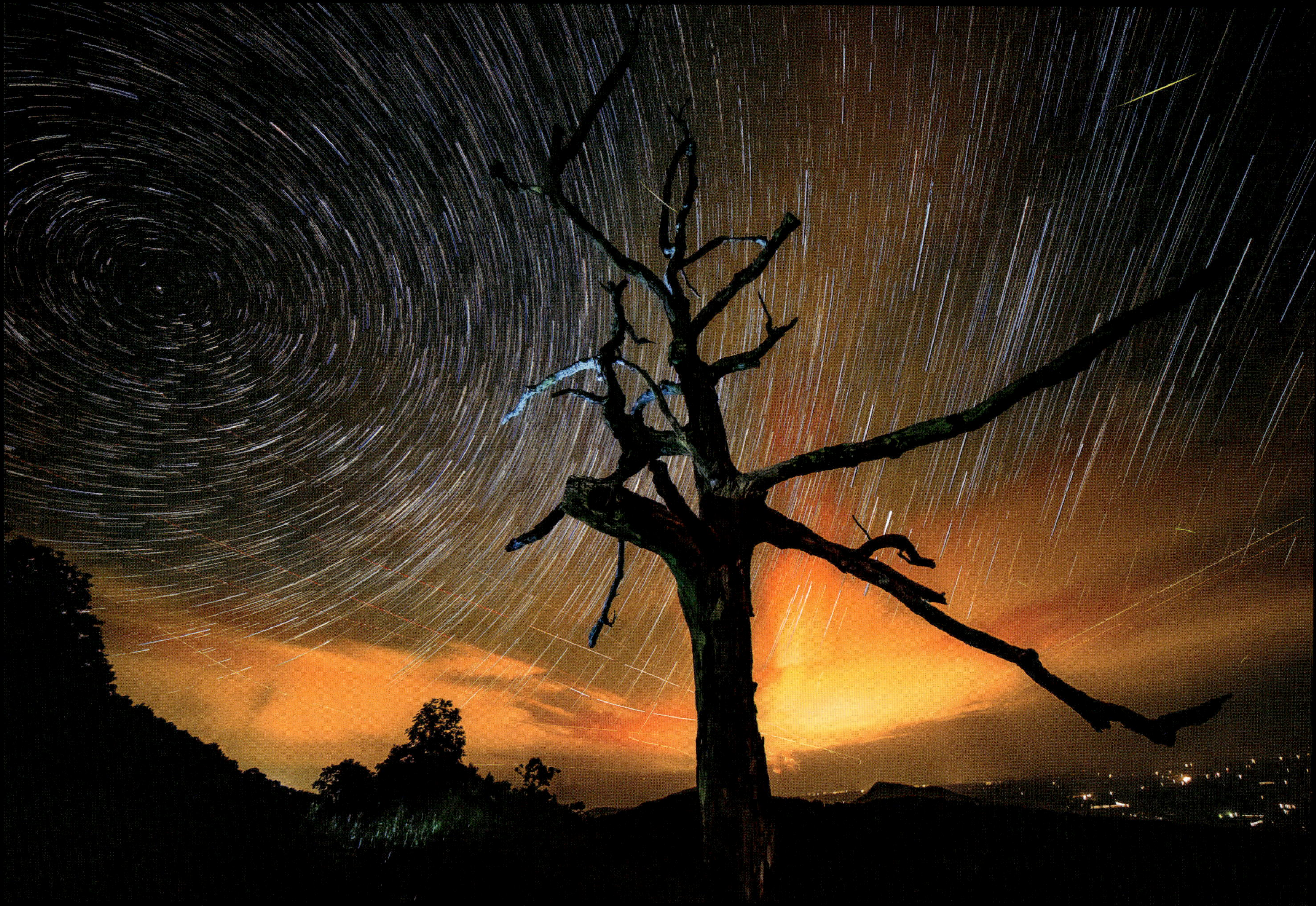

Trees, such as these overlooking the Washington, D.C., area (left) and Los Angeles (right), have been shown to bloom too early due to artificial light.

Shenandoah National Park, Virginia

4

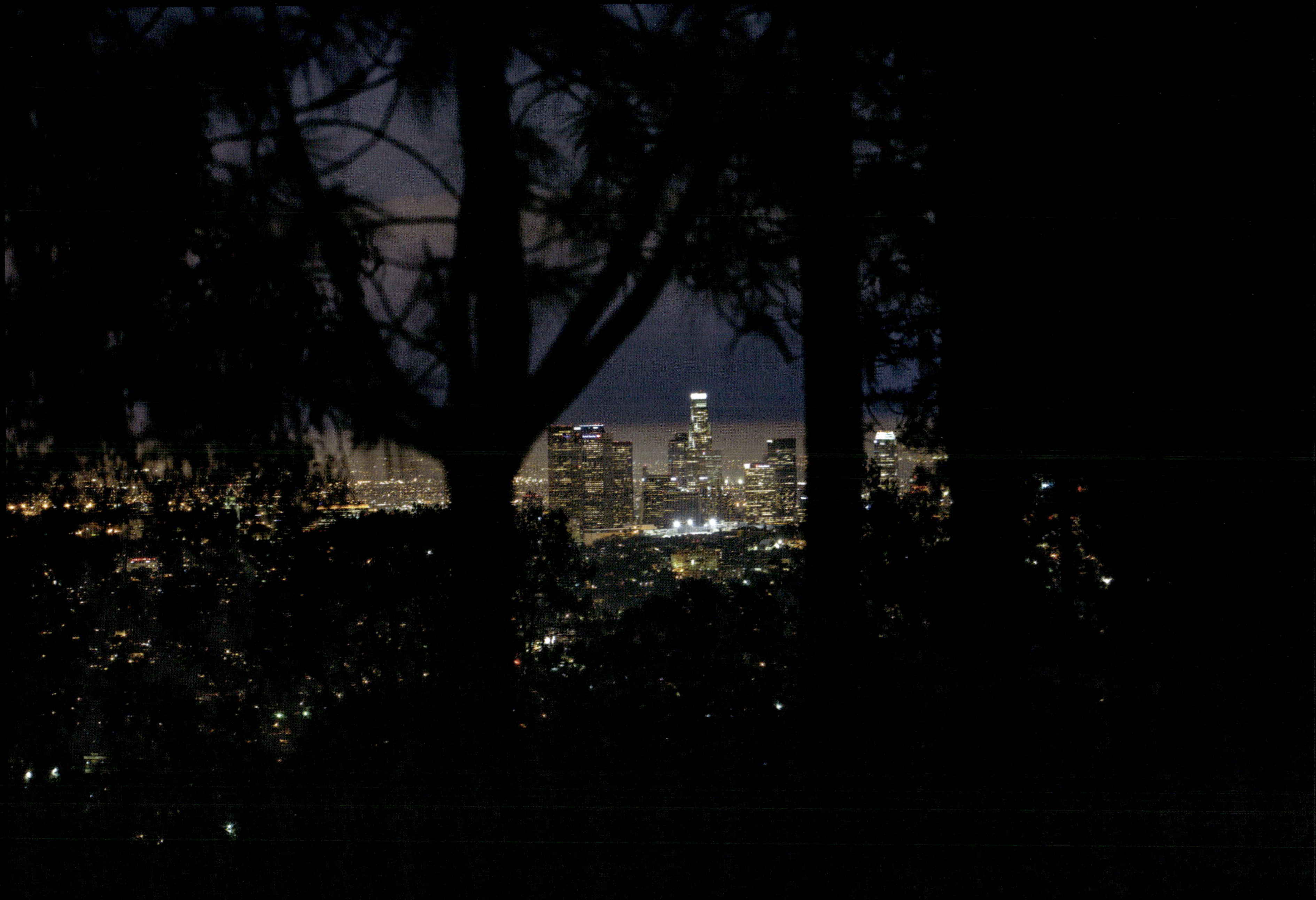

Later-budding trees such as ash trees are most affected by light pollution, which in turn impacts all the animal life that depends on those trees' natural bloom cycle.

Griffith Park overlooking Los Angeles, California

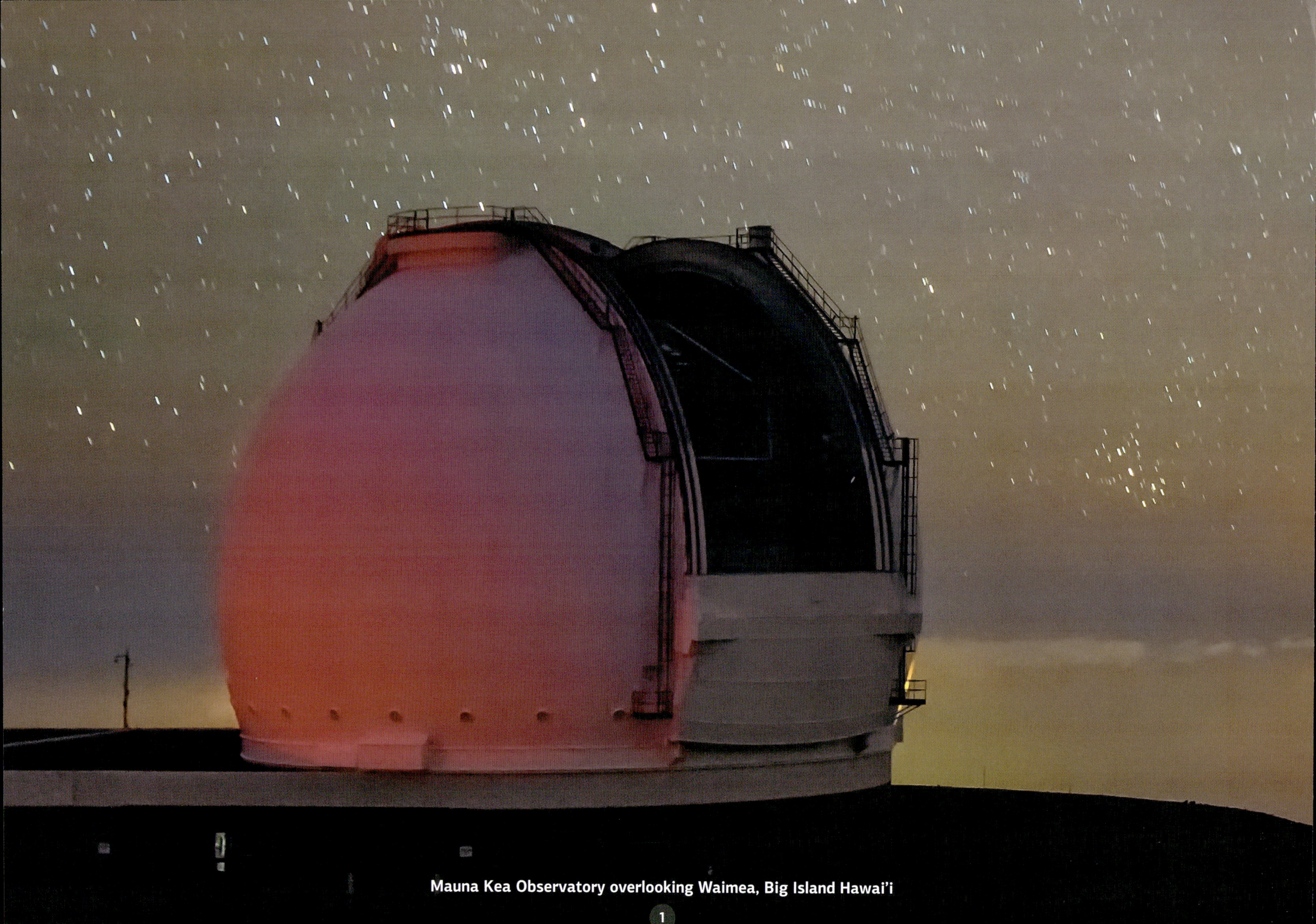

Mauna Kea Observatory overlooking Waimea, Big Island Hawai'i

DARK SKY MOVEMENT

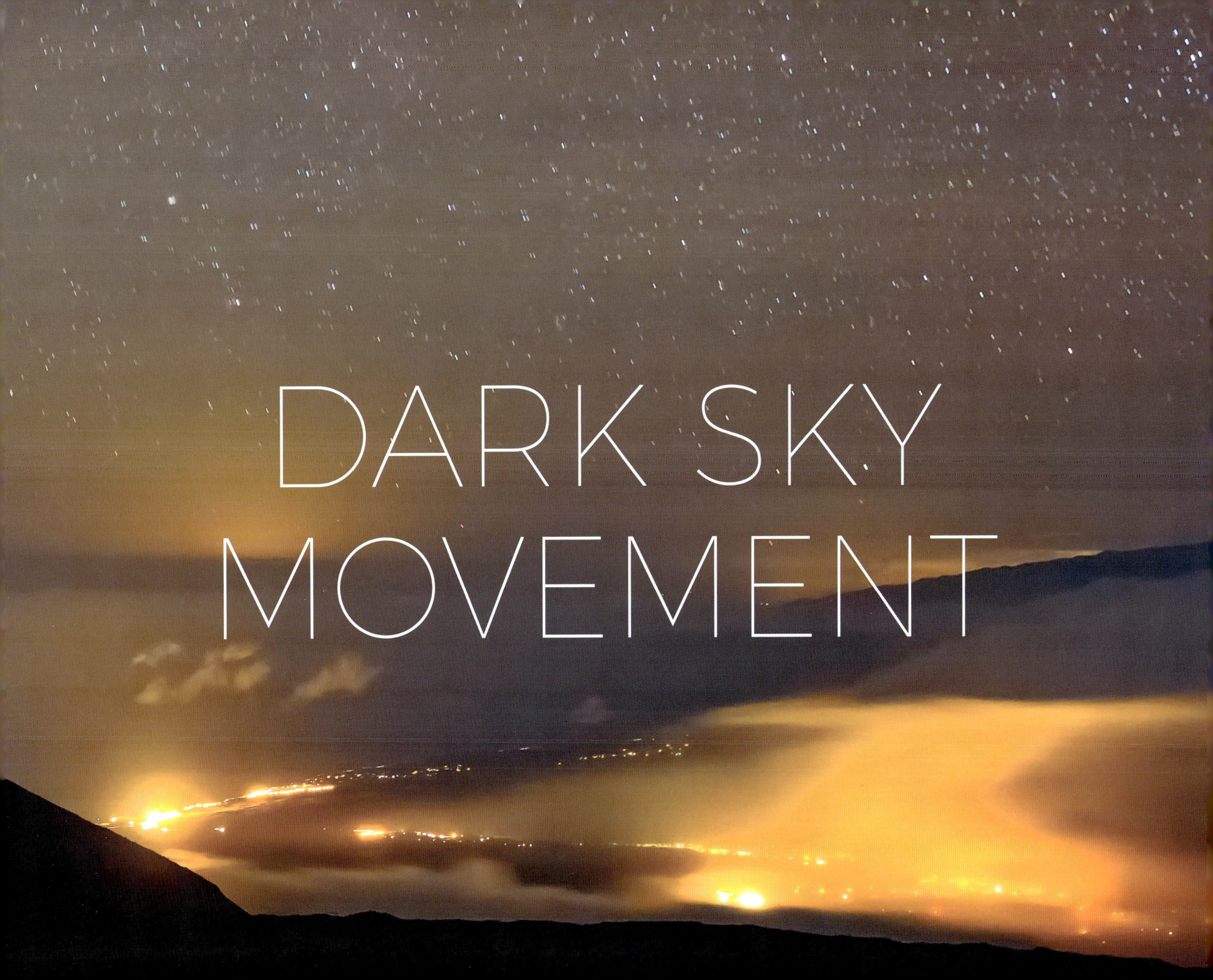

Having an atmosphere is inconvenient.

Well, of course, it's convenient if you want to breathe. That's an advantage. And also if you want to live on a planet with liquid water, and prevent 200-degree temperature swings between day and night. Those all go in the plus column.

But when you're an astronomer, it makes life tougher. For one thing, it blocks a lot of the electromagnetic spectrum: light. Visible light gets through, but infrared, ultraviolet, and more get absorbed on their way down from space. A lot of very interesting objects emit that kind of light, and our atmosphere makes it impossible to see them doing so (but then, ultraviolet, X-rays, and gamma rays are harmful to life, so perhaps there's also a checkmark in the plus column for that, too).

For another, it moves. Little bits of it, parcels a few centimeters across, fly this way and that. Each one of these cells acts like a lens, distorting light coming from space, bending its path this way and that as well. This distorts the apparent shape of the object, making it hard to see fine details.

But there's a third problem, and it may be the toughest of all. Our atmosphere isn't perfectly transparent. If it were, it would be the darkest black you can imagine, and the stars would shine through with clarity and confidence. But it isn't, and they don't. The atmosphere reflects light from below. This is obvious to anyone who has visited a big city; I remember the first time I traveled to Chicago and was very distressed that night to see a blazing orange sky over my head. It was cloudy, which amplified the reflection, but even on clear nights only a handful of stars were visible from my location; just the very few brightest stars, barely able to make themselves known above that raucous background.

That's why astronomers seek remote mountaintops for observatories. The higher you are the less atmosphere there is, and the more light gets down from the heavens. But more importantly the isolation means getting far away from city lights, allowing the faintest astronomical objects to be seen.

Well, at least the fainter ones. Escaping light pollution entirely is nearly impossible in the modern era. Unless, that is, you escape the atmosphere all together. That's a big reason astronomers launch telescopes into space! Hubble Space Telescope is not a particularly large telescope as they go — its 2.4-meter mirror is downright modest compared to the 8 or 10-meter light collectors many ground-based telescopes have these days.

But its big advantage is that it's above the atmosphere, with far darker skies. Faint objects that happen to be cosmically nearby, or luminous ones at the edge of the Universe whose light is dimmed by distance, are more easily within its grasp. Also, without the boiling soup of air above it, it can see finer details than most ground-based optical telescopes.

Of course, lofting a telescope into space adds something of a monetary cost to its construction. The billions of dollars spent on Hubble could have built quite a few Earth-based observatories, but given what it's done for astronomy and space science, the cost was in my opinion worth it. Still, we

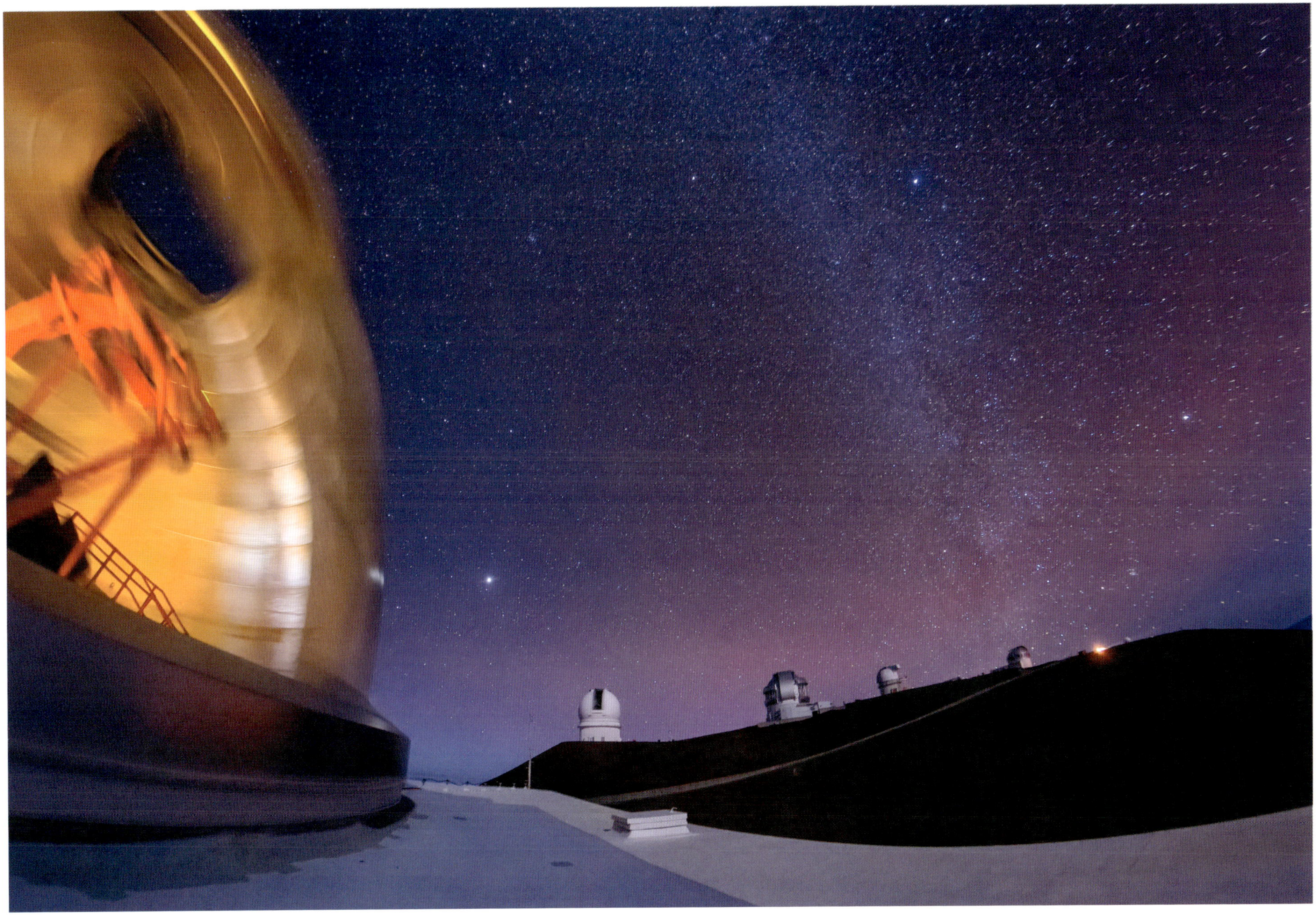

NASA Infrared Telescope, Mauna Kea Observatory, Big Island Hawai'i

1

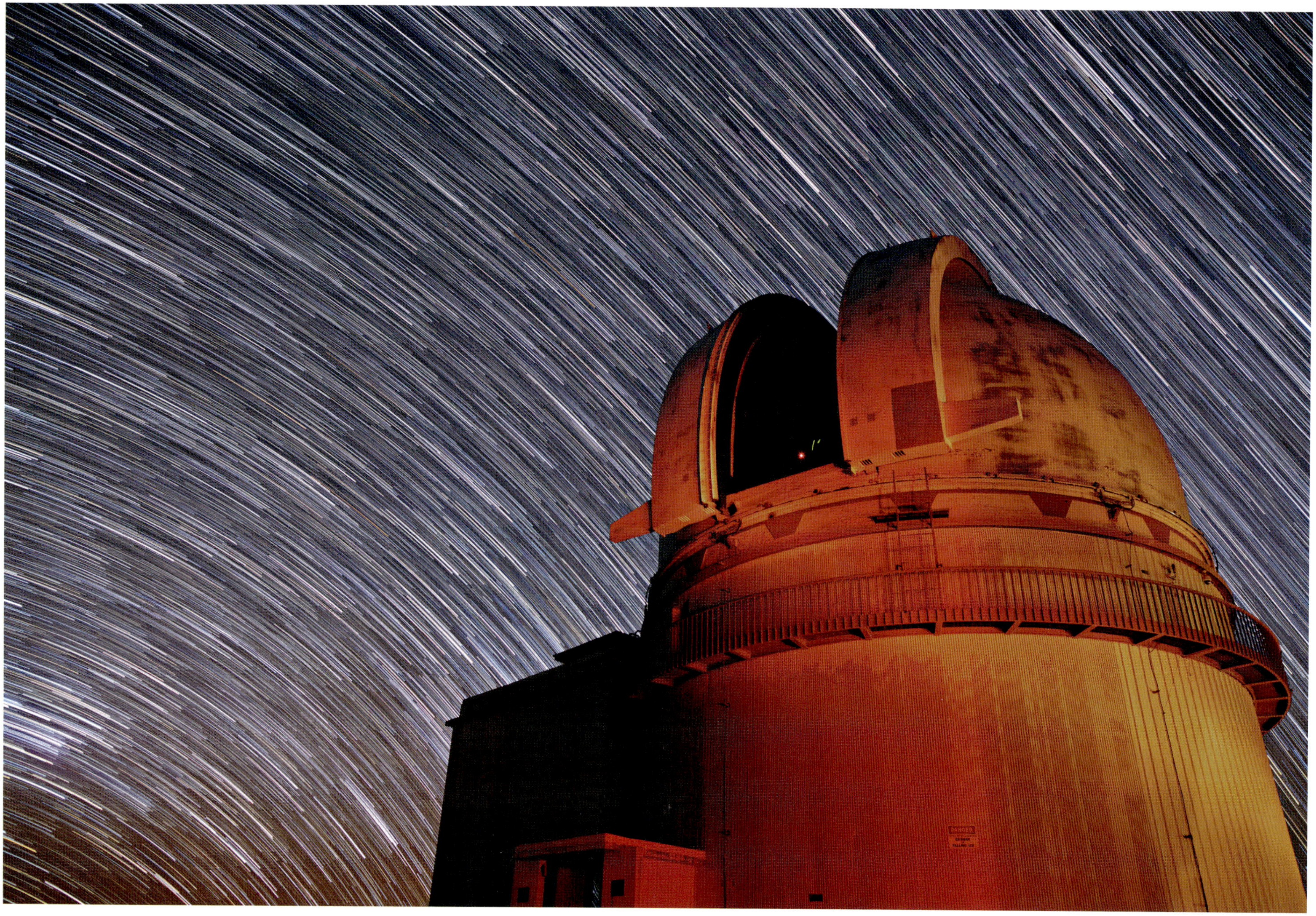

Mauna Kea Observatory, Big Island Hawai'i

1

can't simply build big 'scopes and throw them into space whenever we want; only a few have that luxury. That means light pollution will continue to be an issue for astronomers. What can we do?

There are workarounds. To make images of the sky astronomers use digital detectors much like the cameras in phones (and did you know the tiny light-sensitive computer chip in your phone camera is a direct descendent of ones on Hubble? It's true; the technology used to create bigger, more sensitive, and less expensive detectors for Hubble aided in the eventual ability to mass-produce phone cameras). As light pours in, the camera converts it into electrons, which can then be counted and stored digitally as numbers.

In general, the light coming from the sky is roughly constant over a small area, and that means each pixel in an astronomical image gets roughly the same number value for the sky. A galaxy, for example, will add its own light to that, increasing the number value for those pixels. You can then simply subtract the average value for the sky, leaving just the light from the galaxy in the image. More sophisticated mathematical techniques can be applied as well to more closely approximate the slight change in sky brightness over the area of the image, better than subtracting it away.

While this is a practical technique to lower the effects of light pollution, it's thorny, and can be difficult to pursue in practice. It can leave artifacts in the image, which is unpleasant on the artistic side, and untenable on the scientific one. I saw one such photo that shows the Milky Way in it, even though it was taken from Los Angeles! That was a remarkable achievement, but it still left me more than a bit melancholy. We shouldn't need to go to such heroic efforts to see the galaxy in which we live. And you need not live in a major metropolis to have the Milky Way hidden from you: a study done in 2016 mapped light pollution across the planet, and found that 80% of Americans have skies too bright to see the Milky Way, and overall on Earth our home galaxy is blotted out to 1/3 of all people on Earth.

That's well over two billion people who can't see the amassed glow of a hundred billion stars. And while that might be a difficult situation for professional astronomers' science, for the public at large it's a tragedy of the soul.

There is something deeply profound about standing under a velvet sky, gleaming stars spread across it. Even before we understood the true nature of the stars, ancient people gazed upon them and wondered. These people saw themselves reflected in the skies, assigning stories to the stars that were morality tales of their own lives. Eventually the stars were given even more influence, becoming the homes of the gods, or able to influence our own lives depending on when and where we were born. Astrology may be just so much nonsense, but the very fact that people would ascribe so much power to the stars speaks of their intrinsic effect on us.

And now we do understand them, which adds so much to their appeal. Vega is blue, Antares red, which is due to their different temperatures...and that itself is tied to their masses, their relative ages, their chemical composition. Just seeing the stars was not enough for science, we sought to know them as well.

Science has helped us unlock the mysteries of the Universe, and has also provided a decent post facto solution to the problems of light pollution when it comes to astrophotography and astronomical observations. But that's missing the point. Seeing the stars, being able to go outside and experience them, was the first step to all this. That is becoming harder, as the stars are diminished one by one as a result of our own carelessness. Studies have shown that poor illumination, besides increasing wasted light thrown into the sky, also wastes money, and has a negative impact on wildlife as well as our own safety when walking at night.

But there is a deeper impact, a damage to our souls. When we lose the stars, we lose ourselves. And that is a darkness I do not want to see.

- Phil Plait, "The Bad Astronomer"

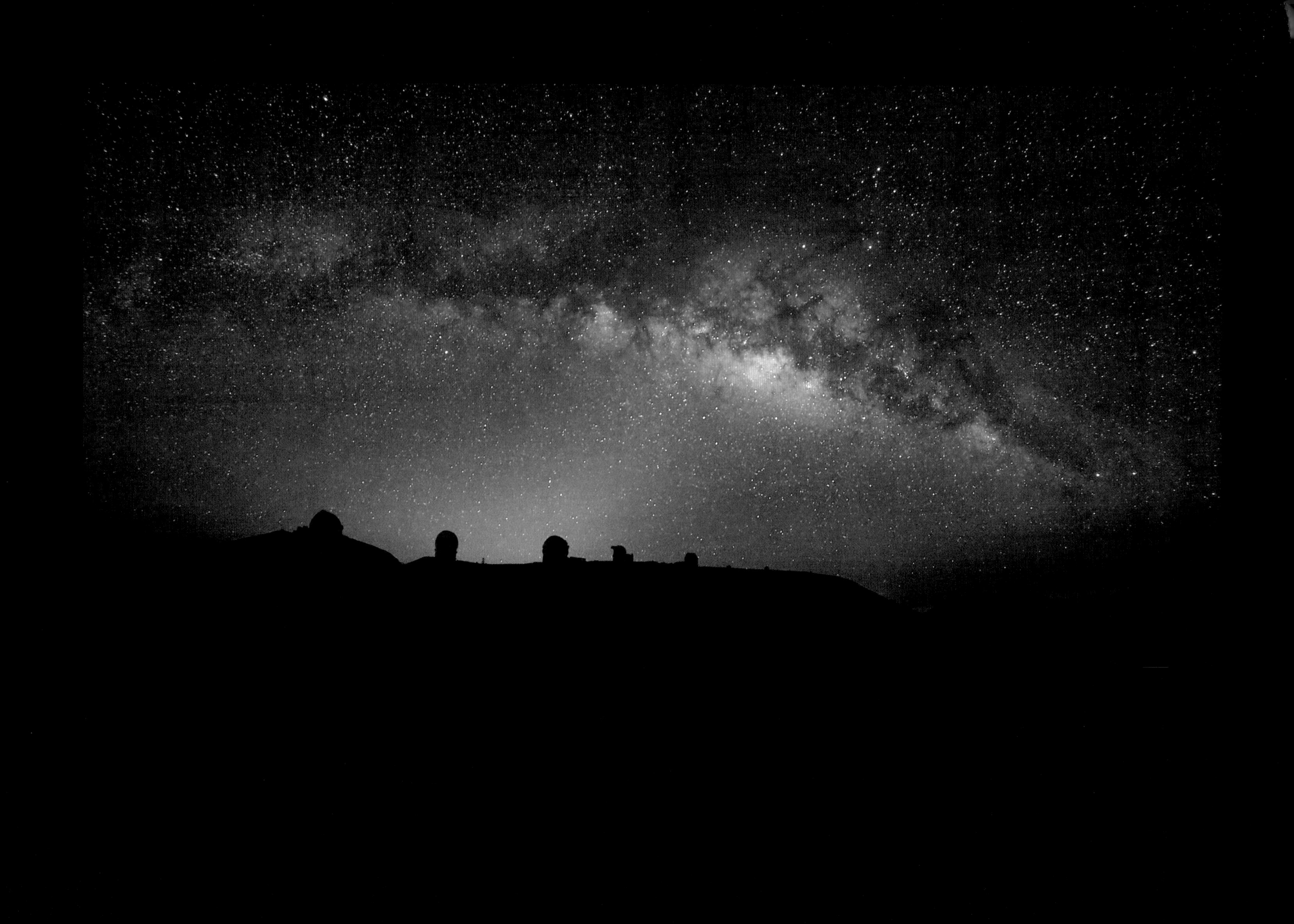

Mauna Kea Observatory, Big Island Hawai'i

1

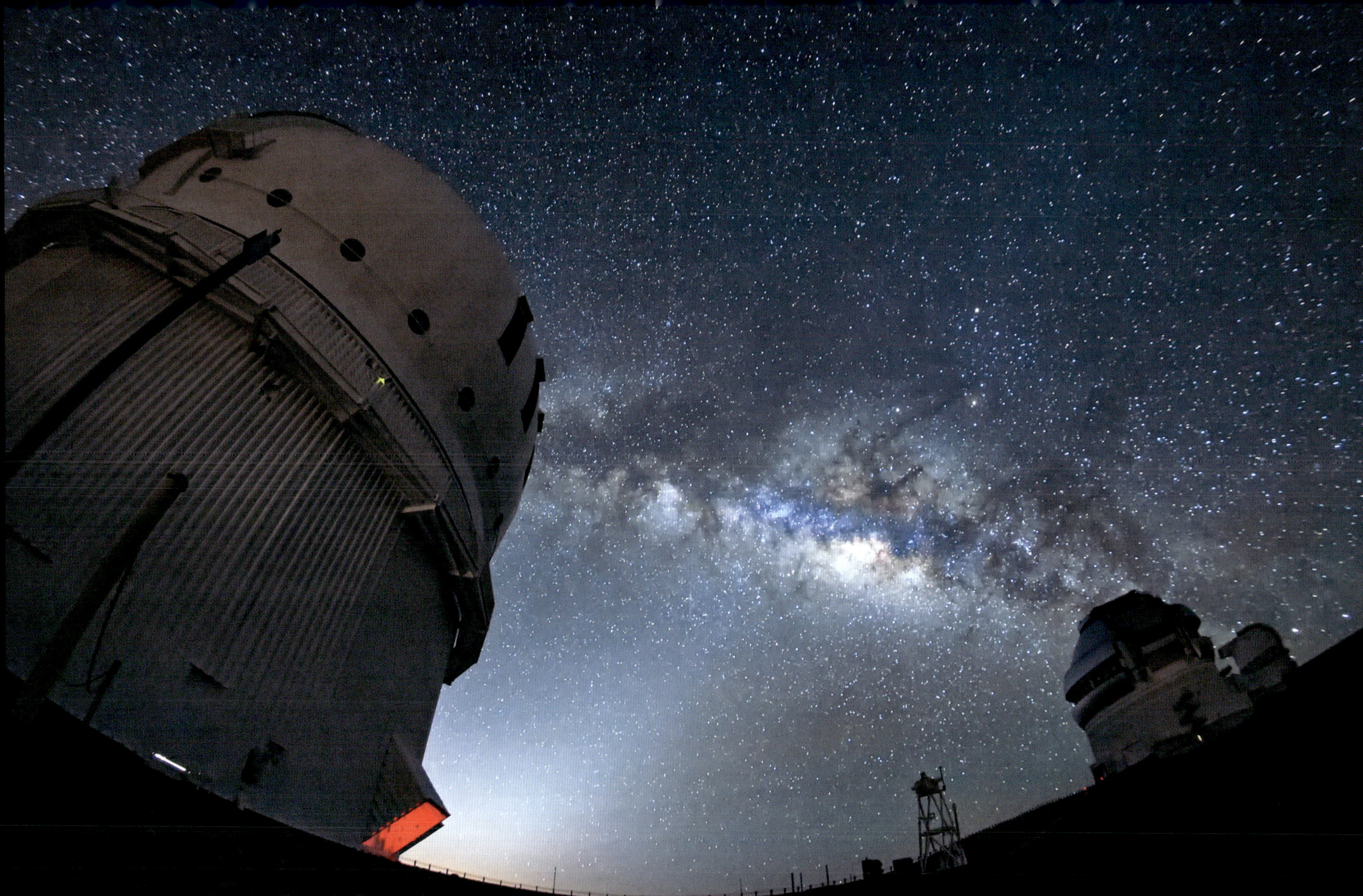

From the very earliest days of electric street lighting, astronomers were the first humans to notice the effects of skyglow. Observational astronomy, disrupted by artificial light, began to move farther and farther away from cities, chasing ever-receding pristine night skies.

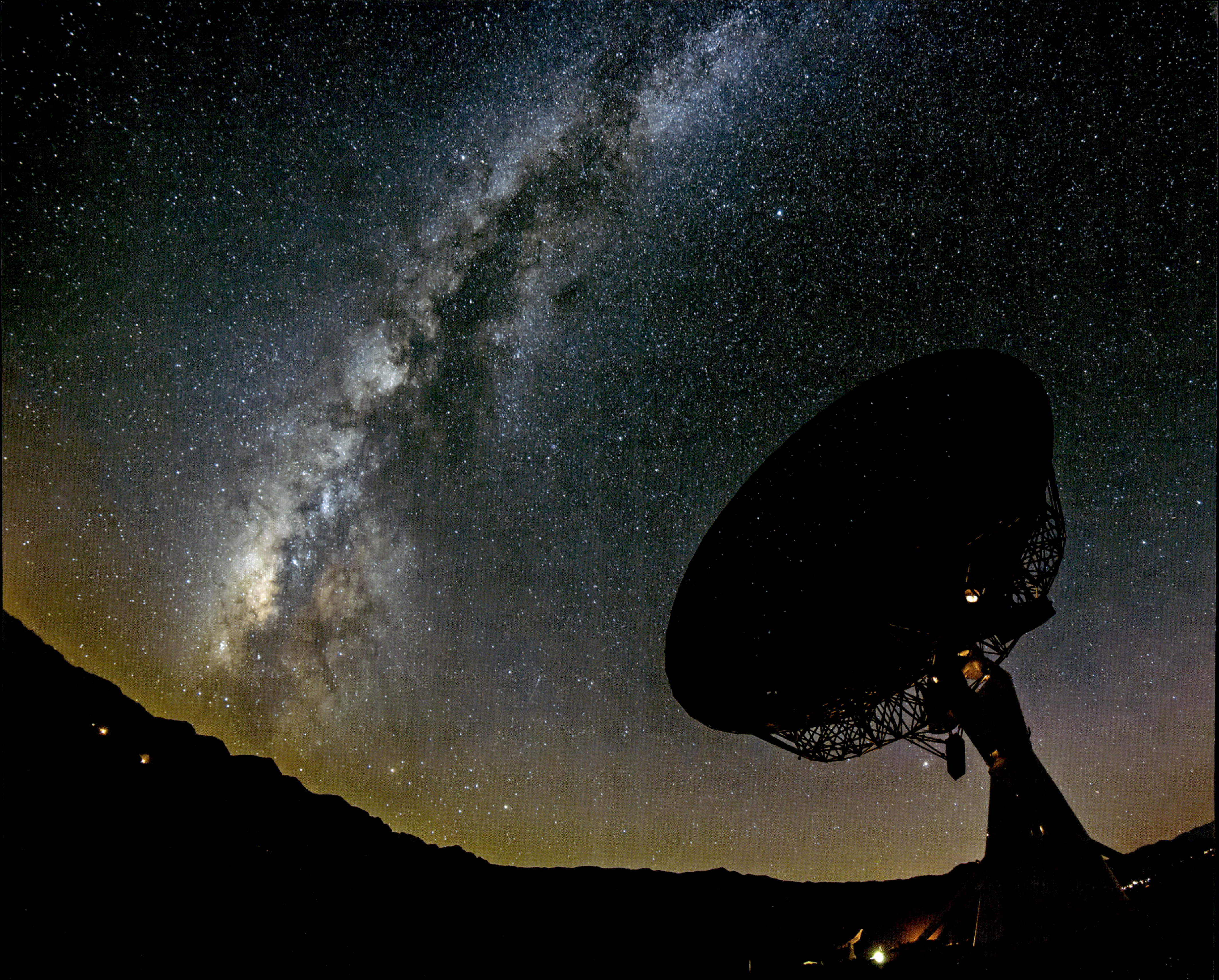

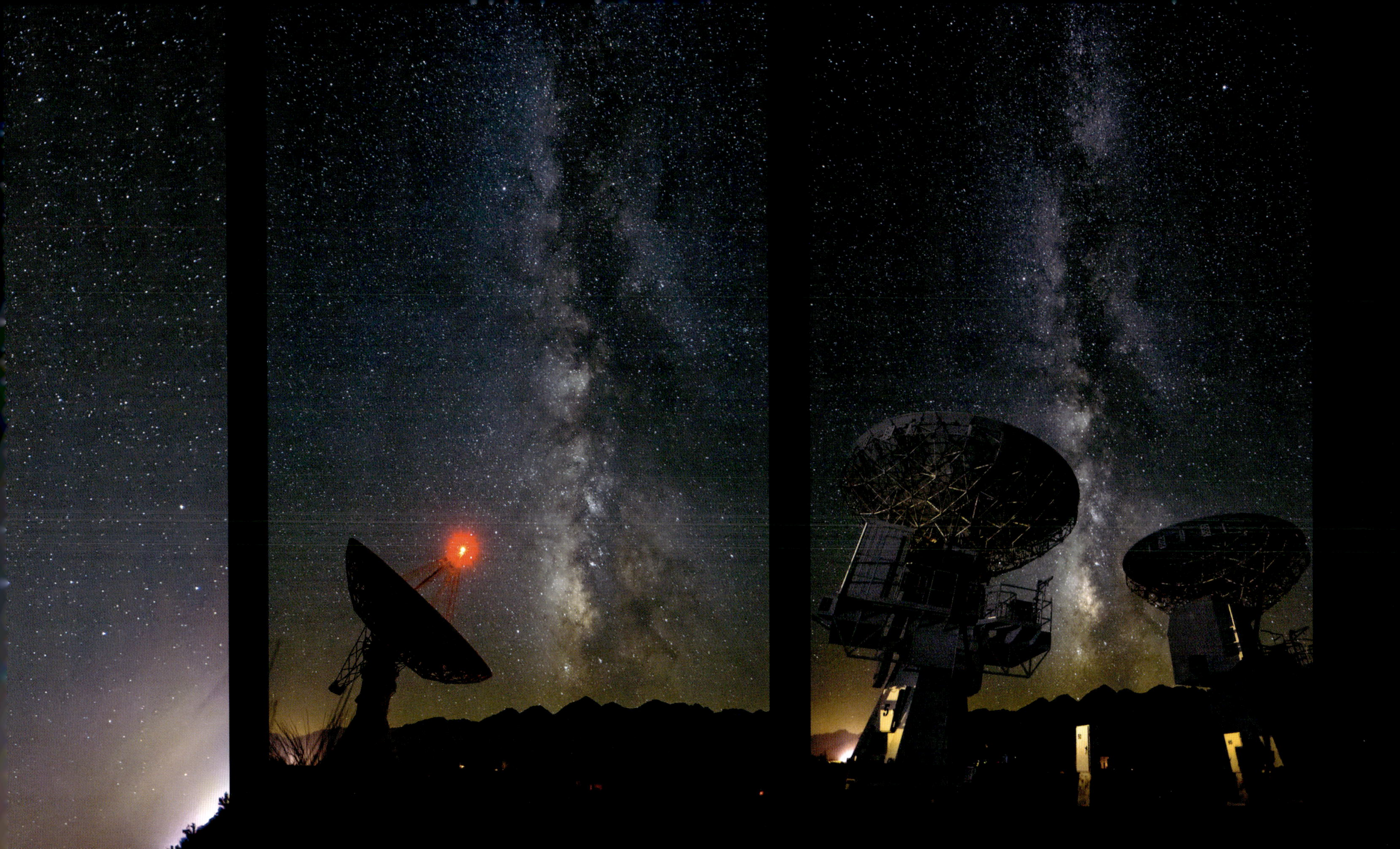

Observational astronomers were not the only ones responding to the growing amount of technological disruption. Radio observatories such as Owens Valley were constructed in remote parts of the United States to escape man-made obstructions like radio transmissions, cell phones, and wireless Internet signals.

Caltech Owens Valley Radio Observatory, California

1

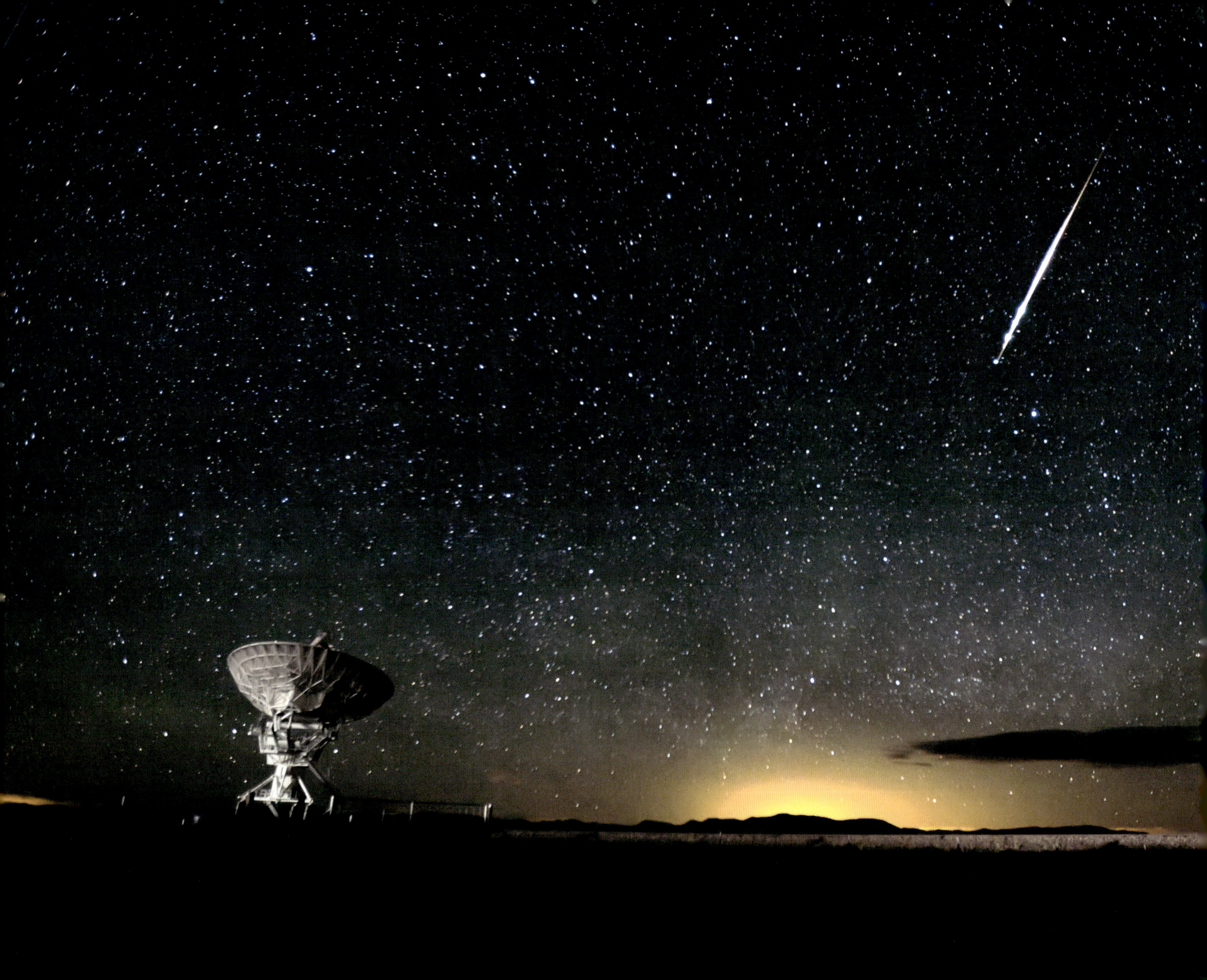

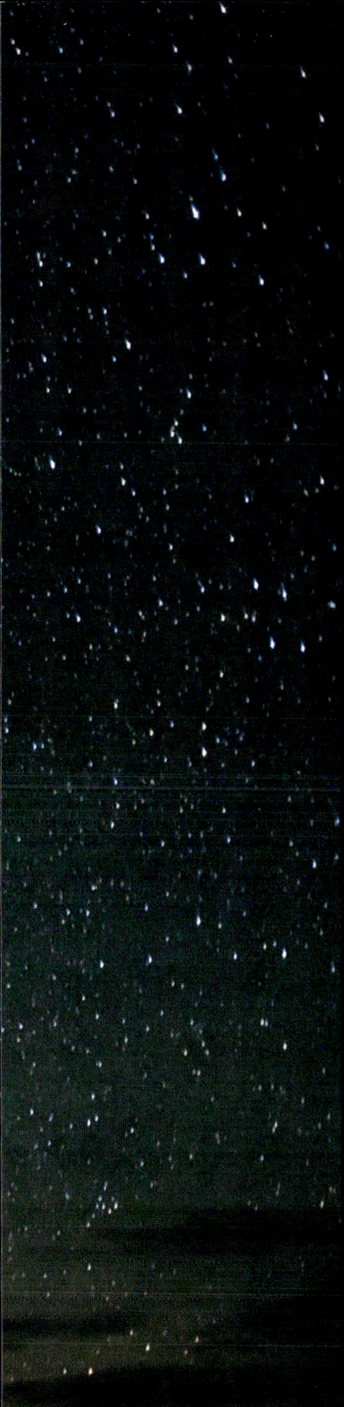

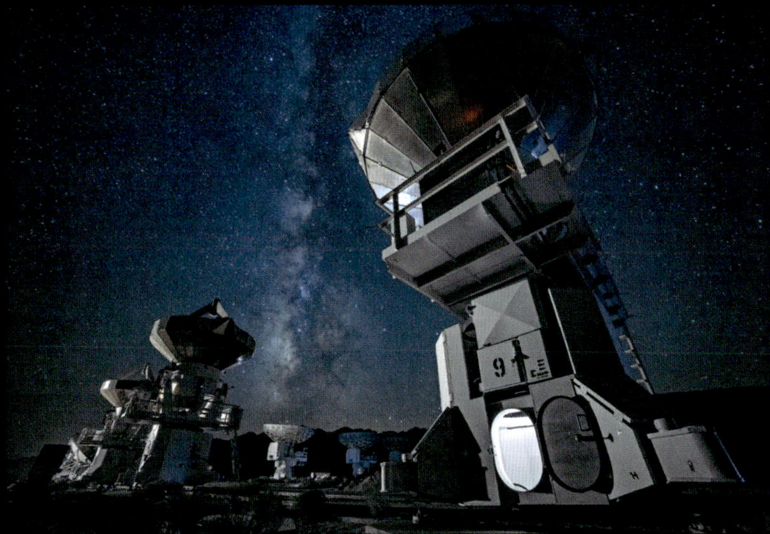

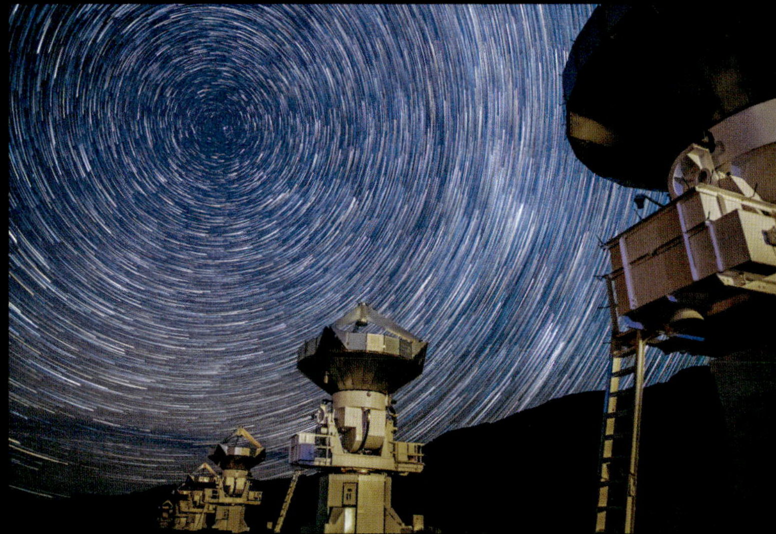

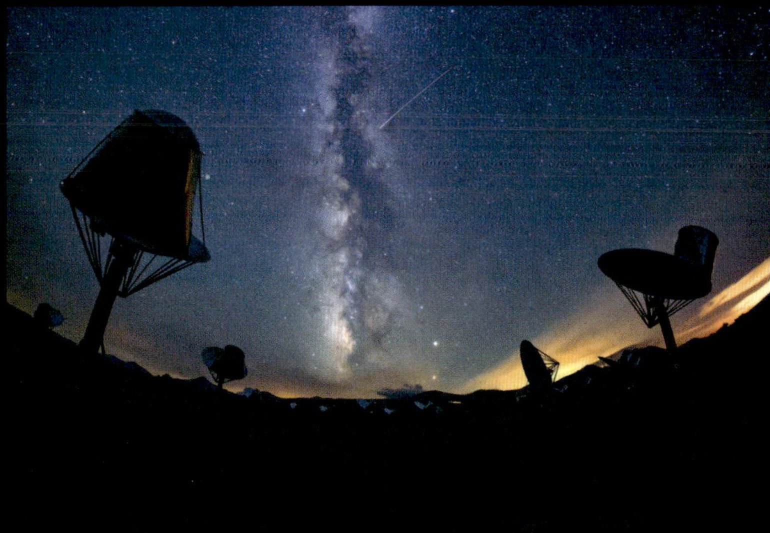

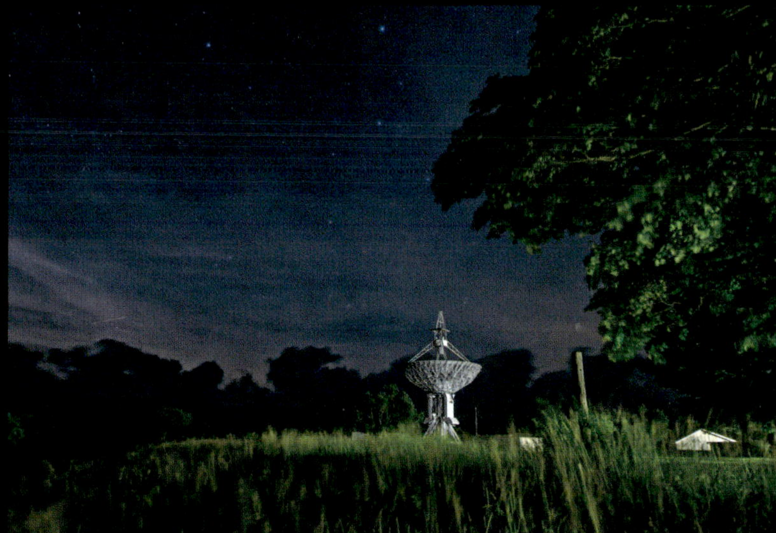

However, when it became clear that simply distancing themselves from expanding cities would not stop the inevitable destruction of all dark skies, both professional and amateur astronomers began to fight back.

TOP: **Caltech Owens Valley Radio Observatory, California**

BOTTOM LEFT: **: SETI Allen Telescope Array, Hat Creek Observatory, California**

BOTTOM RIGHT: **Green Bank National Radio Observatory, West Virginia**

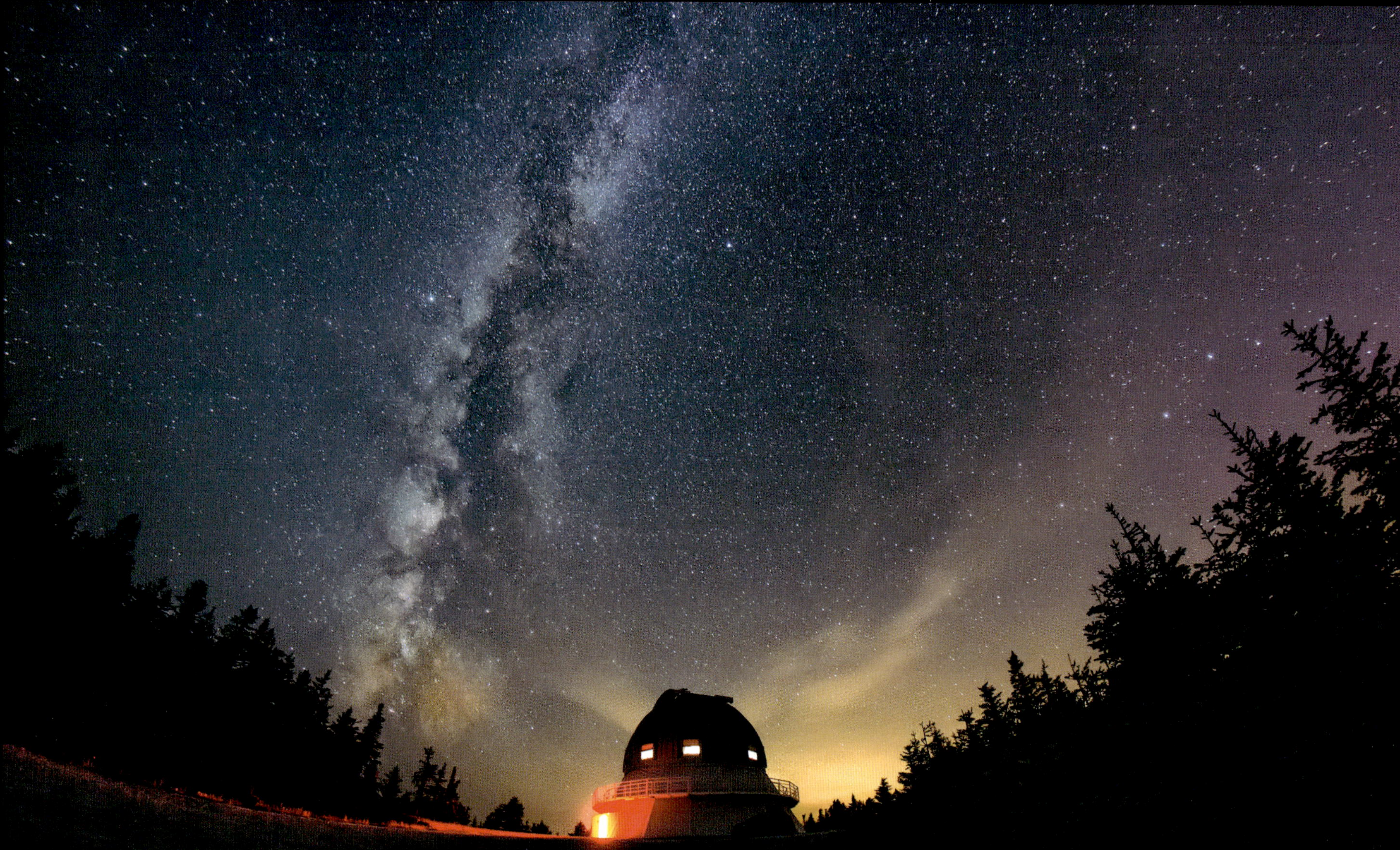

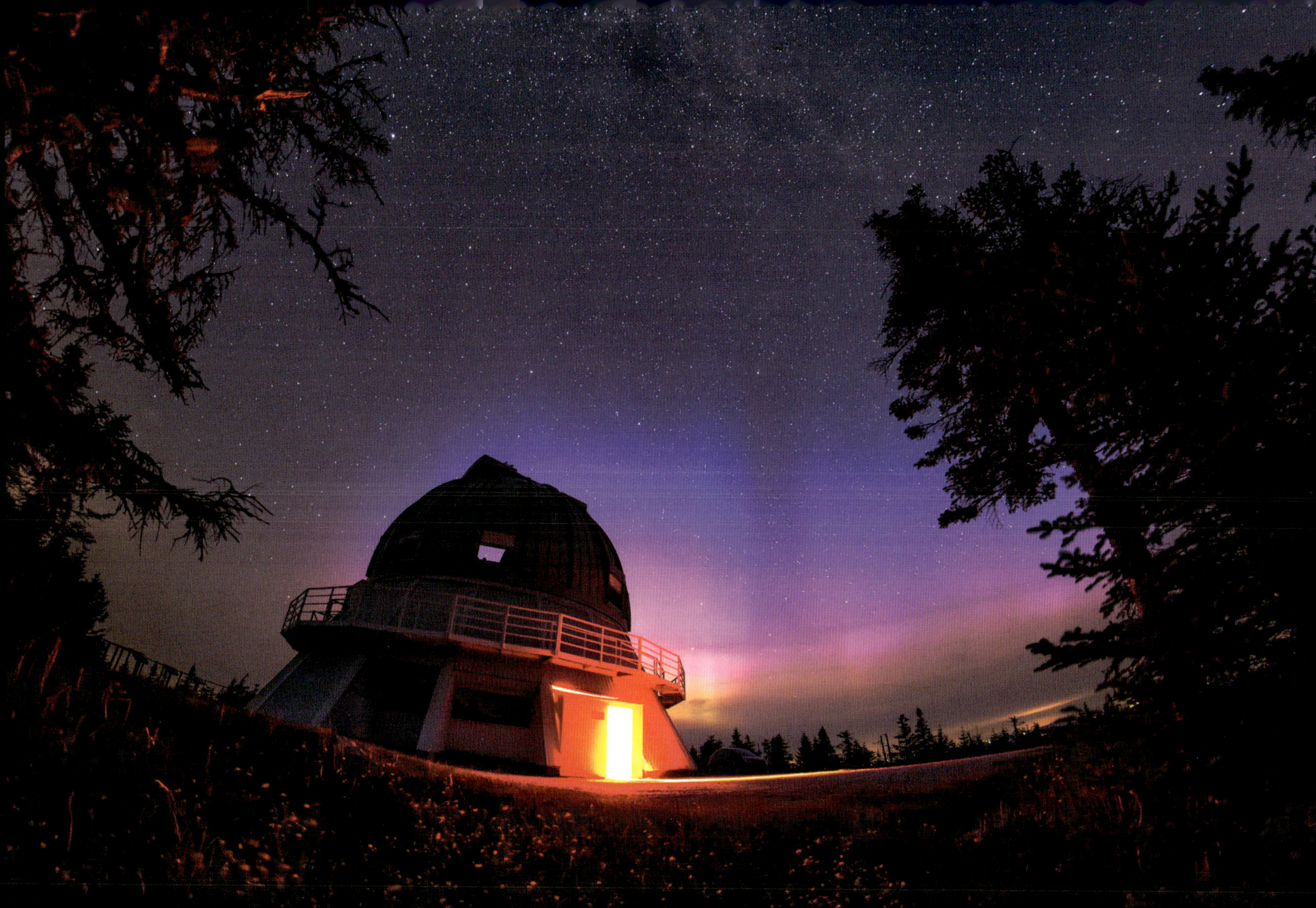

In 1988, a group of Tucson-based astronomers formed the International Dark-Sky Association (IDA), a non-profit with the sole purpose of raising awareness about light pollution and fighting for the preservation of dark skies. In 2008, while lobbying all surrounding towns to change their lighting restrictions, they helped protect a remote observational astronomy site in Québec, Mont-Mégantic, forming the First International Dark-Sky Reserve.

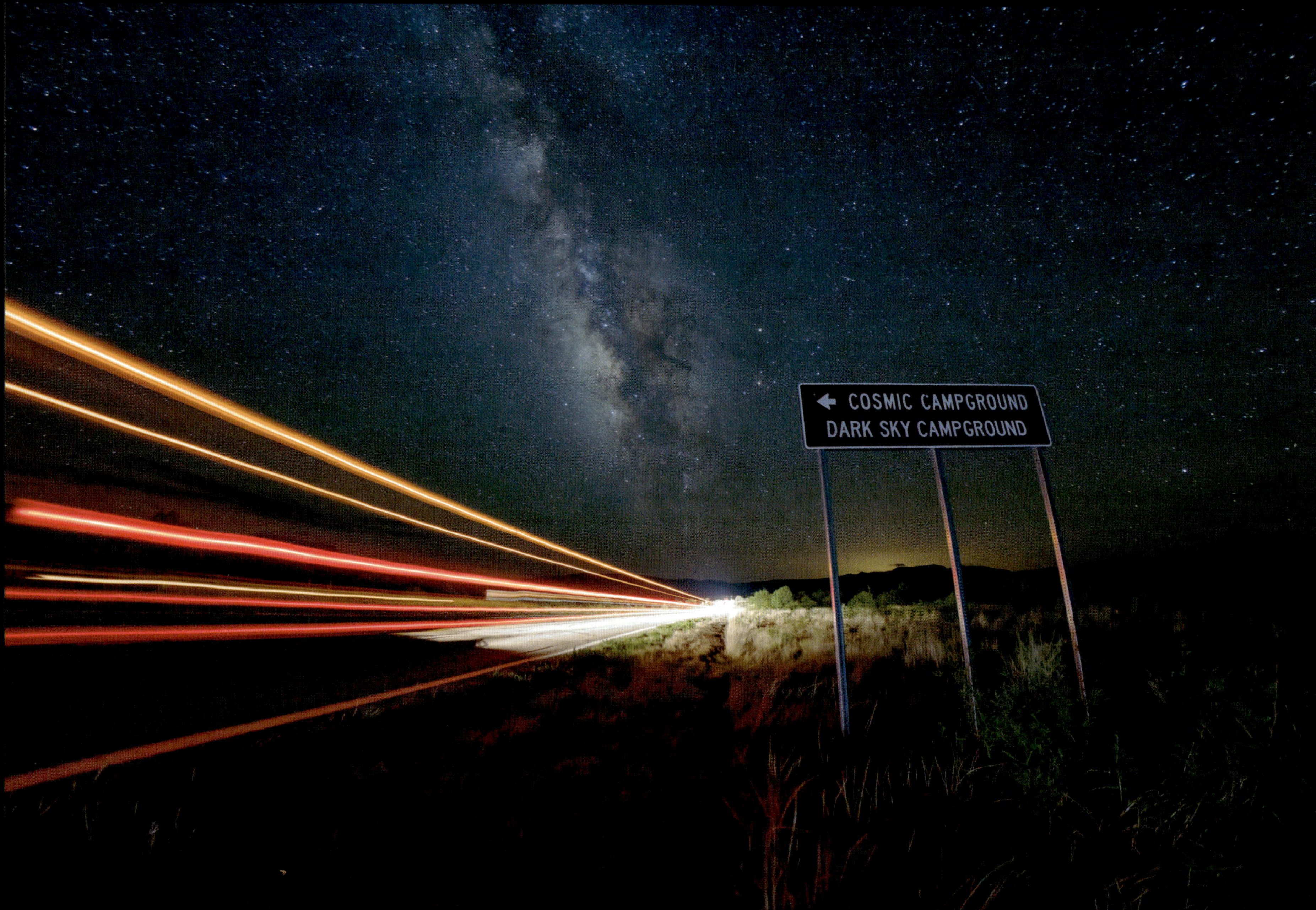

Amateur astronomers also responded to the increase in light pollution by establishing remote campgrounds where they would be allowed to performtelescope observations. In 2015, one such location, New Mexico's Cosmic Campground, became the IDA's First International Dark-Sky Sanctuary.

First International Dark-Sky Sanctuary, Cosmic Campground, New Mexico

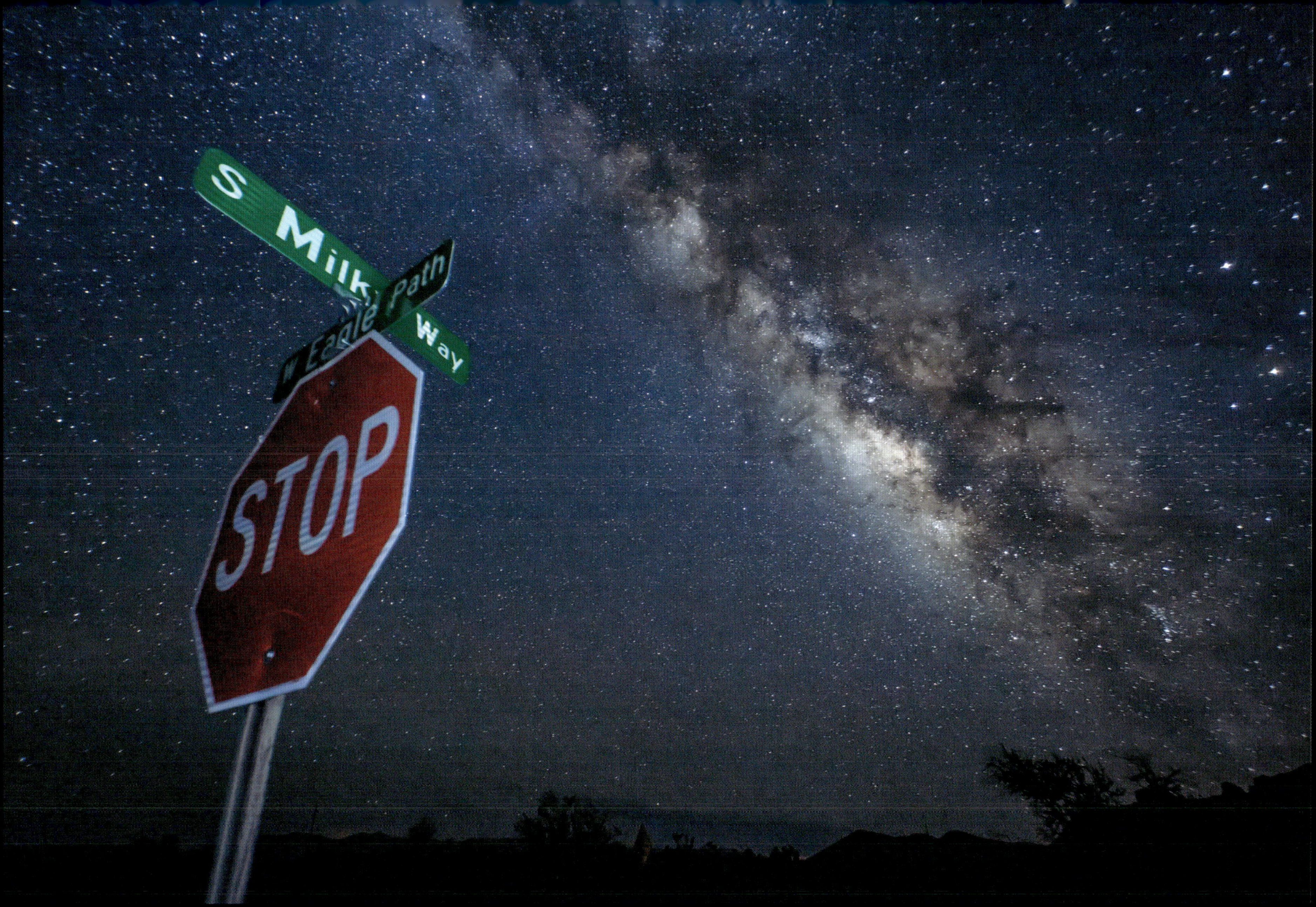

Some amateur astronomers went so far as to create their own towns, such as Arizona's Sky Village, where residents can observe largely untarnished night skies throughout most of the year due to its location in the dry Chiricahua Desert and no significant light pollution in a hundred-mile radius.

Arizona Sky Village, Portal, Arizona

1

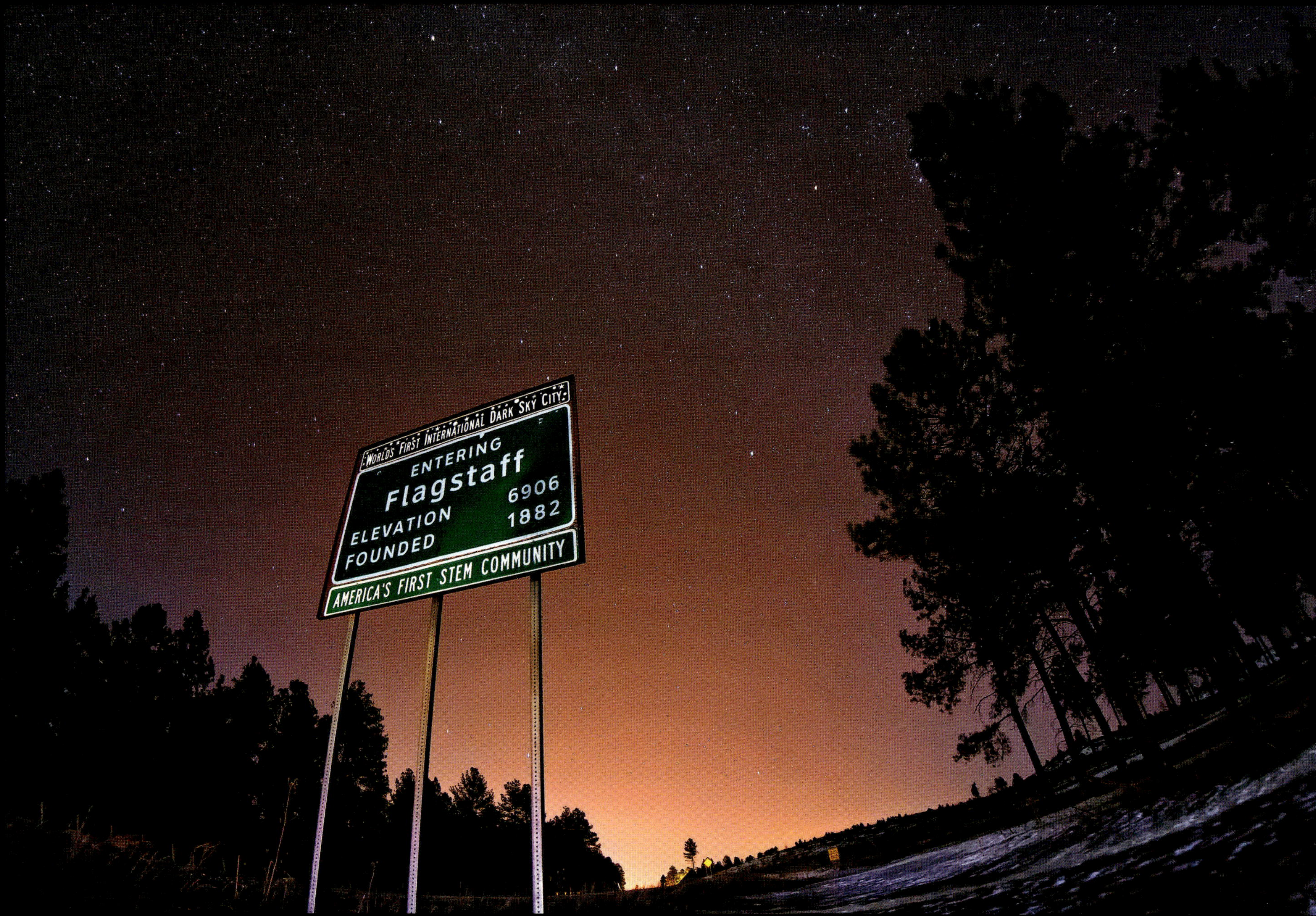
WORLD'S FIRST INTERNATIONAL DARK SKY CITY
ENTERING
Flagstaff
ELEVATION 6906
FOUNDED 1882
AMERICA'S FIRST STEM COMMUNITY

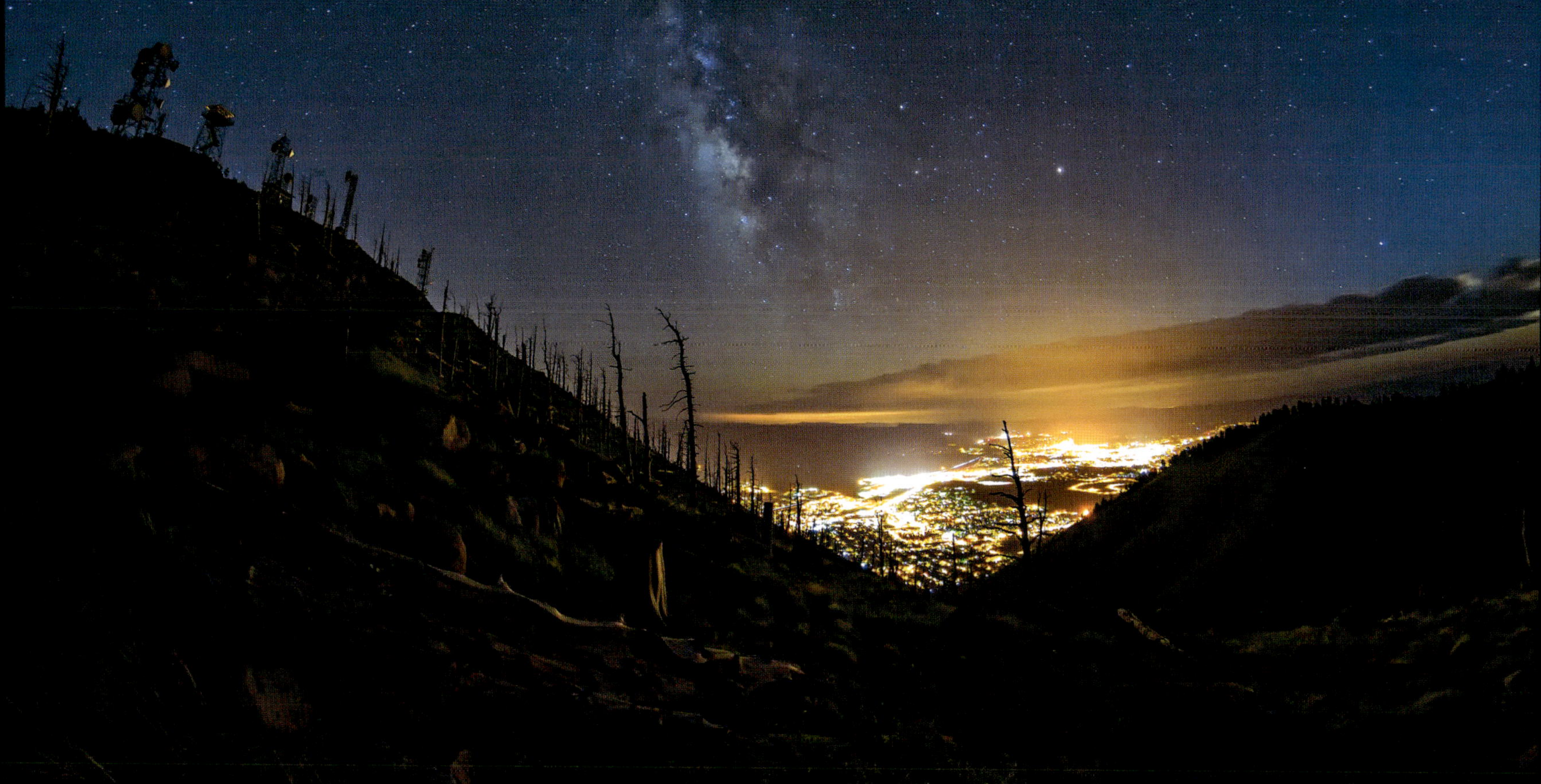

The fight for dark skies began long before IDA was formed. In 1958, the small mountain town of Flagstaff, Arizona, became the first municipality in history to enact a light pollution control ordinance. Flagstaff is home to the Lowell Observatory, where Pluto was discovered, the early Martian surface research was conducted, and the first evidence of Dark Matter was registered. Flagstaff's Northern Arizona University Astronomic Research Observatory was first to conduct an infrared observation of the sky from Earth's surface.

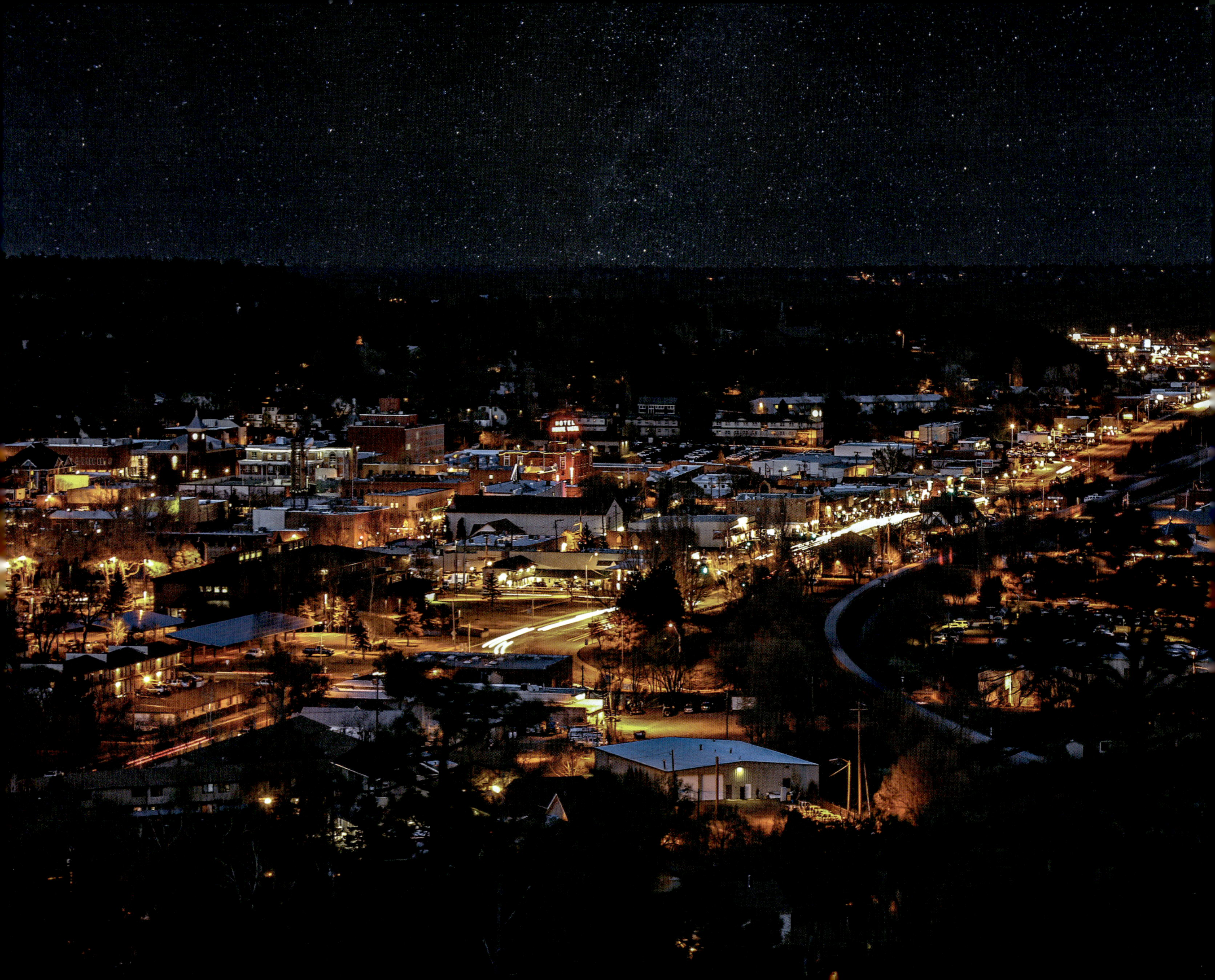
HOTEL

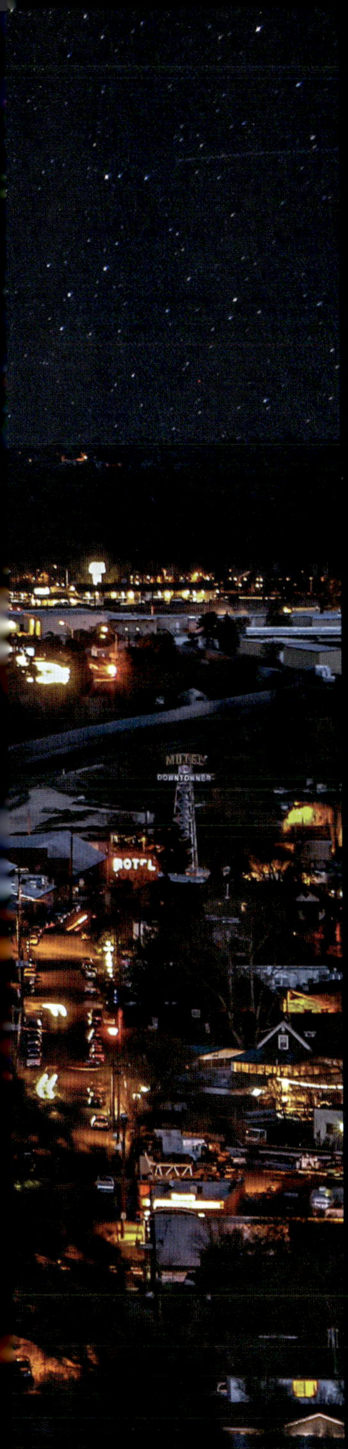

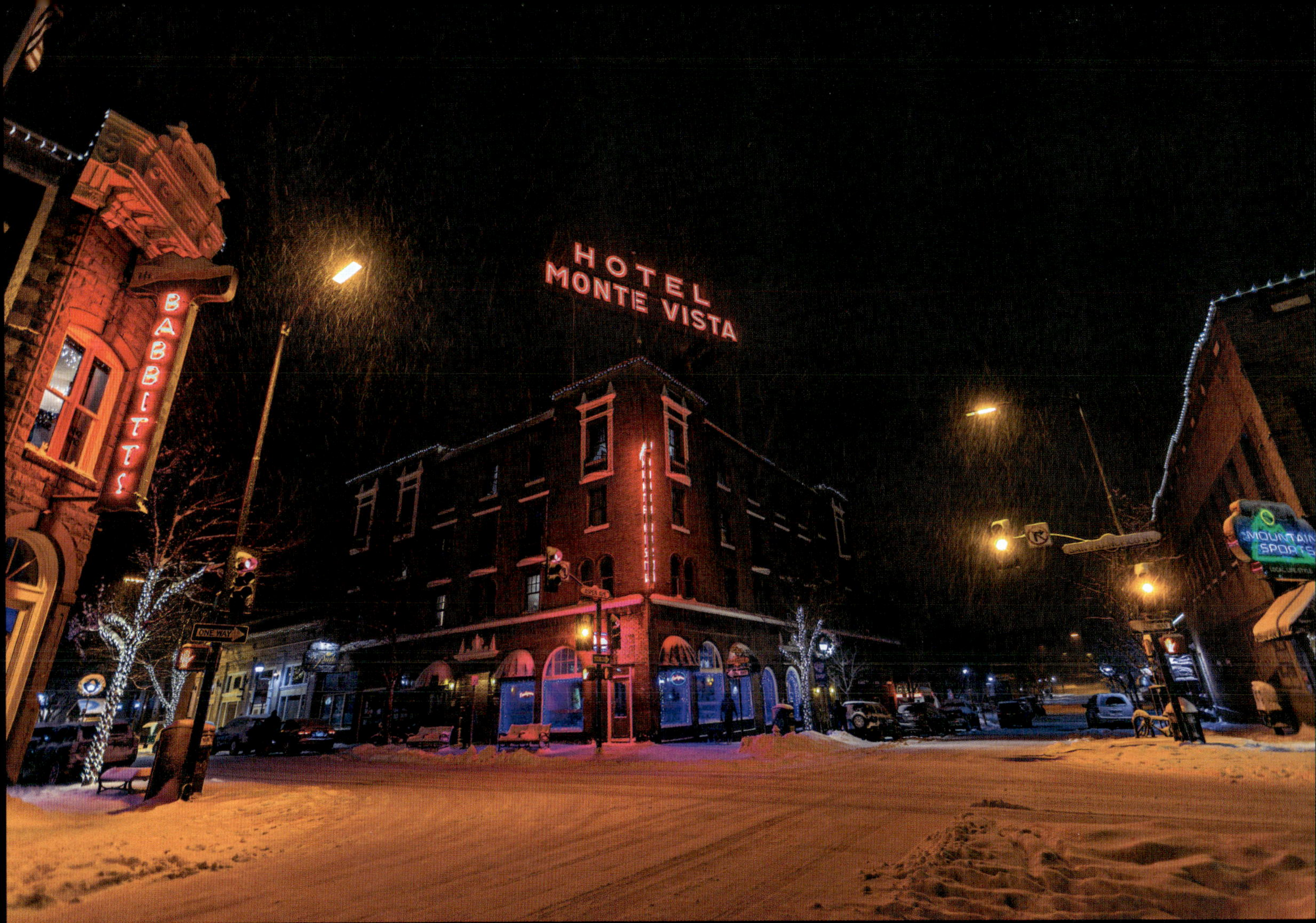

Flagstaff's light control ordinance prohibited "the use of commercial searchlights within the city limits." Punishments for violating the ordinance included "imprisonment in the city jail not to exceed 90 days."

Protection of the night sky continued for decades as Flagstaff grew in population. Coconino County, where Flagstaff is located, enacted a lighting code in 1973, covering the second-largest county in the United States. In 1981, the county took it a step further, banning all billboards and the accompanying lighting.

Finally, in recognition of its pioneering role in protecting the night sky, Flagstaff was designated the World's First International Dark-Sky Community by the IDA in 2001.

Downtown Flagstaff, Arizona

6

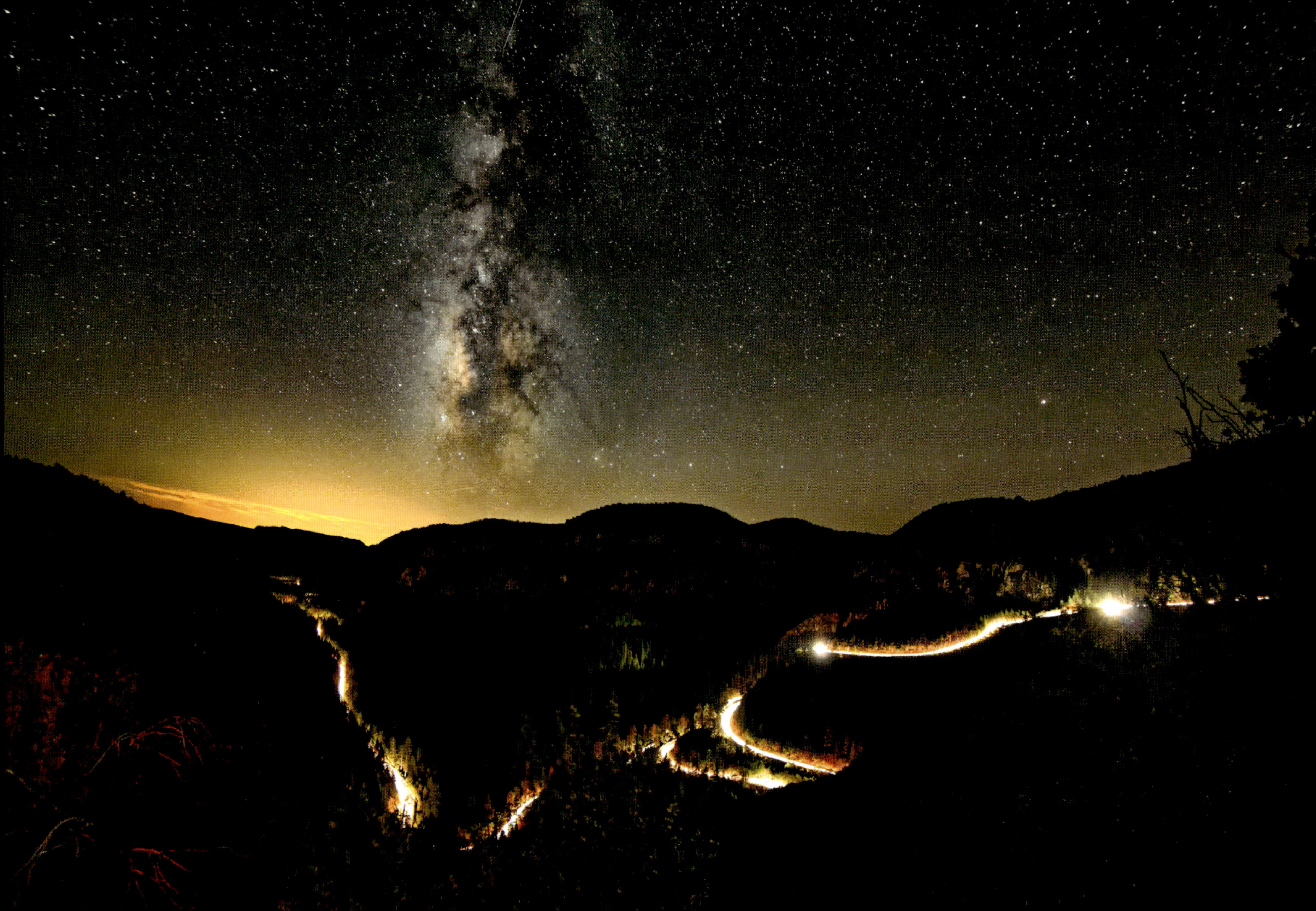

Oak Creek Canyon, facing the Eighth International Dark-Sky Community, Sedona, Arizona

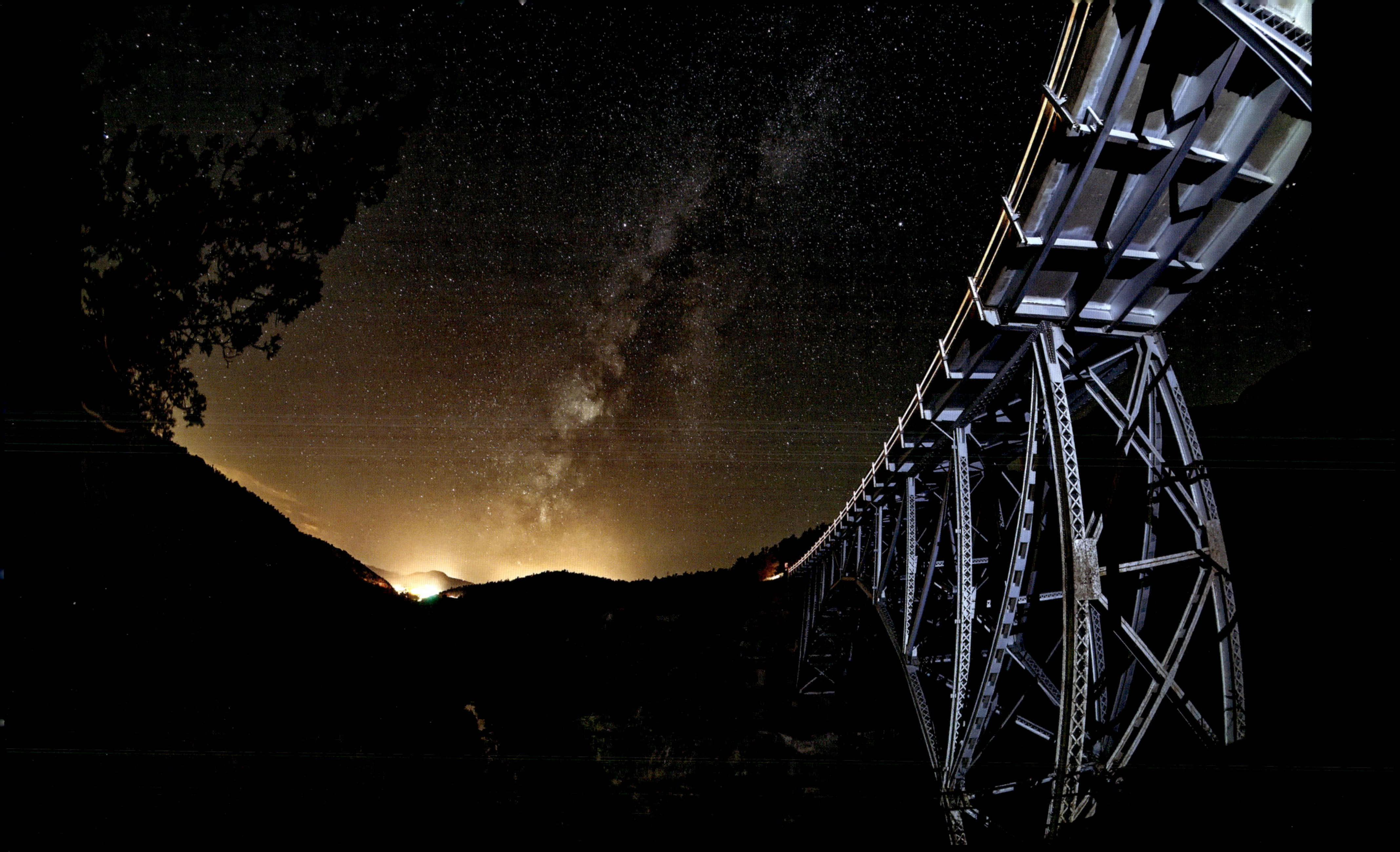

Flagstaff was not alone; other municipalities followed. Just 30 miles south, down the Oak Creek Canyon, the city of Sedona was designated the Eighth International Dark-Sky Community in 2014, upon instituting a light pollution control ordinance. Today, those who travel through the Oak Creek Canyon, which joins the two cities, can observe the Milky Way with their own eyes.

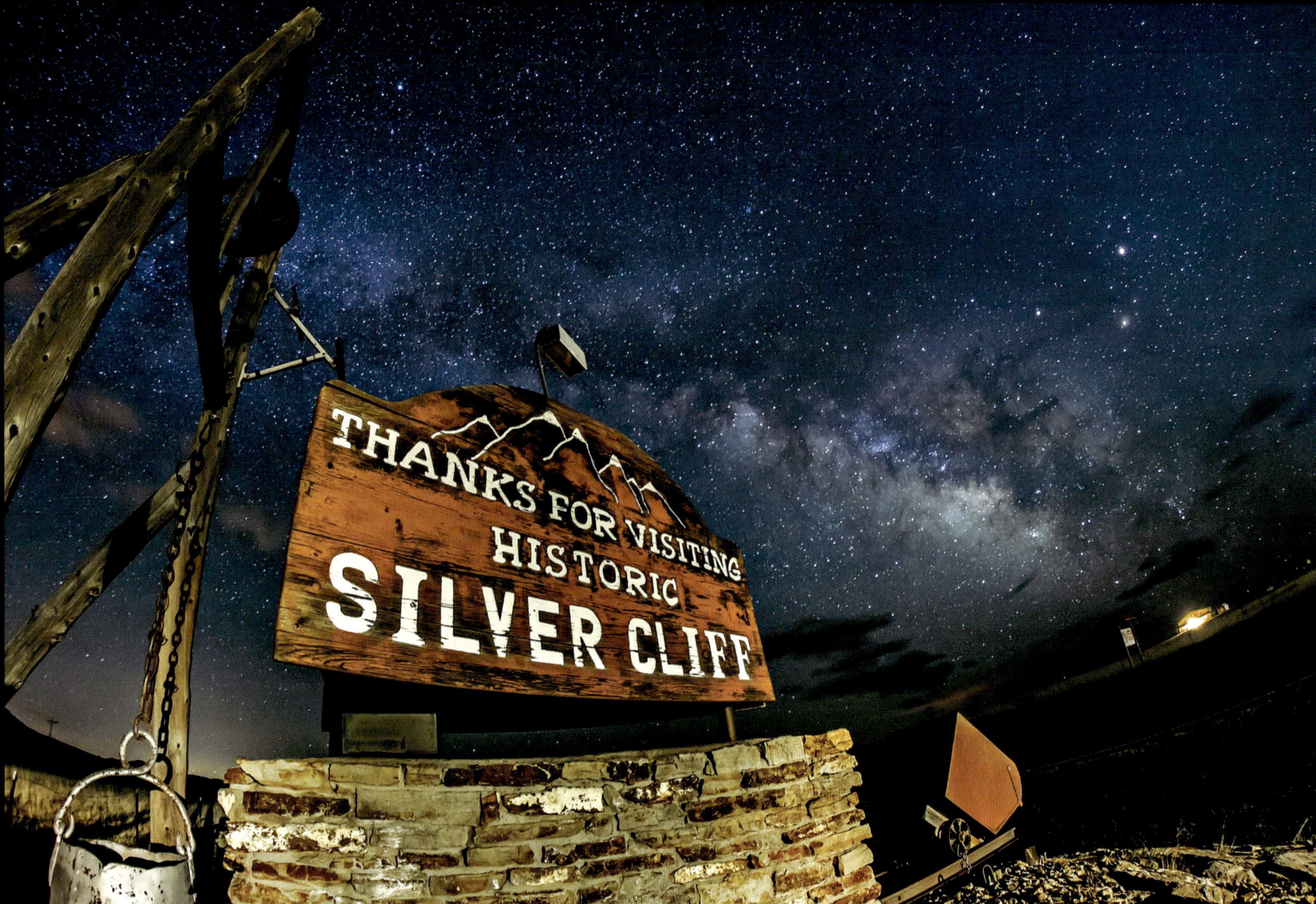

Ninth International Dark-Sky Community, Westcliffe/Silver Cliff, Colorado

3

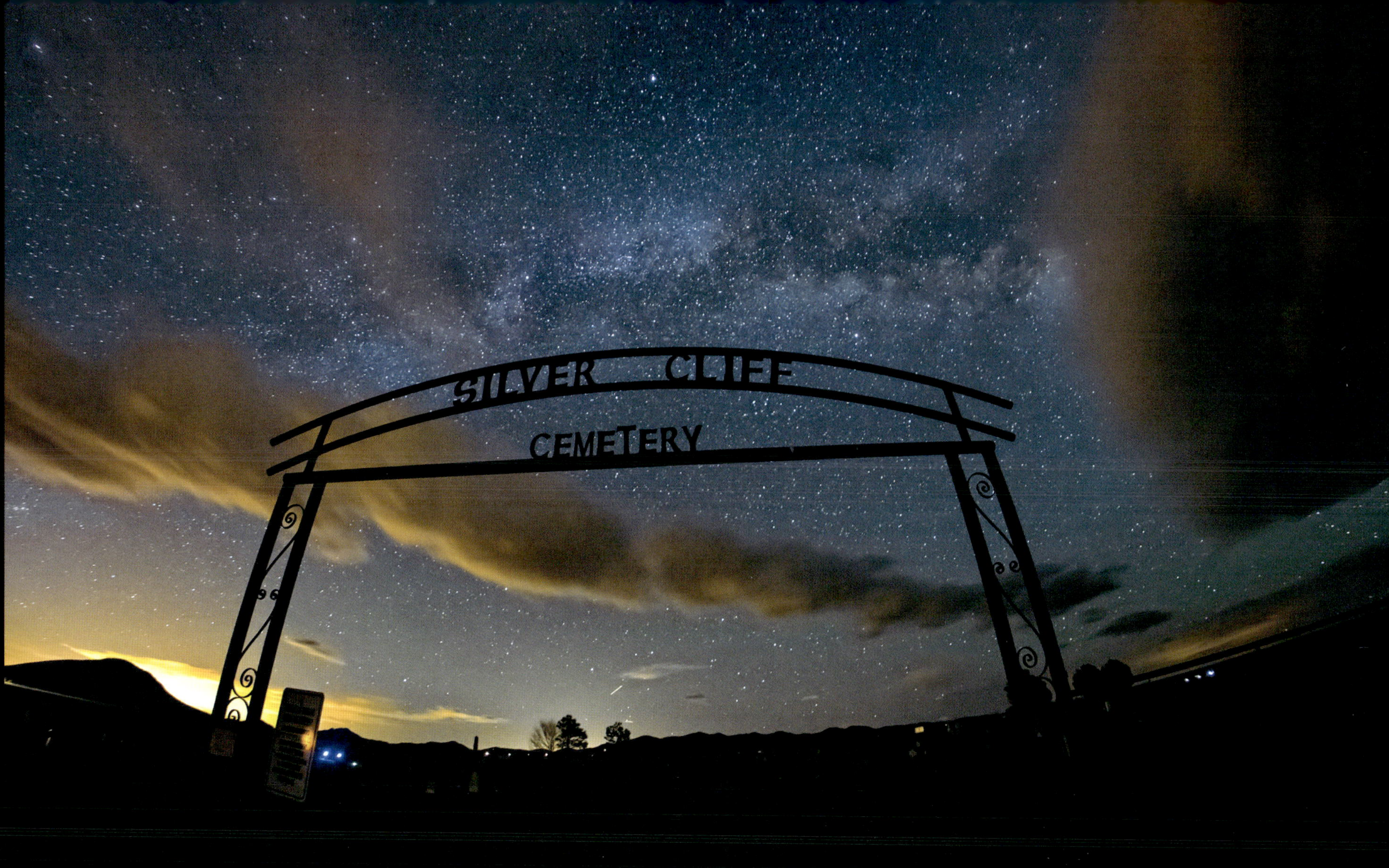

Places such as the Ninth International Dark-Sky Community, Westcliffe/Silver Cliff, acted to protect their local night light phenomena such as the Silver Cliff Cemetery's "dancing blue lights," which are seen on occasion around the cemetery. According to reports, the lights resemble blue lanterns or white spheres, and are said to float through the cemetery and bounce on headstones.

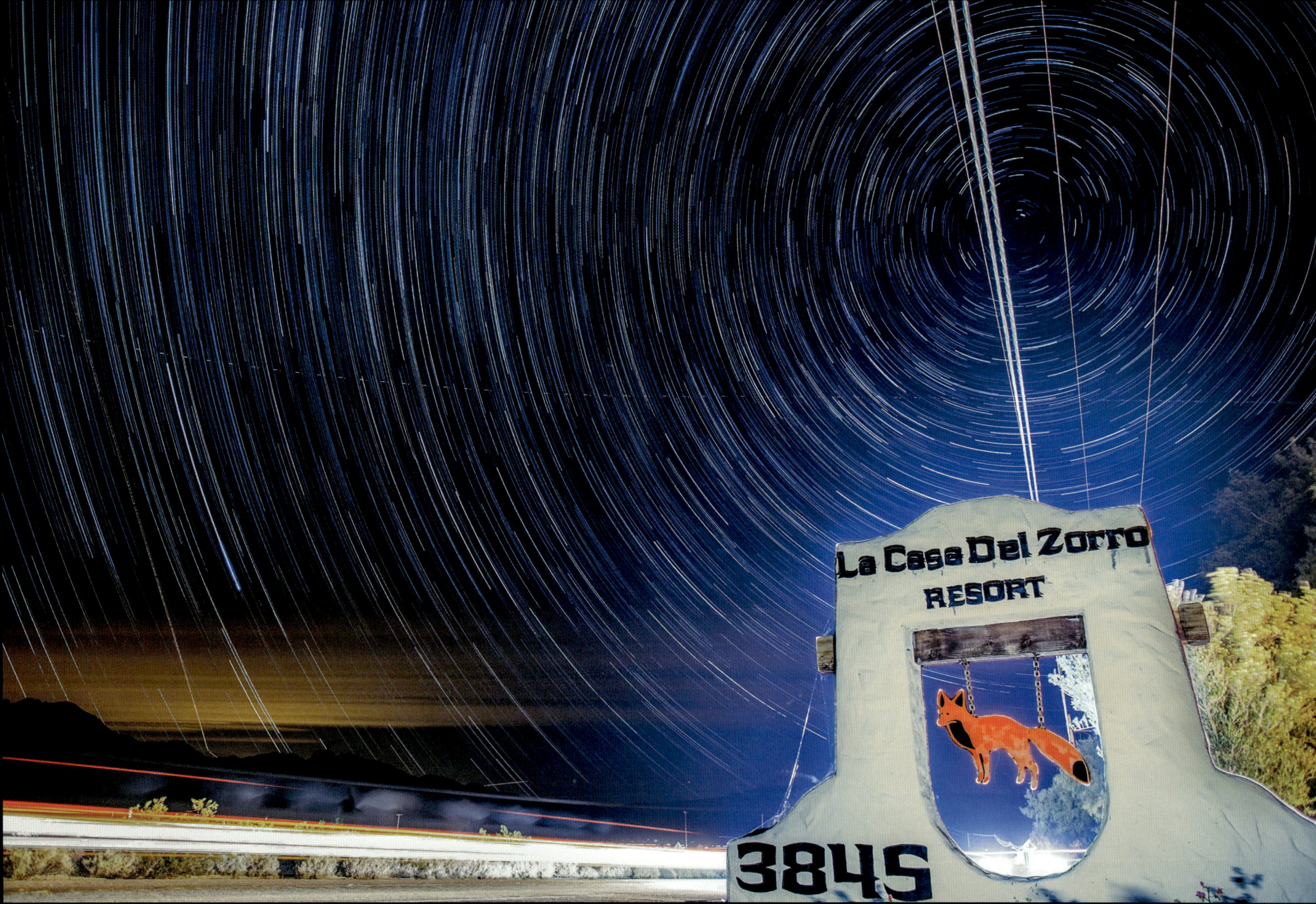

Second International Dark-Sky Community, Borrego Springs, California

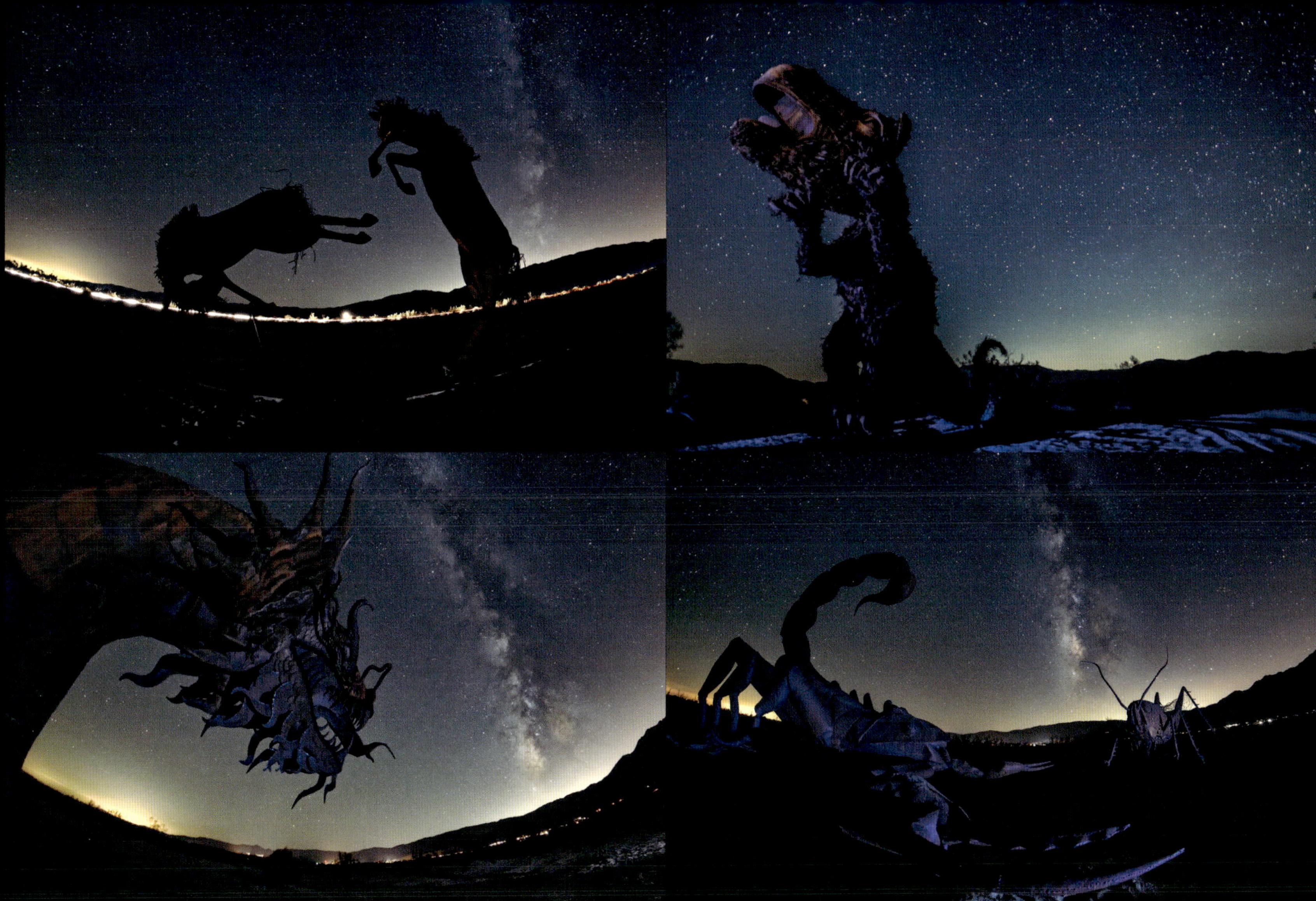

We don't leave a light on for anyone. We even tell people to turn out their porch lights. We put shields on our lights so they don't shine up — and we use amber hues and lower-wattage bulbs. We manage mankind's light so it doesn't negatively impact the night sky. If you come here, you won't be able to see your hand in front of your face — but you'll see the Milky Way like you've never seen it before.

- Linda Haddock, director of the Chamber of Commerce of Borrego Springs, a town just a few hours away from Los Angeles, which was designated the Second International Dark-Sky Community in 2009

Half The Park Is After Dark

By Tyler Nordgren

On a lovely May night in 2005 I did as millions of visitors to America's national parks do each year, I went to an evening ranger talk. That night in an outdoor amphitheater in Yosemite National Park, the park ranger showed us pictures of planets and moons, and Hubble Space Telescope photos of galaxies containing a hundred billion stars. As each slide appeared the crowd ooh'd and aah'd at sights the average person has never seen before. I'd seen them all, of course, because I'm an astronomer.

I've spent my life traveling to far-off observatories, and at each of these places with names like Palomar and Mauna Kea a million stars stretch from horizon to horizon every moonless night of the year. I love the night and what it reveals about the universe in which we live. As a result, I knew why I was there in that audience that night, what I didn't at that time understand was why everyone else was there. But as the show continued, and evening twilight gave way to fully dark night, the stars came out and I at last understood. The sight of a star-filled sky is as rare today for most people as the views of waterfalls and wildlife that bring visitors to the parks by day.

Current estimates, based on light-pollution measurements, are that 60% of US and European residents no longer live where the Milky Way is even faintly visible. The World Health Organization estimates that 50% of the world's population now lives in urban areas, meaning that 50% of the children born this year will probably never see the Milky Way in their lives.

That's a sobering thought. We have no problem imagining a road trip or a family vacation to visit the Grand Canyon; there is only one such place so of course we should have to travel to see it. But the Milky Way used to be something everyone could see from wherever they lived. Now it too is something that can only be seen by a conscious effort, but only by those who know it's still there. Our loss of a natural night sky has been so gradual that most of us hardly know what we are missing.

But the National Park Service has begun to realize that a natural night sky is an important resource, worthy of protection. Surveys of park visitors conducted for a number of western parks reveal that a starry night sky is as important to people as wildflowers and wildlife. It's not hard to understand why when you consider that for the vast majority of people a typical sky is now a featureless orange glow. We still talk about "sleeping under the stars" when we think about camping and our mental image is of a tent under a starlit canopy.

Attendance numbers reveal astronomy programs are typically one of the top ranger programs parks offer. In addition, over the last two decades, more and more parks around the country have begun to offer "Night Sky Parties" and "Astronomy Festivals."

Bryce Canyon National Park has been offering these since the 1990s. What began with an astronomy volunteer and an interested ranger or two offering a talk and a look through a telescope is now a three-day festival of talks, model rockets, scale model solar system walks, night sky constellation tours, and dozens of telescopes set up by local astronomy clubs specifically for the public to see the stars and Milky Way overhead in the midst of one of Utah's great dark-sky regions.

In 2010 I stood on the rim of Bryce Canyon chatting with former NASA astronaut, Story Musgrave, who was responsible for fixing the Hubble Space Telescope

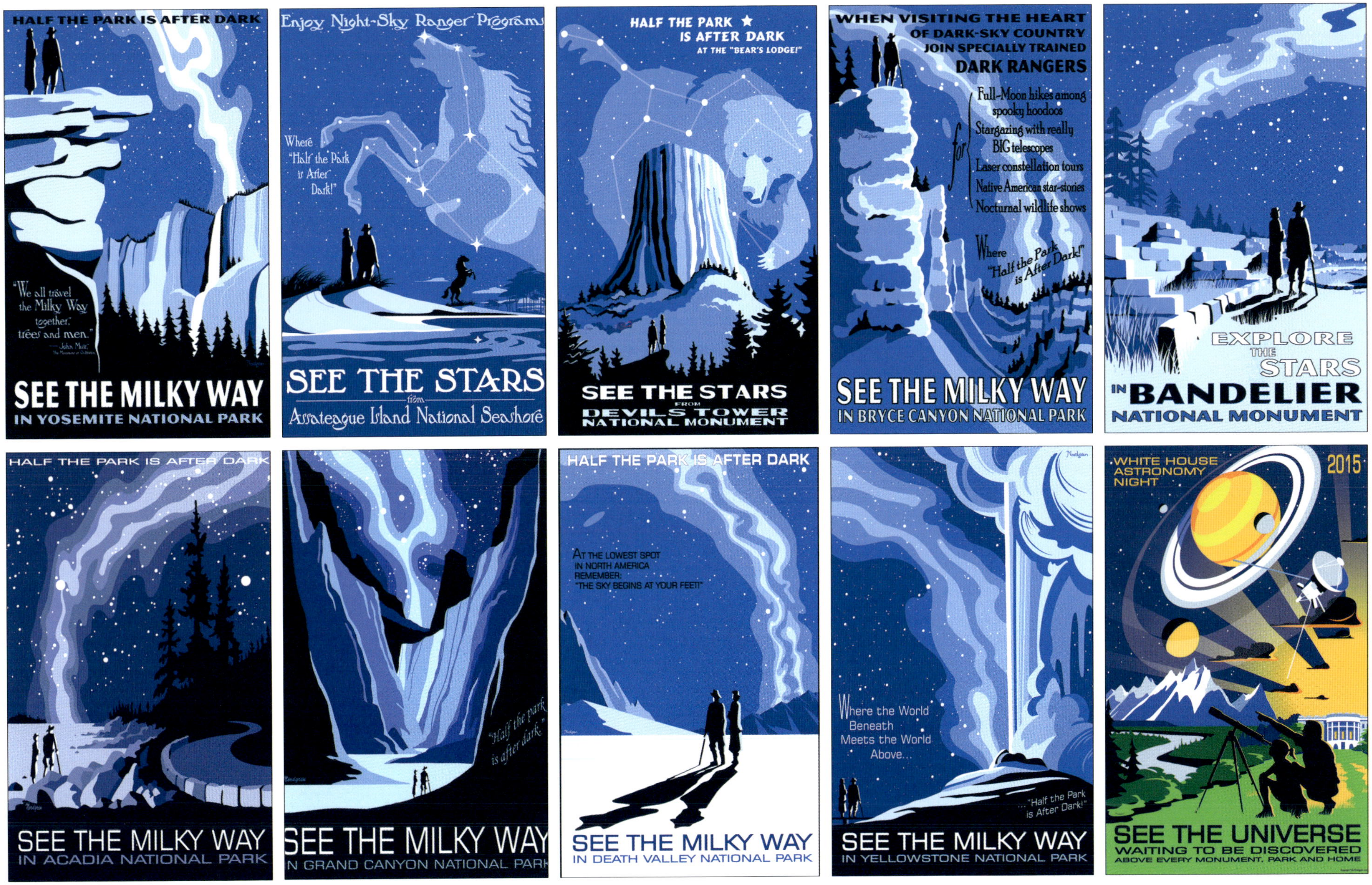

Tyler Nordgren National Park Posters

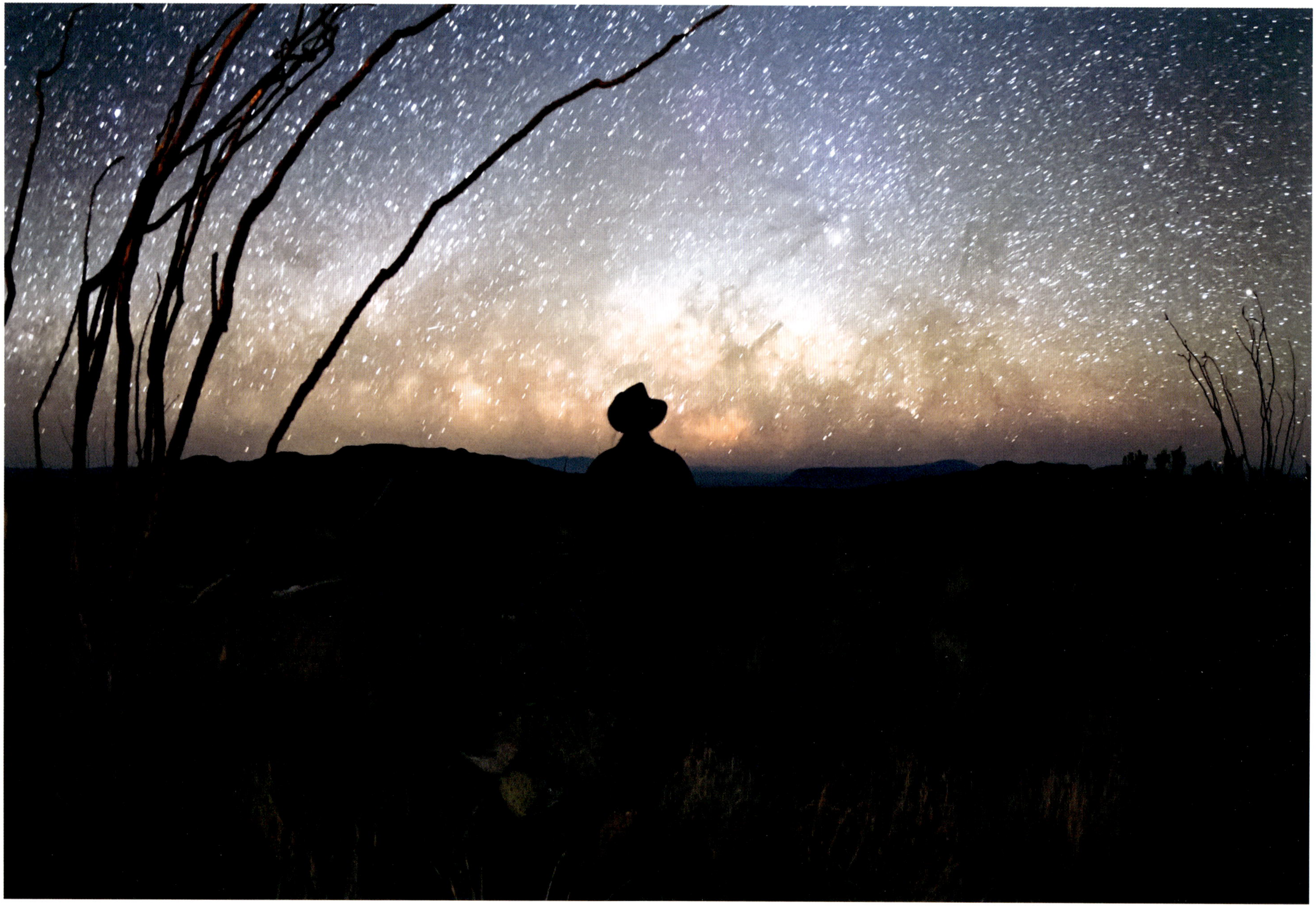

Tyler Nordgren self-portrait as he observes the rare 360-degree view of the rise of the Milky Way at Big Bend, one of the darkest National Parks in the country.

Tenth International Dark-Sky Park, Big Bend National Park, Texas

in orbit in the early 1990s. We talked about the beauty of the parks and our love of the night sky and I posed a question that a lighting engineer had once asked me. When we have space telescopes like Hubble, where every school child can see galaxies at the edge of the universe on their computer monitors, why do they still need to be able to see the stars? Musgrave looked out over the multi-hued hoodoos that stretched for as far as we could see and said, "We have pictures of all of this on the internet, so why should anyone still come here?"

The National Park Service was created in 1916 to protect and preserve the scenery, natural, and historical objects of the parks for the enjoyment of future generations. For much of the last century these natural scenes did not include the sky above the horizon, and definitely not the sky at night. As the 20th century gave way to the 21st, park rangers at Death Valley National Park and Pinnacles National Monument (now National Park) in California noticed that the skies above their parks were getting noticeably brighter. They set out to measure how bright this polluting light was getting and if they could determine where it was coming from. Over the next decade this Night Sky Team, a small group of dedicated park rangers and astronomers, measured the darkness of parks all over the country and began to bring the message of our nation's night skies as a natural resource in need of protecting.

In 2007, on the basis of these observations, the International Dark-Sky Association, in conjunction with the National Park Service, declared Natural Bridges National Monument in southern Utah, the world's first International Dark-Sky Park. Standing beneath the enormous sandstone arcs visitors to this tiny monument experience the night sky the way every generation of human beings once saw it. There, the center of our galaxy is the brightest thing the human eye sees (once the Sun and Moon have set) and the light of a billion stars casts shadows across the ground. It's a magical experience and one more and more people are experiencing each year. In the years since Natural Bridges' designation visitors who before would simply drive by or spend no more than an hour looking around now make plans to stay past sunset.

And other parks and the communities that surround them have begun taking note. Death Valley, Big Bend, Great Basin, and Grand Canyon National Parks have all been designated international dark-sky parks, with more applying for designation every year. As it becomes increasingly difficult to see the stars elsewhere, astronomical tourism is growing in popularity and communities looking to support local businesses in faraway rural regions (like Bar Harbor, Maine, outside Acadia National Park) realize that it's not a matter of stars versus safety when it comes to limiting light pollution, or even stars versus dollars when it comes to business. But rather we all win when local communities help protect the night sky and give the public a place to see the beauty the parks preserve by both day and night.

Half the Park is After Dark is a phrase I first thought of in Big Bend in far-distant West Texas along the Rio Grande. There, over two years after my eye-opening experience in Yosemite, I spent five nights camped beside my car in the back country of this most isolated park. At night, there is no trace that the electric lightbulb has been invented and the stars visible from horizon to horizon give an actual sense of three-dimensional distance to the cosmos.

Near midnight on a stunningly clear night in March, I beheld a sight I had never seen before and have never seen since. The Milky Way rose over a full 180 degrees of horizon at once. Only in the most southern parts of the US does this phenomenon ever happen. And as I stood there staring off towards the southeast horizon I saw the center of the galaxy rise with its arms wrapped around me, circling the horizon. At that moment my face was lit by the light of a billion suns, and at that instant I knew where I was in the Galaxy.

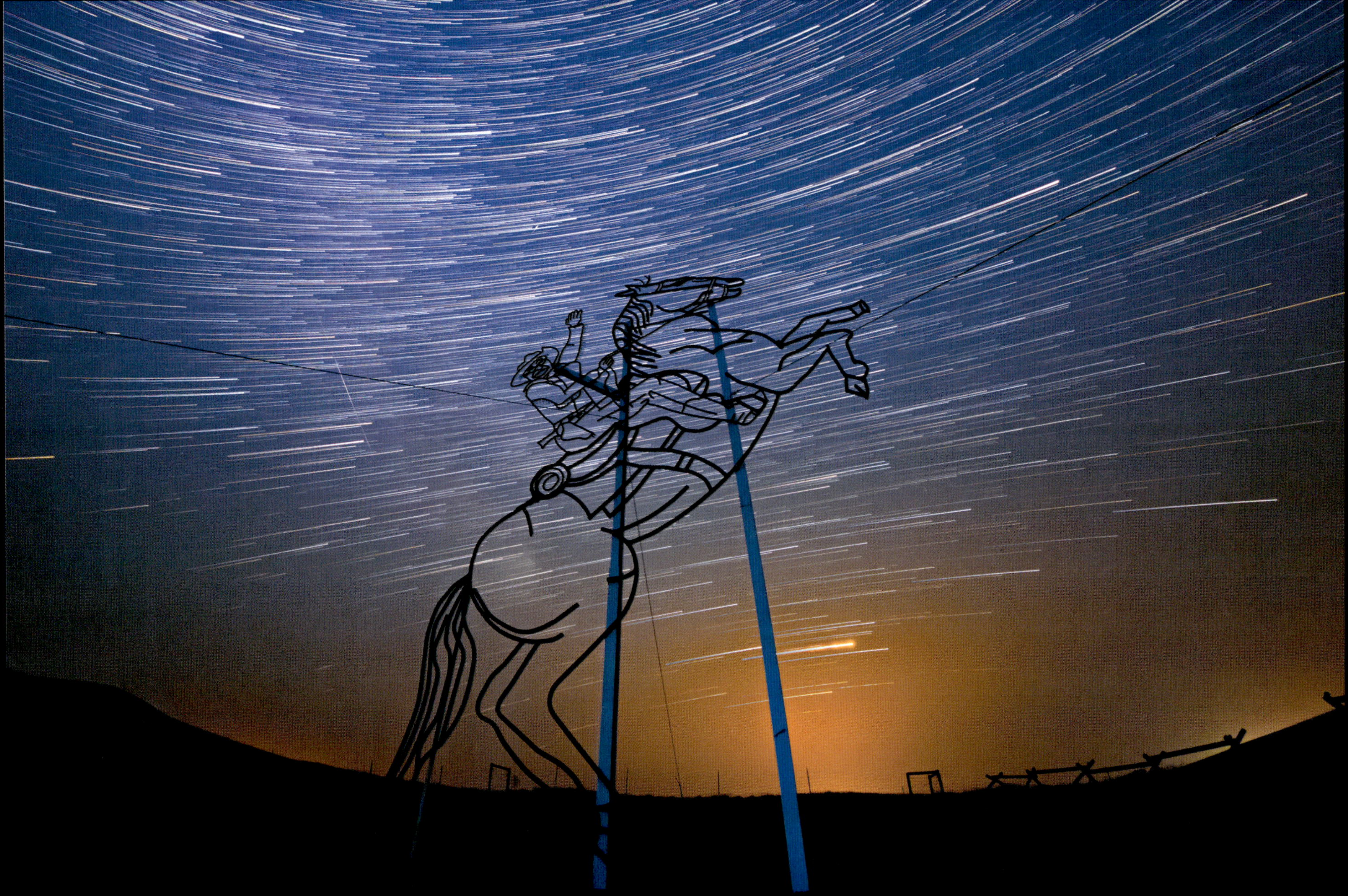

Considered one of the most important figures in National Park conservation, President Teddy Roosevelt presided over the then-largest expansion of protected federal lands in history.

Teddy Roosevelt "The Conservation President" Statue, Enchanted Highway, North Dakota

3

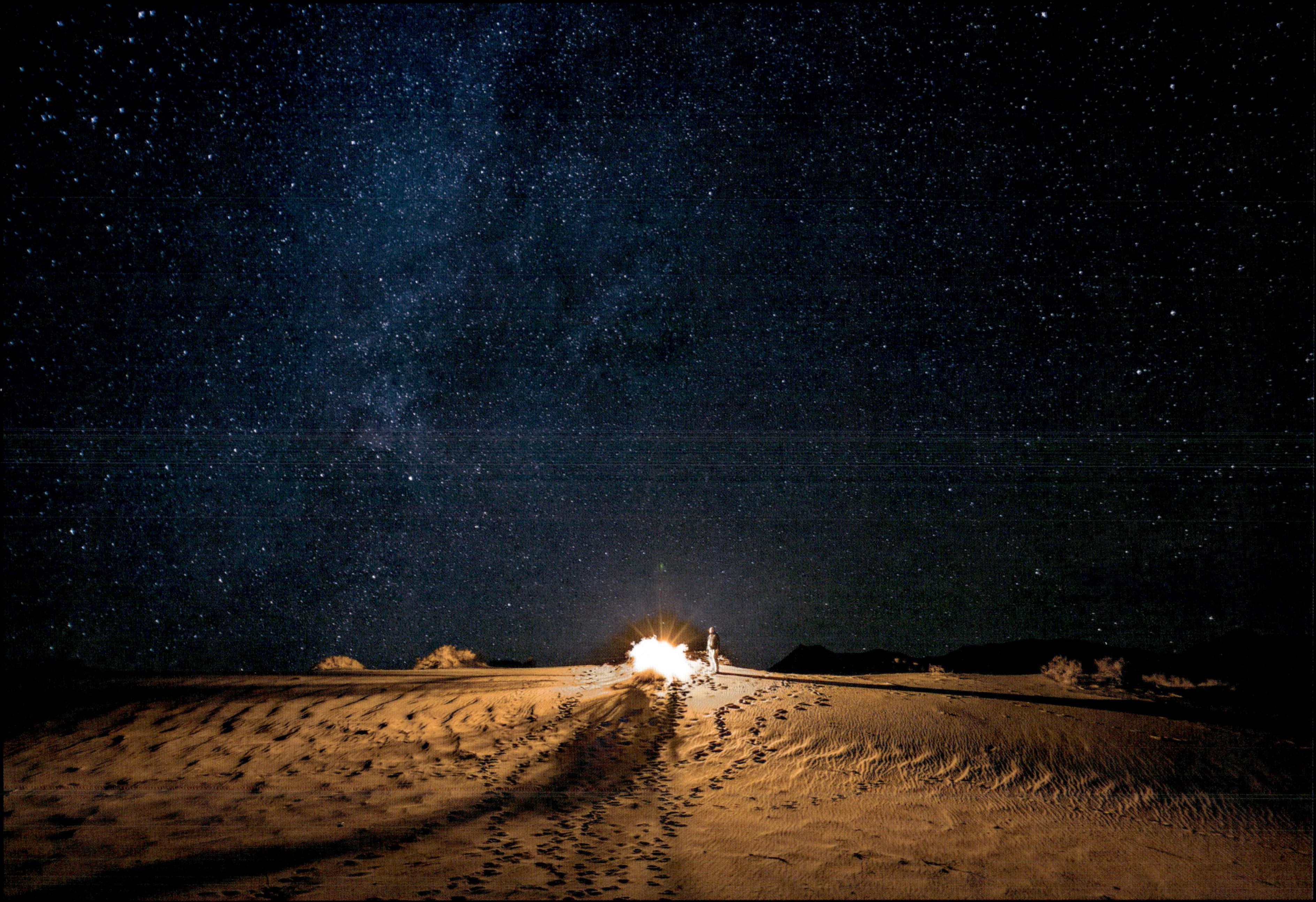

Considered one of the best dark-sky observation areas in the world, Death Valley National Park benefits from mainly clear skies, little air pollution, and very low humidity.

Eureka Dunes, Death Valley National Park, California

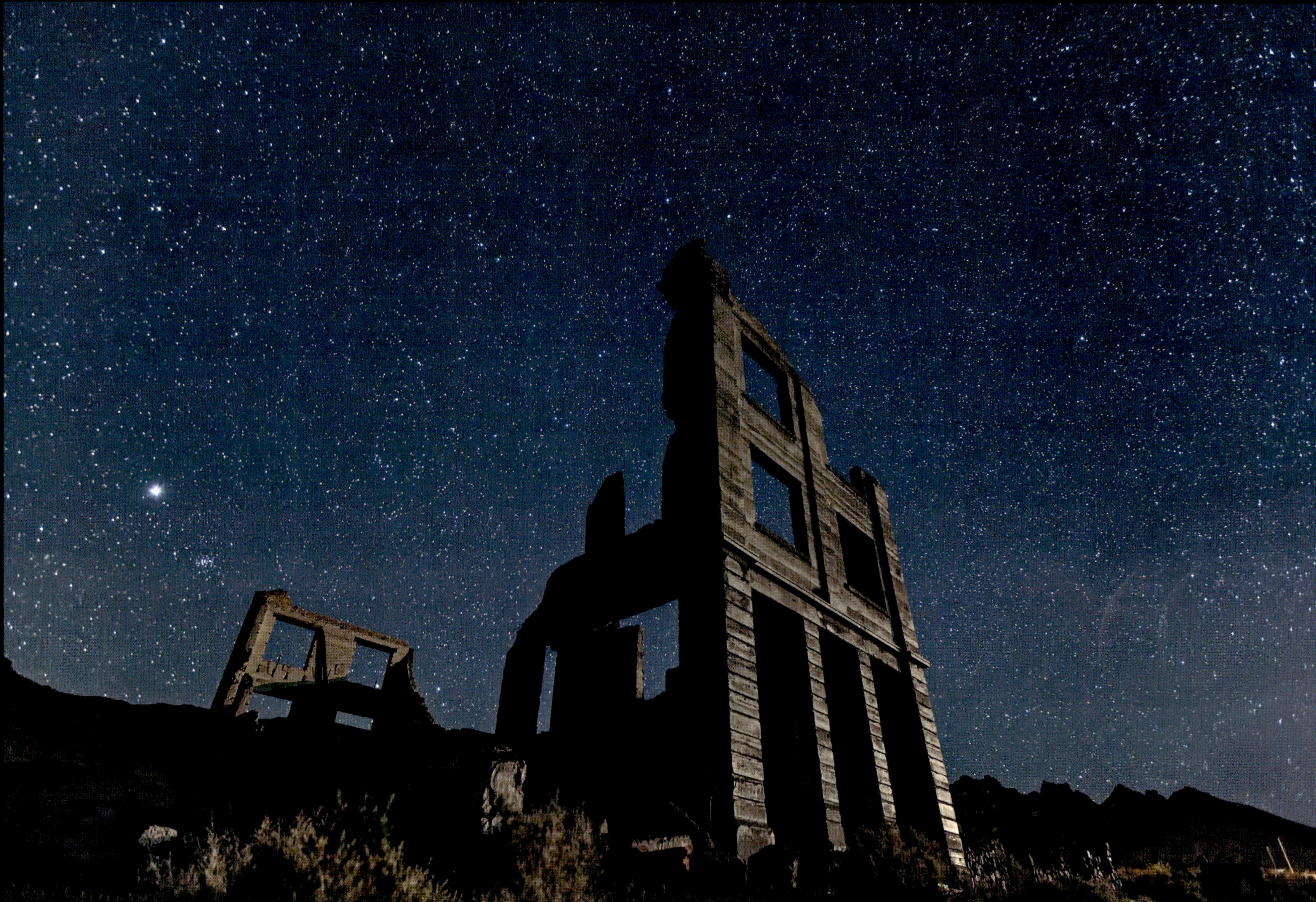

One of the other key reasons why Death Valley maintains great dark skies is that few people remain permanently situated there,

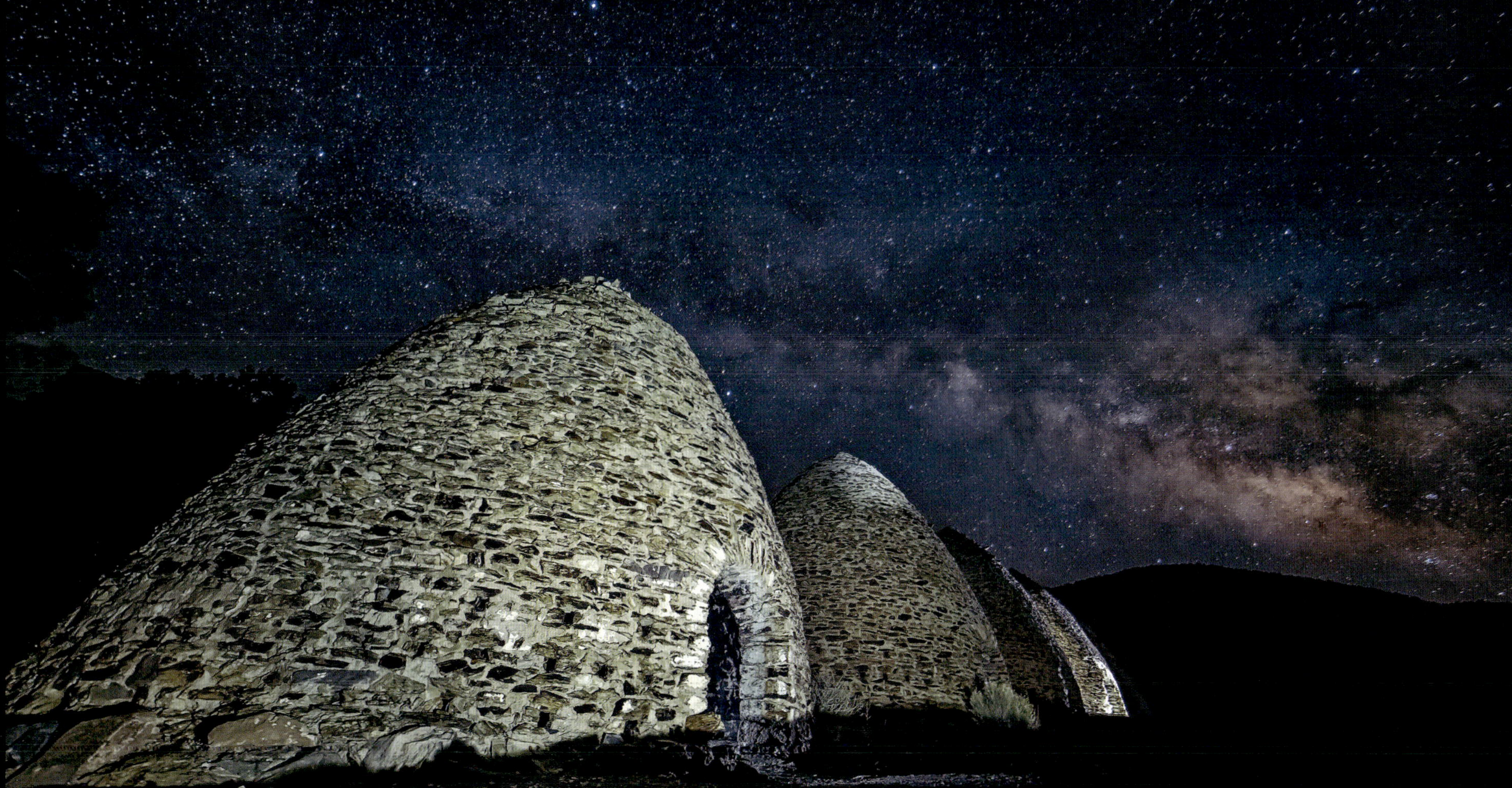

Death Valley National Park was designated the Eleventh International Dark-Sky Park by the IDA in 2013.

Charcoal Kilns, Death Valley National Park, California

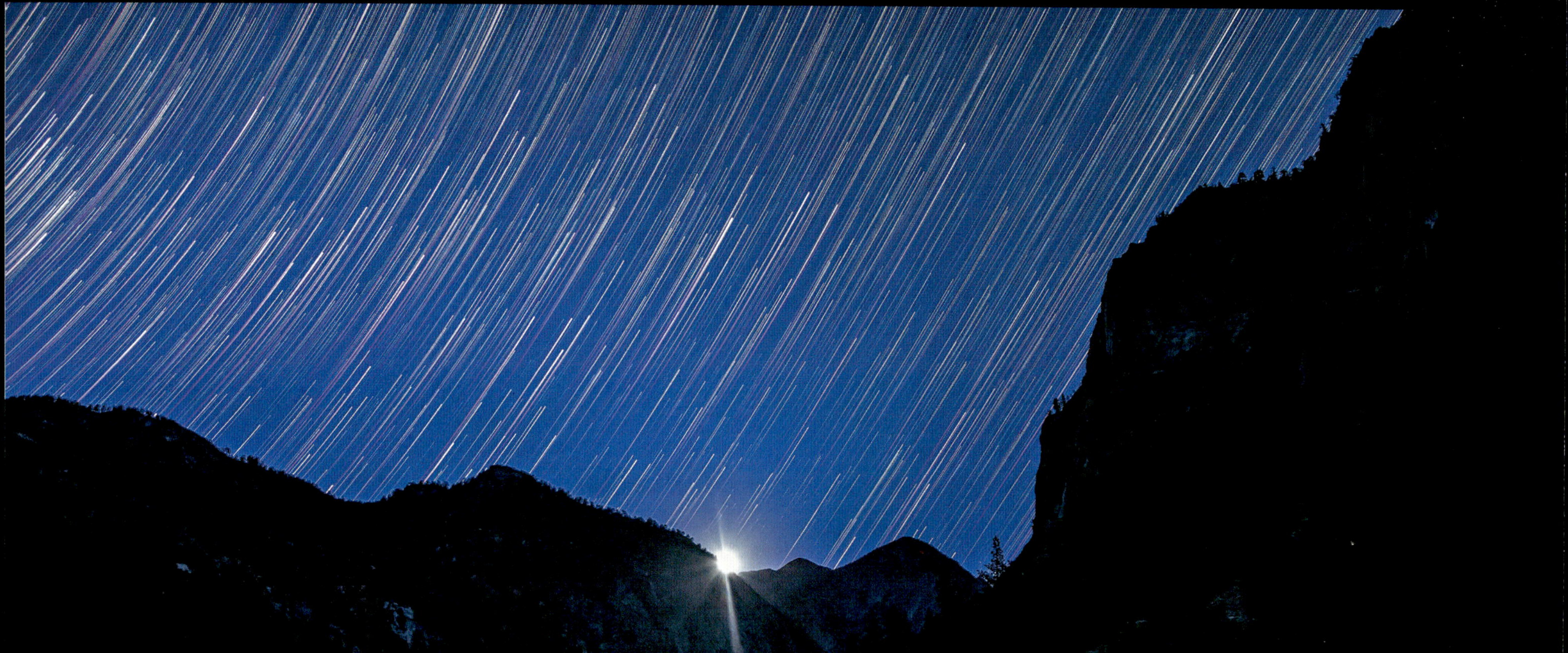

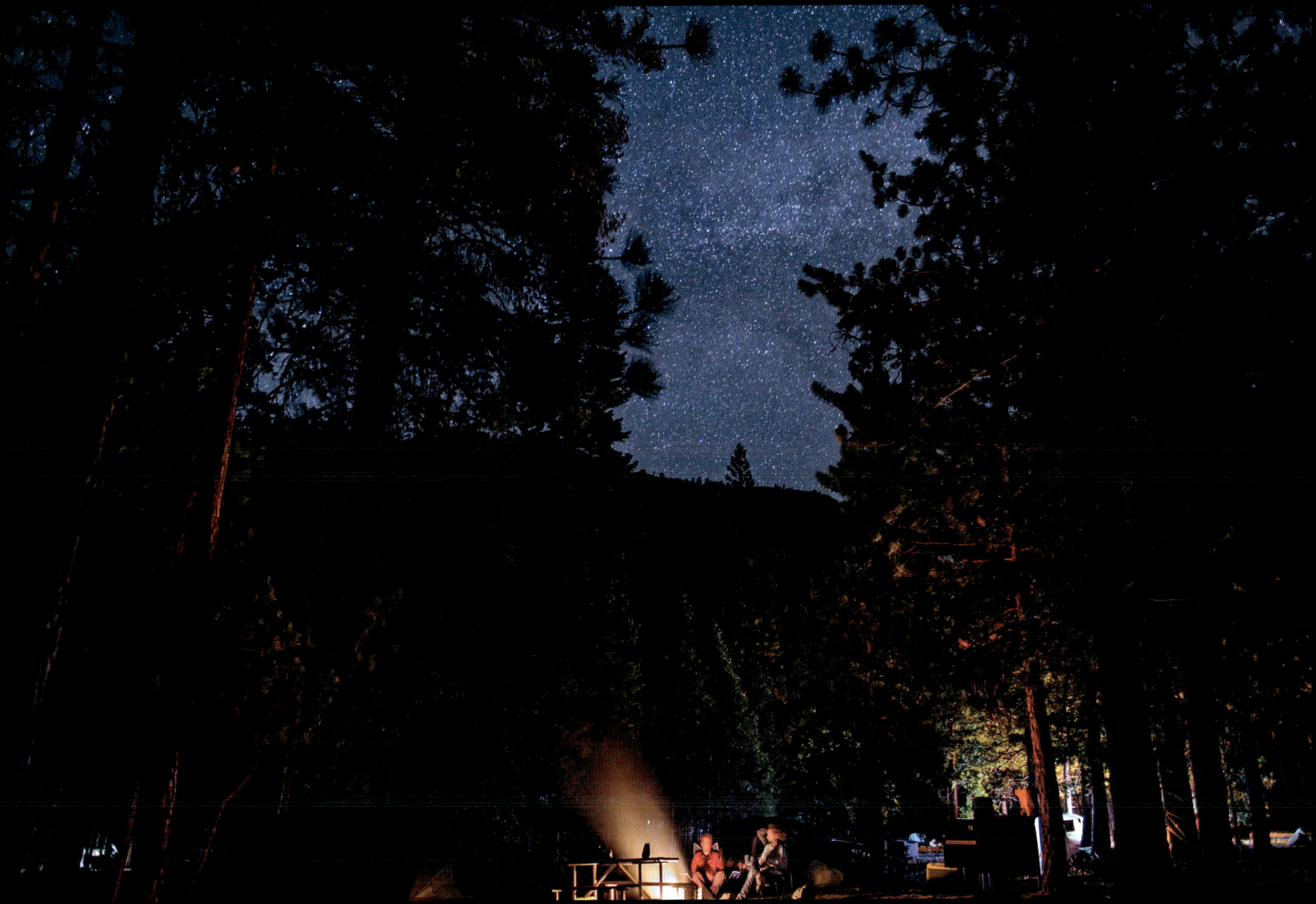

Just west of Death Valley sits Kings Canyon National Park, where residents of heavily light-polluted cities such as Los Angeles, San Francisco, Fresno, and Sacramento have escaped for years in order to enjoy the starry skies above Canyon View Campground.

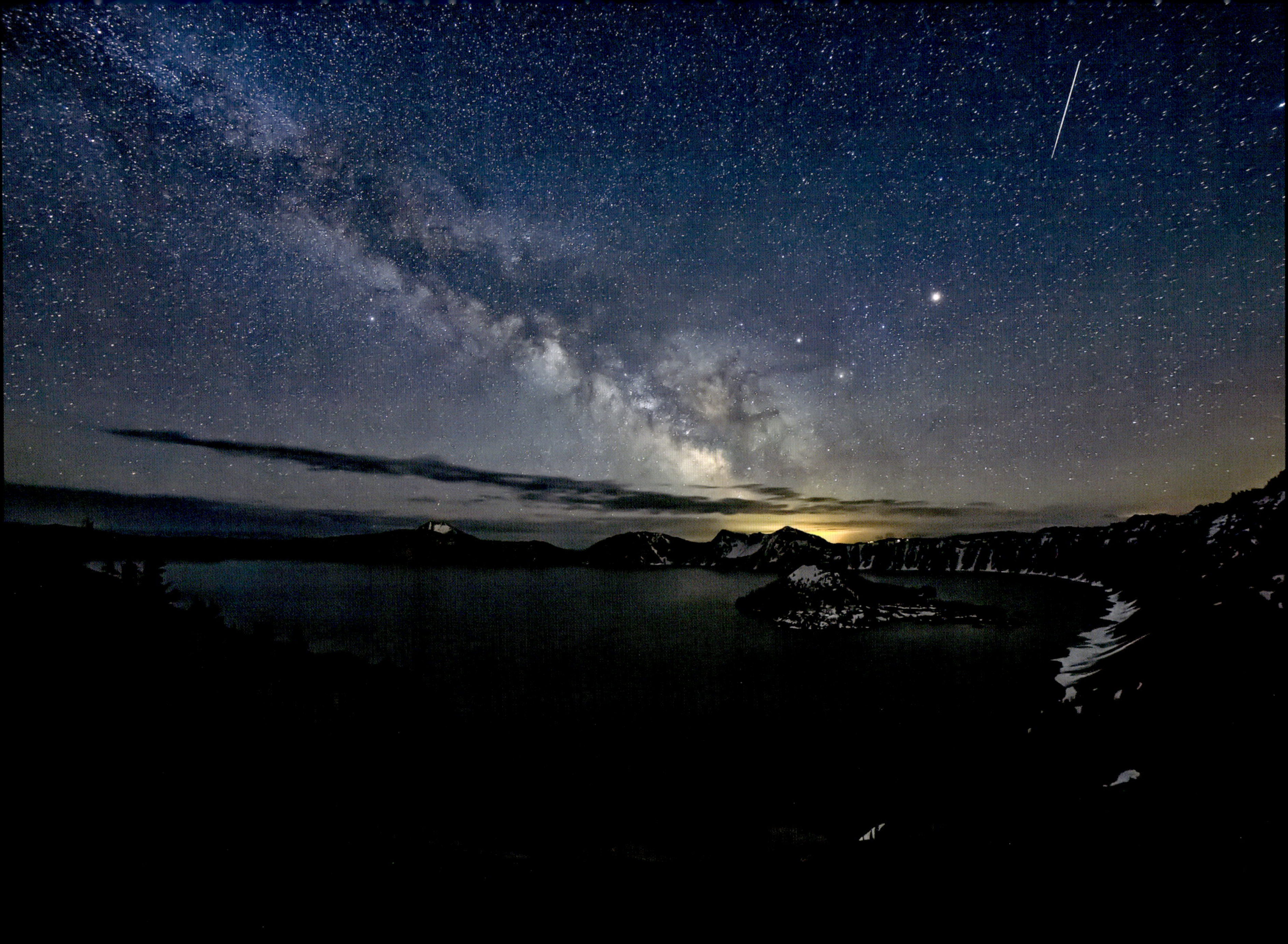

Crater Lake National Park, Oregon

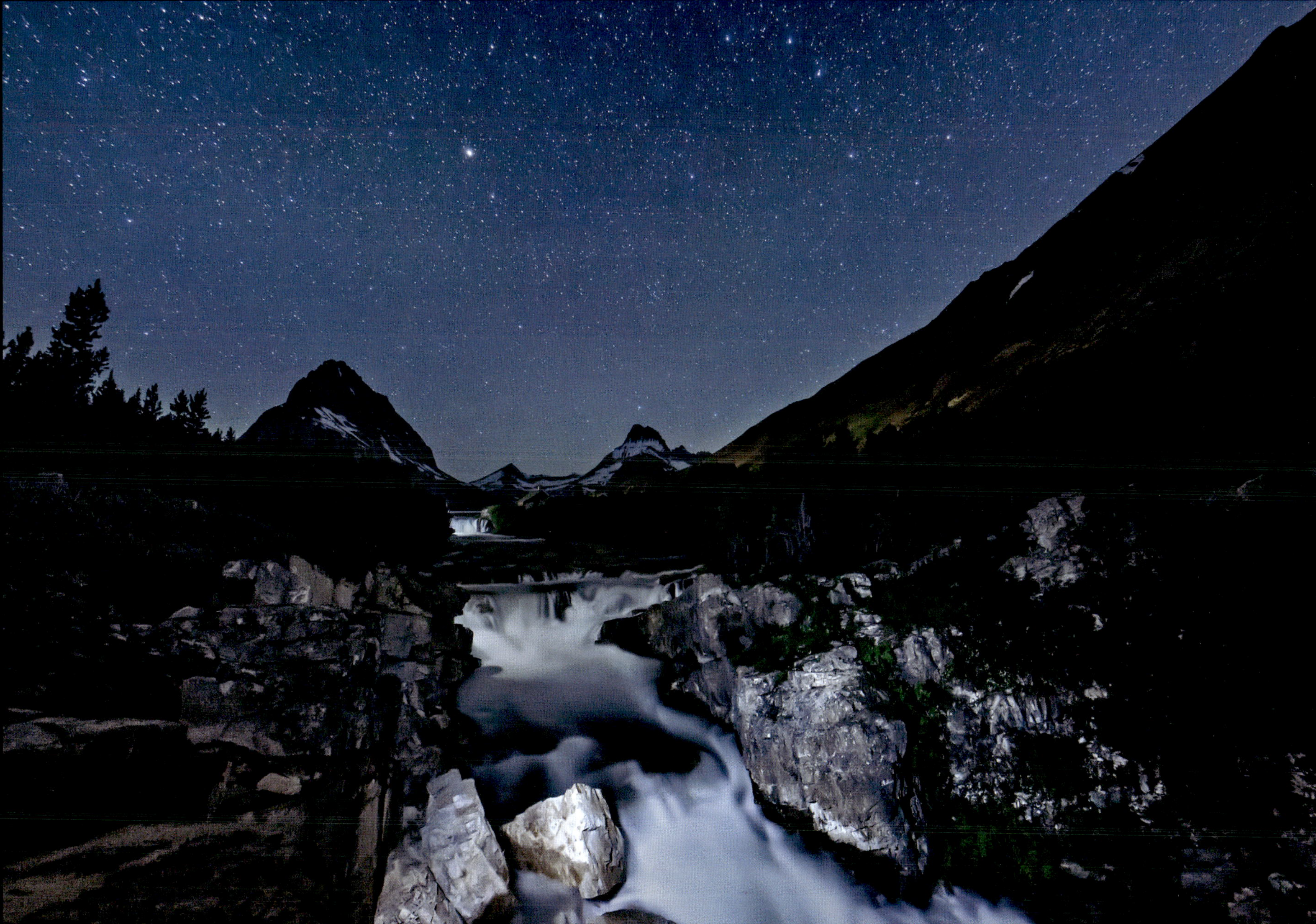

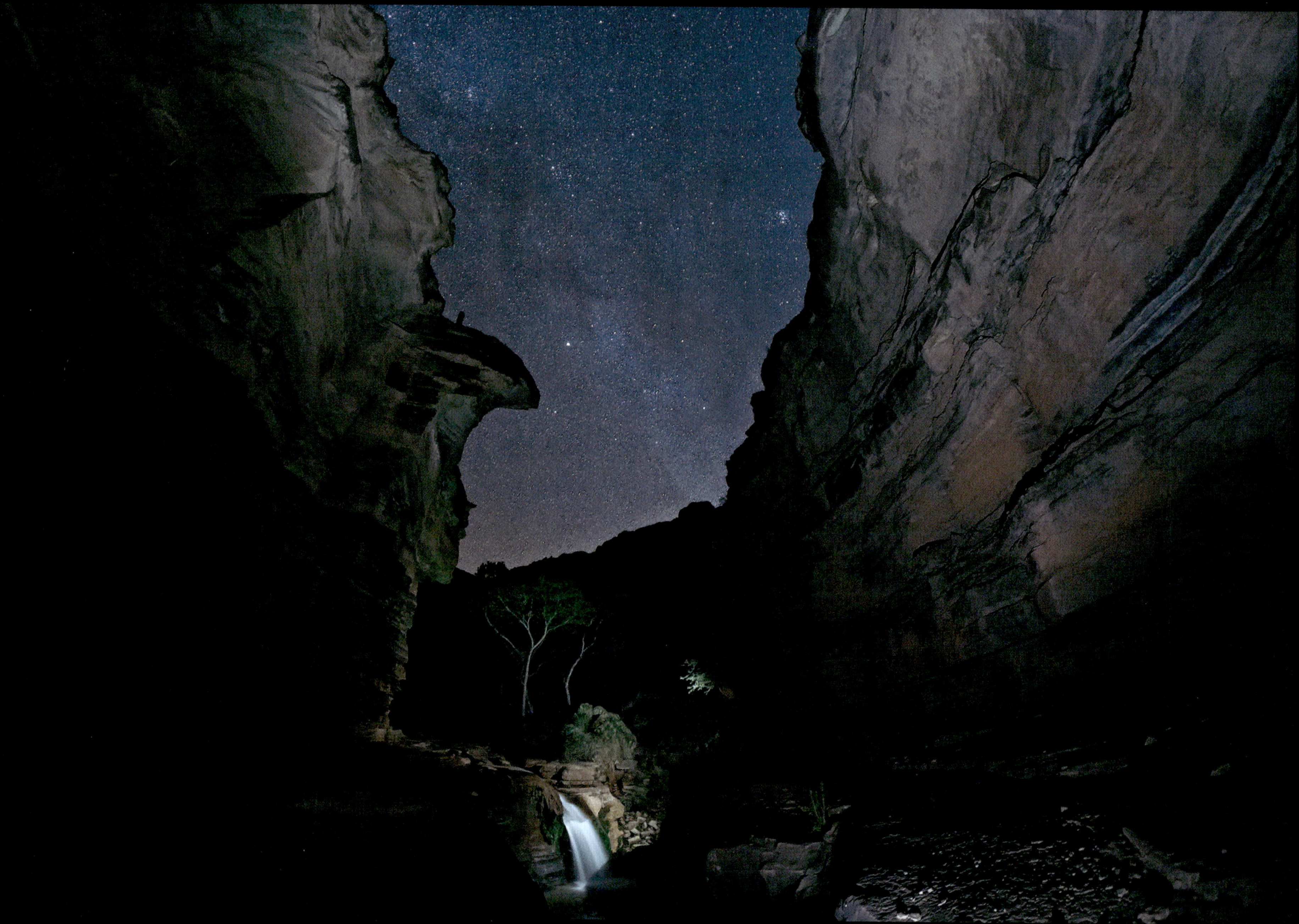

Deer Creek, Grand Canyon National Park, Arizona

1

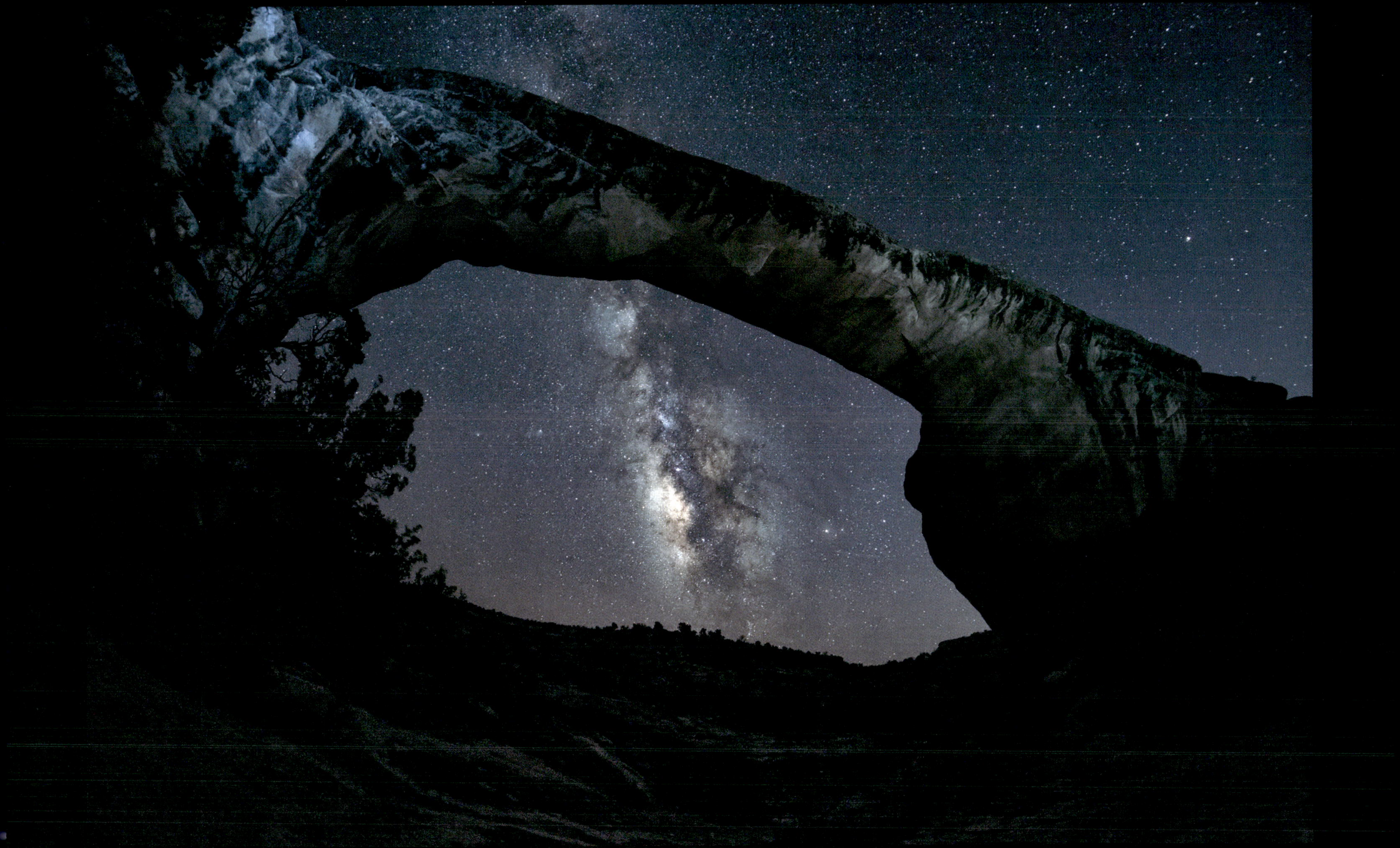

First International Dark-Sky Park, Natural Bridges National Monument, Utah

1

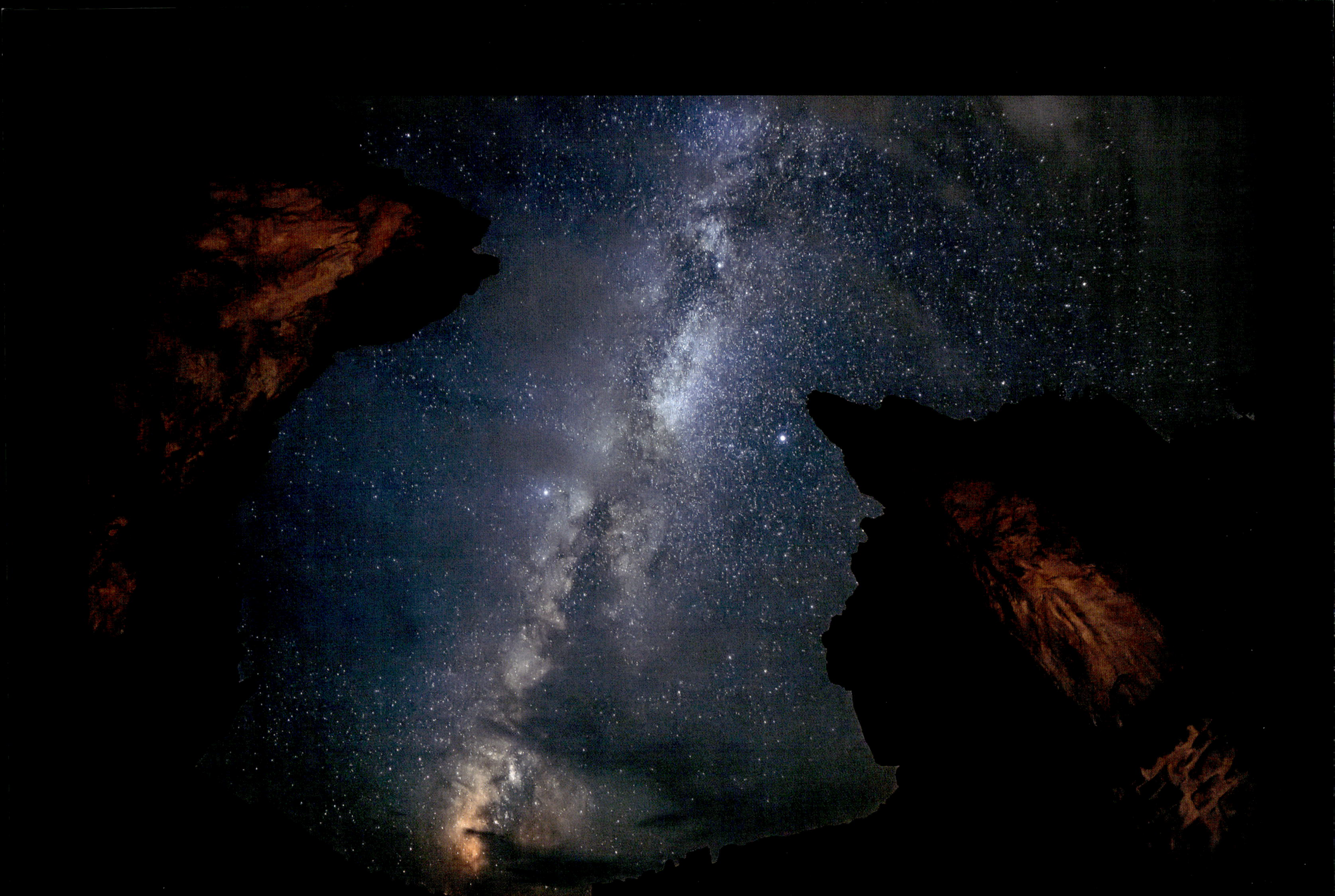

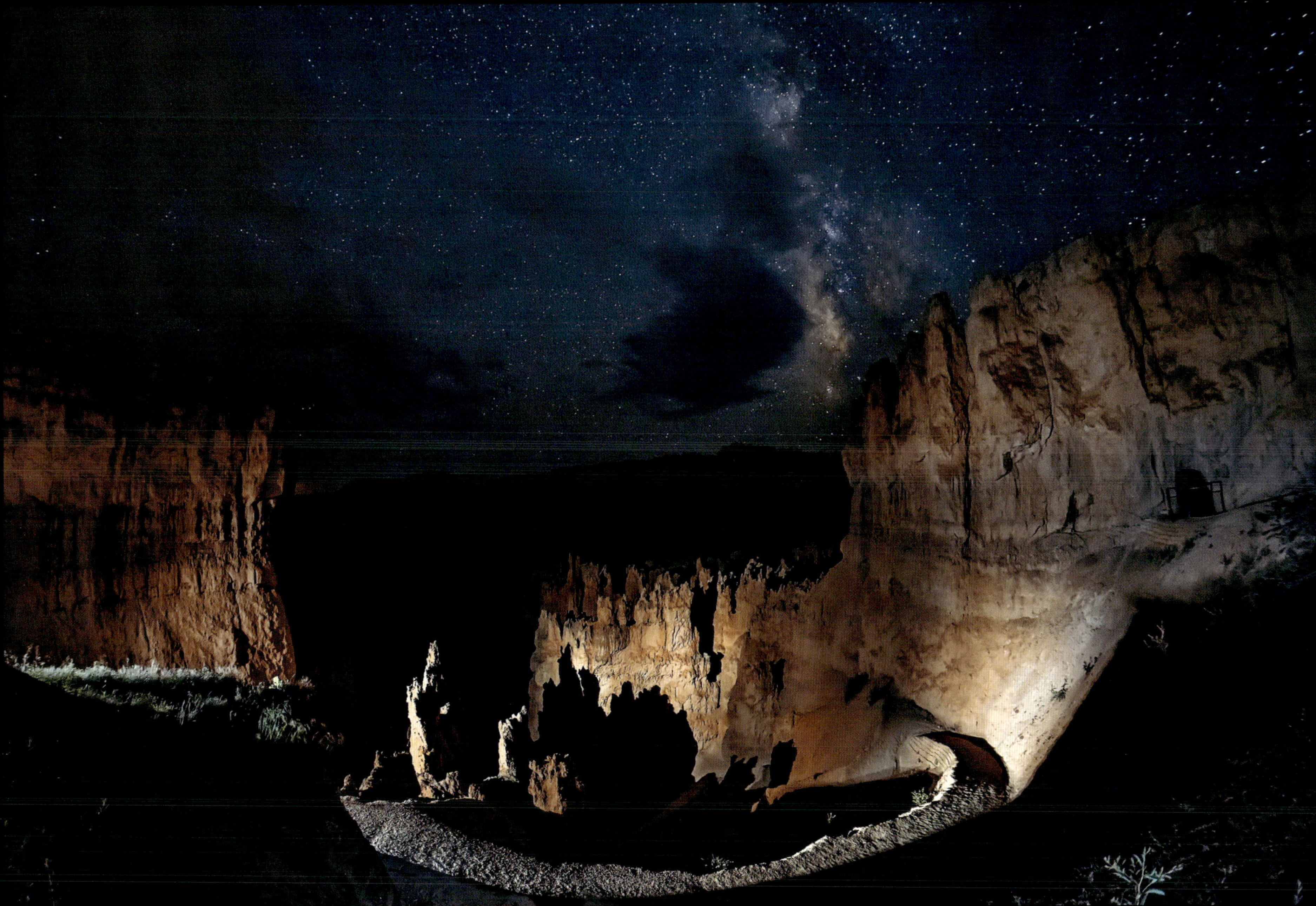

Today, the National Park Service protects some of the most pristine dark sky areas in the country. The National Park Service Natural Sounds and Night Skies Division measures light pollution in parks around the country and develops strategies to curb the impacts of skyglow.

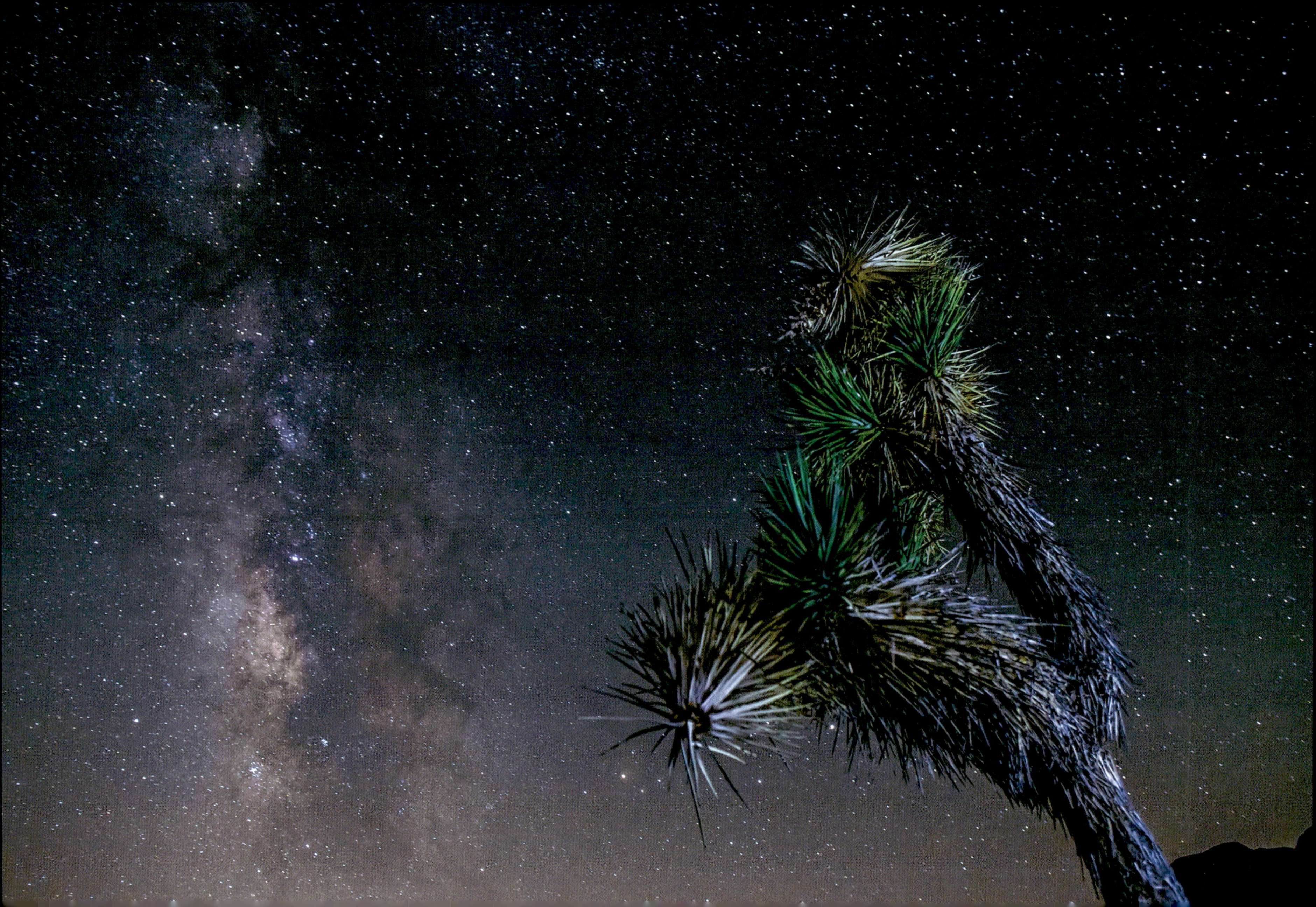

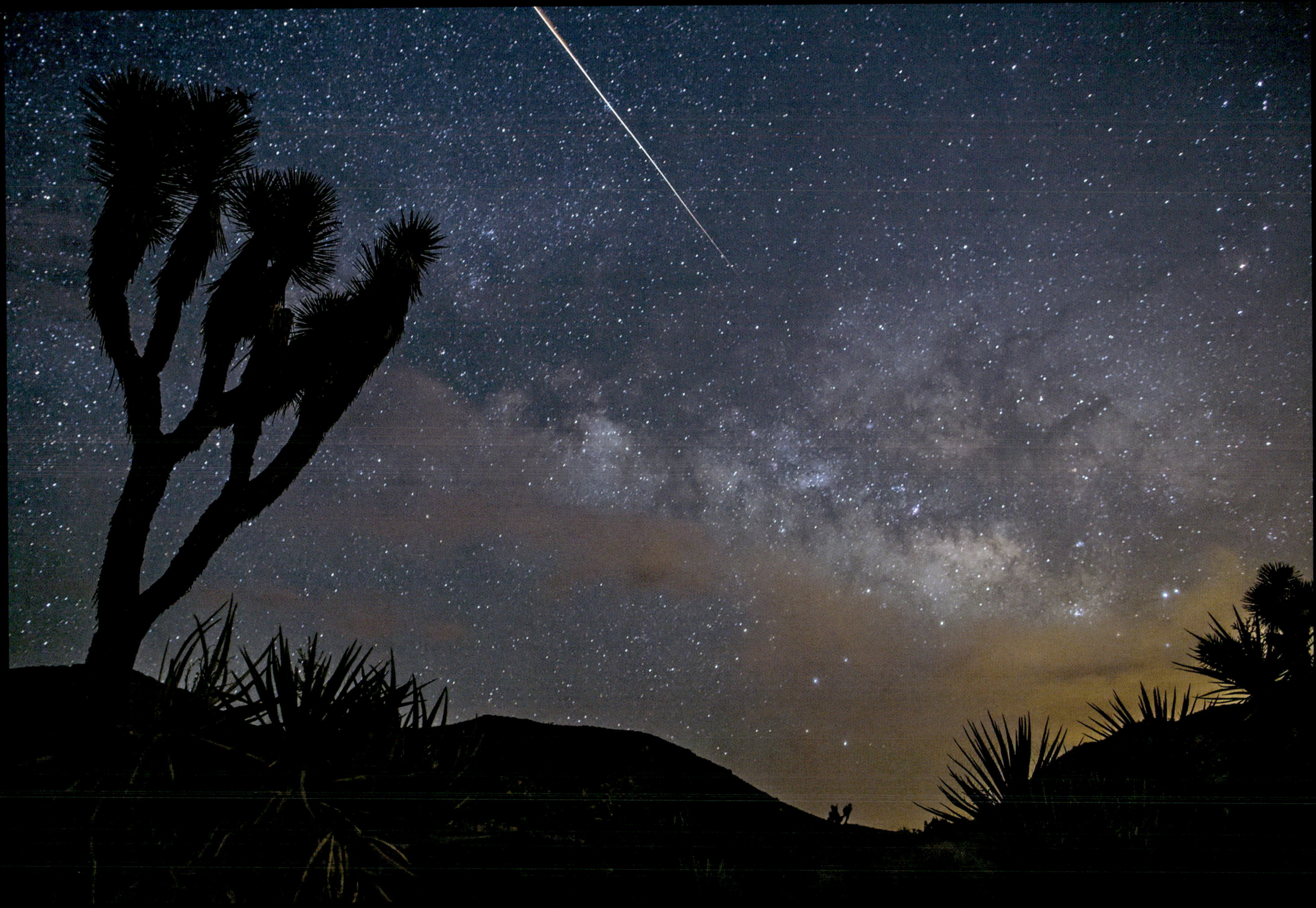

Despite National Park efforts to preserve nature for the enjoyment of the people, the use of artificial light both within and outside the boundaries of the National Parks threatens the quality of night skies. Joshua Tree is one of such parks affected by the light from neighboring Los Angeles and Palm Springs.

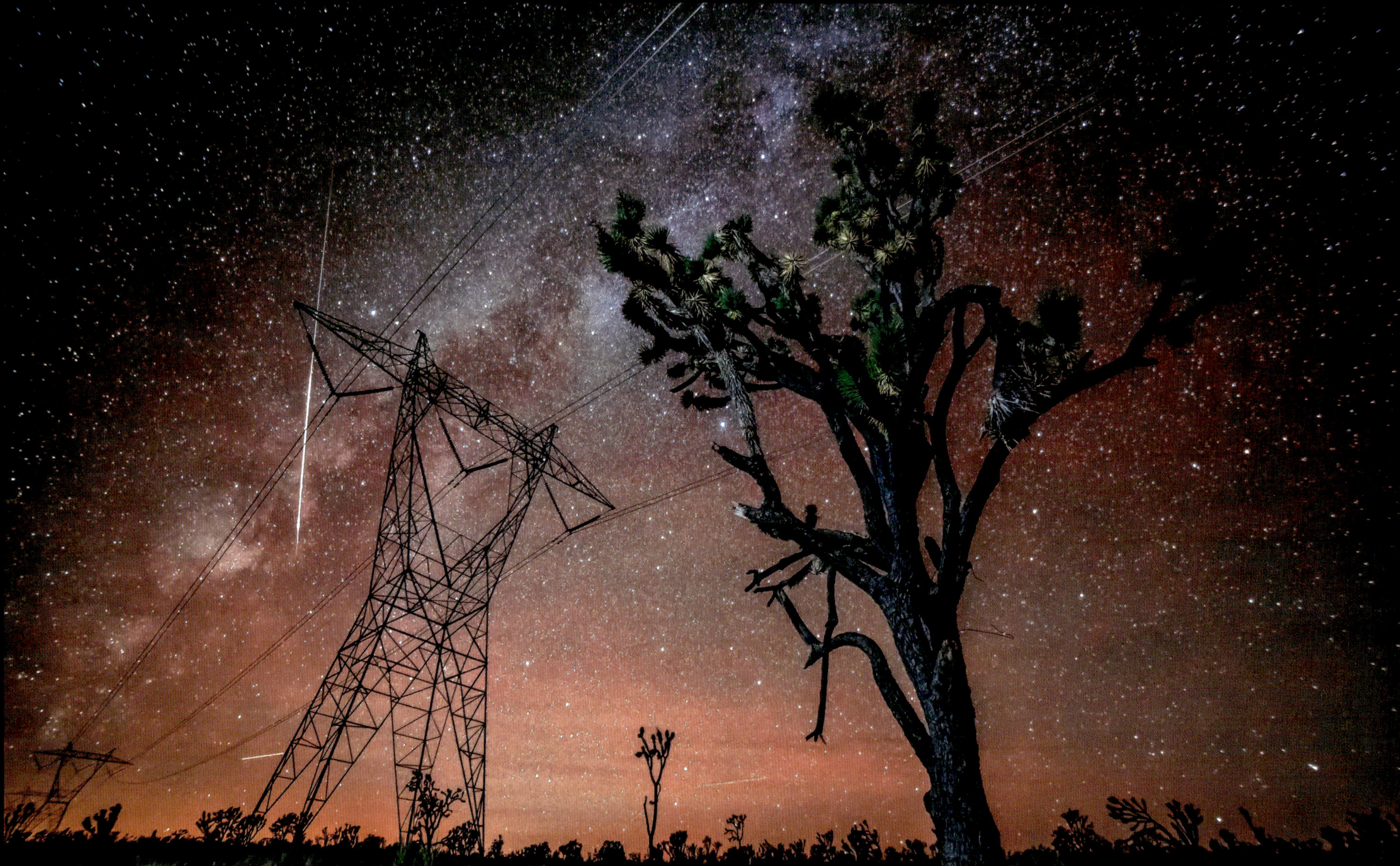

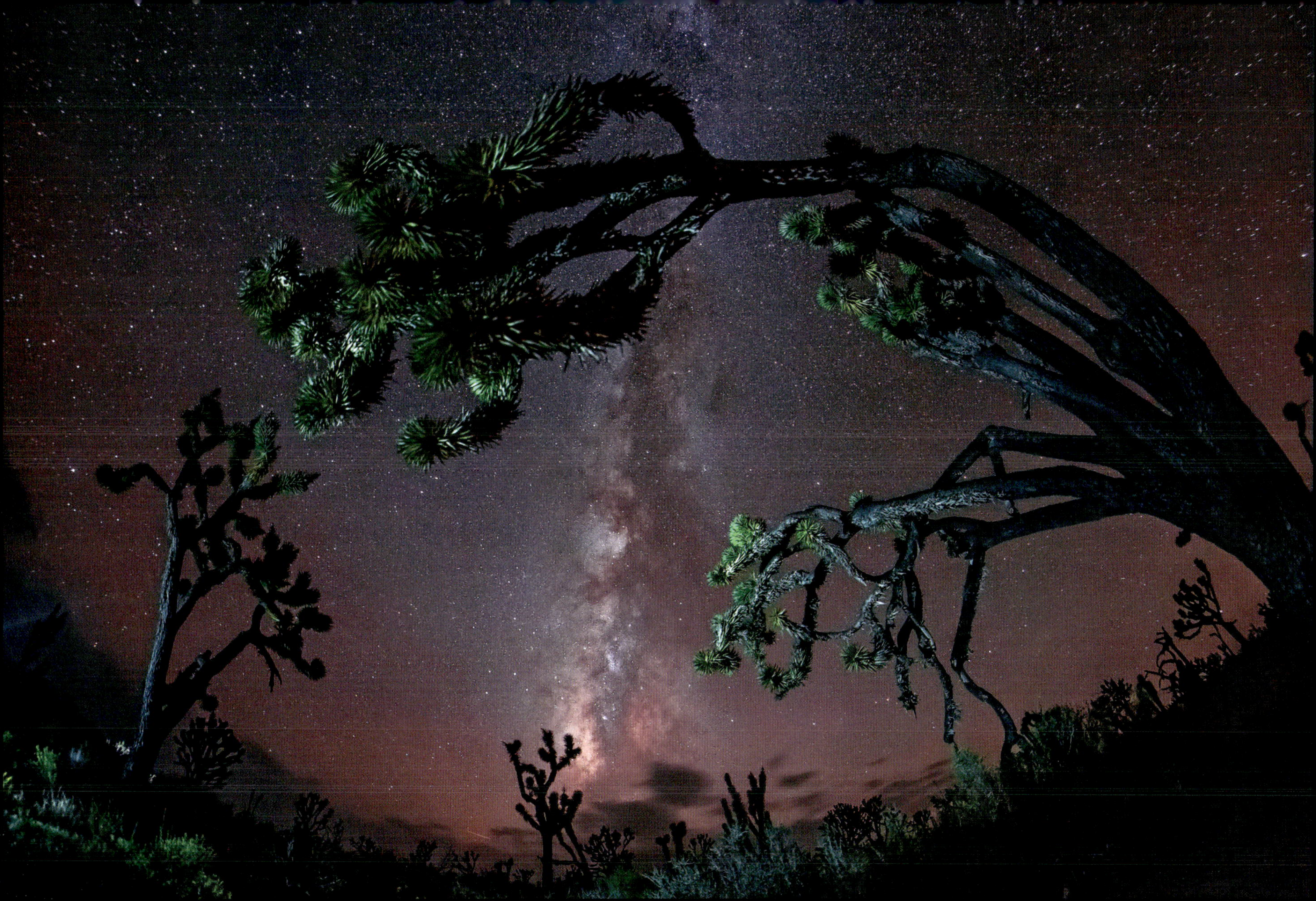

Being stuck between the Los Angeles area and Las Vegas keeps even the remotest parts of the Mojave National Preserve from maintaining clear night skies.

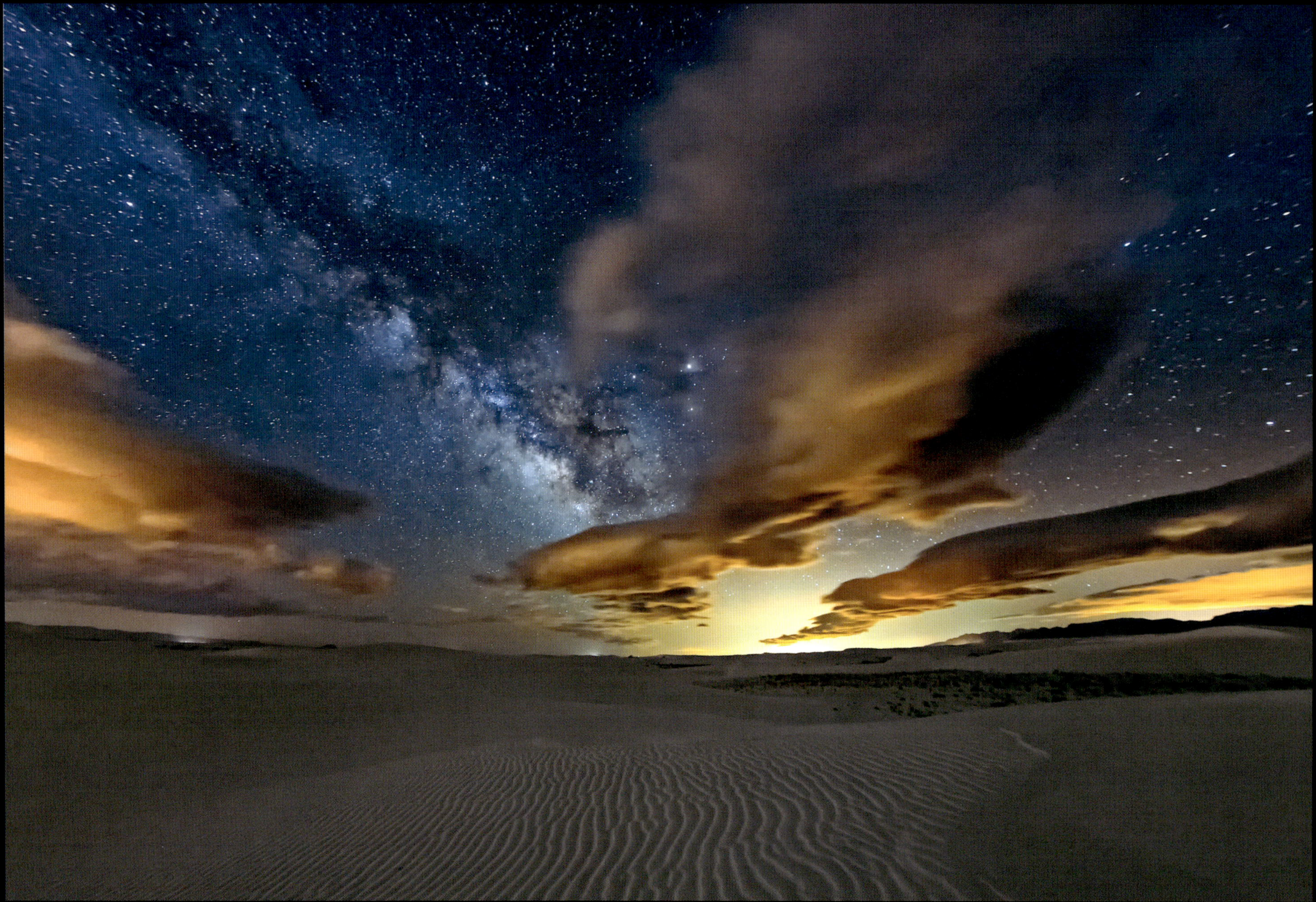

Much of New Mexico's White Sands National Monument suffers from light pollution coming from nearby cities Alamogordo and Las Cruces.

White Sands National Monument, New Mexico

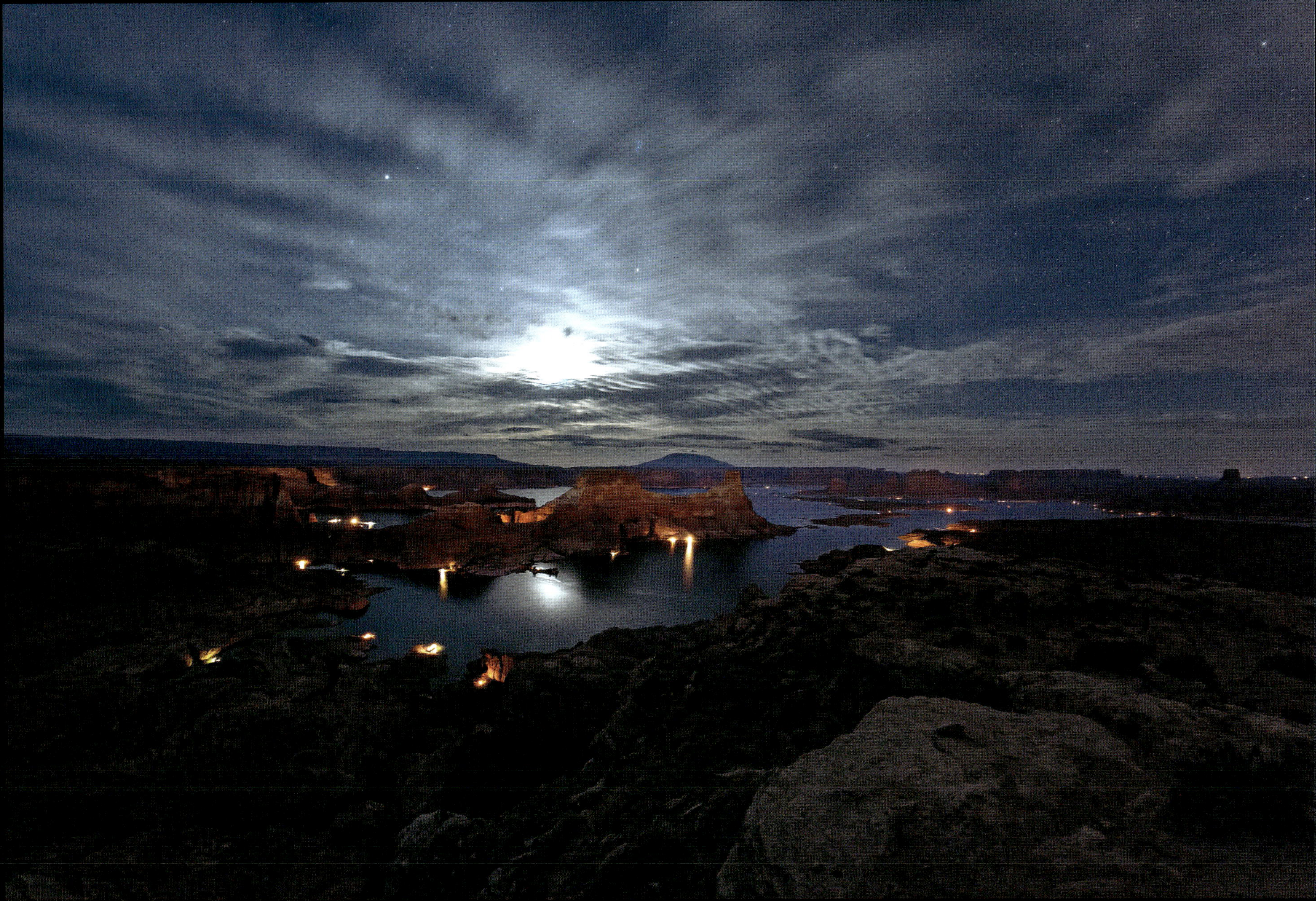

Some of the remotest parts of Lake Powell are marred by artificial light
on the lake itself or from the nearby town of Page, Arizona.

Alstrom Point, Glen Canyon National Recreation Area, Utah

2

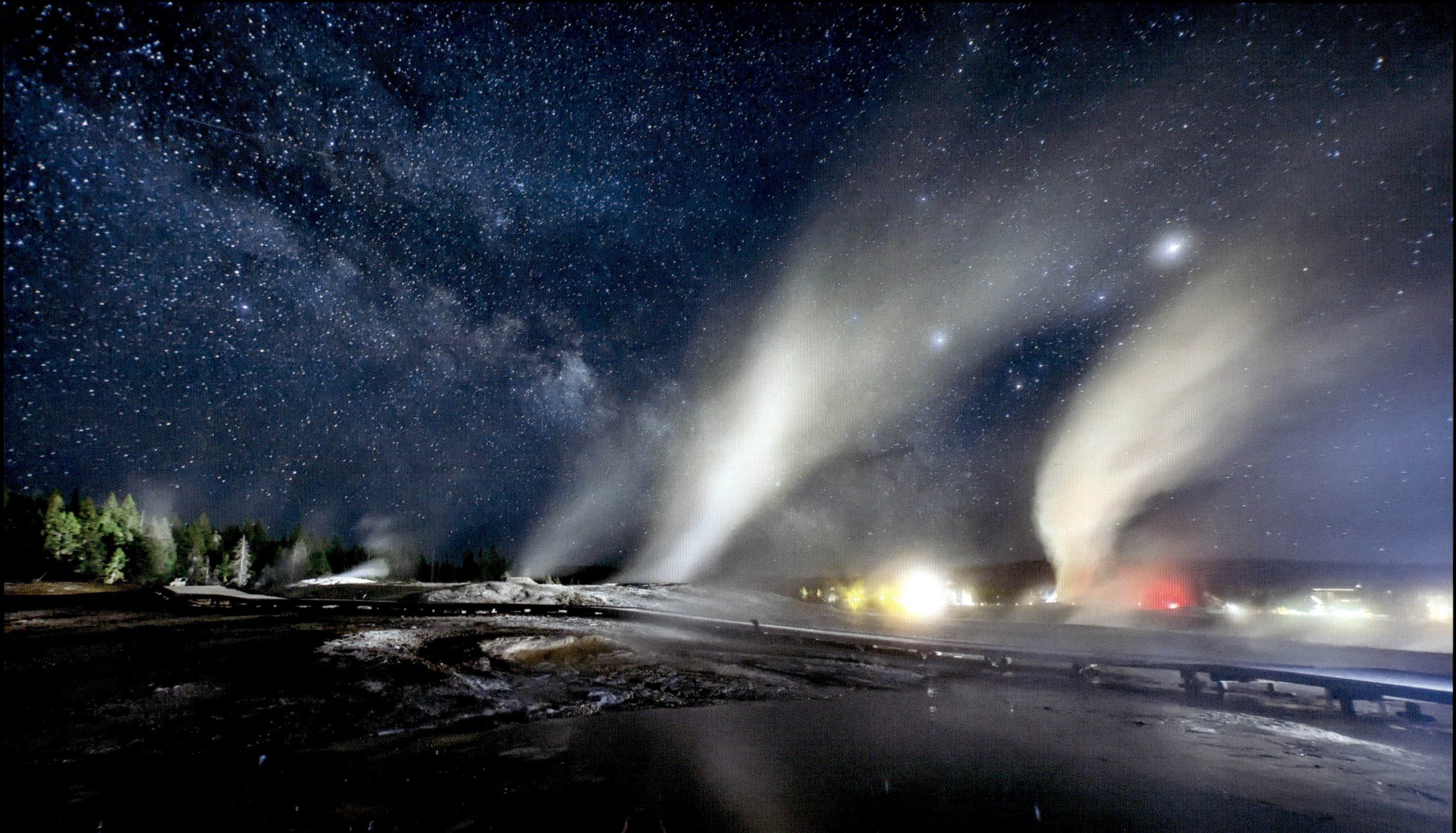

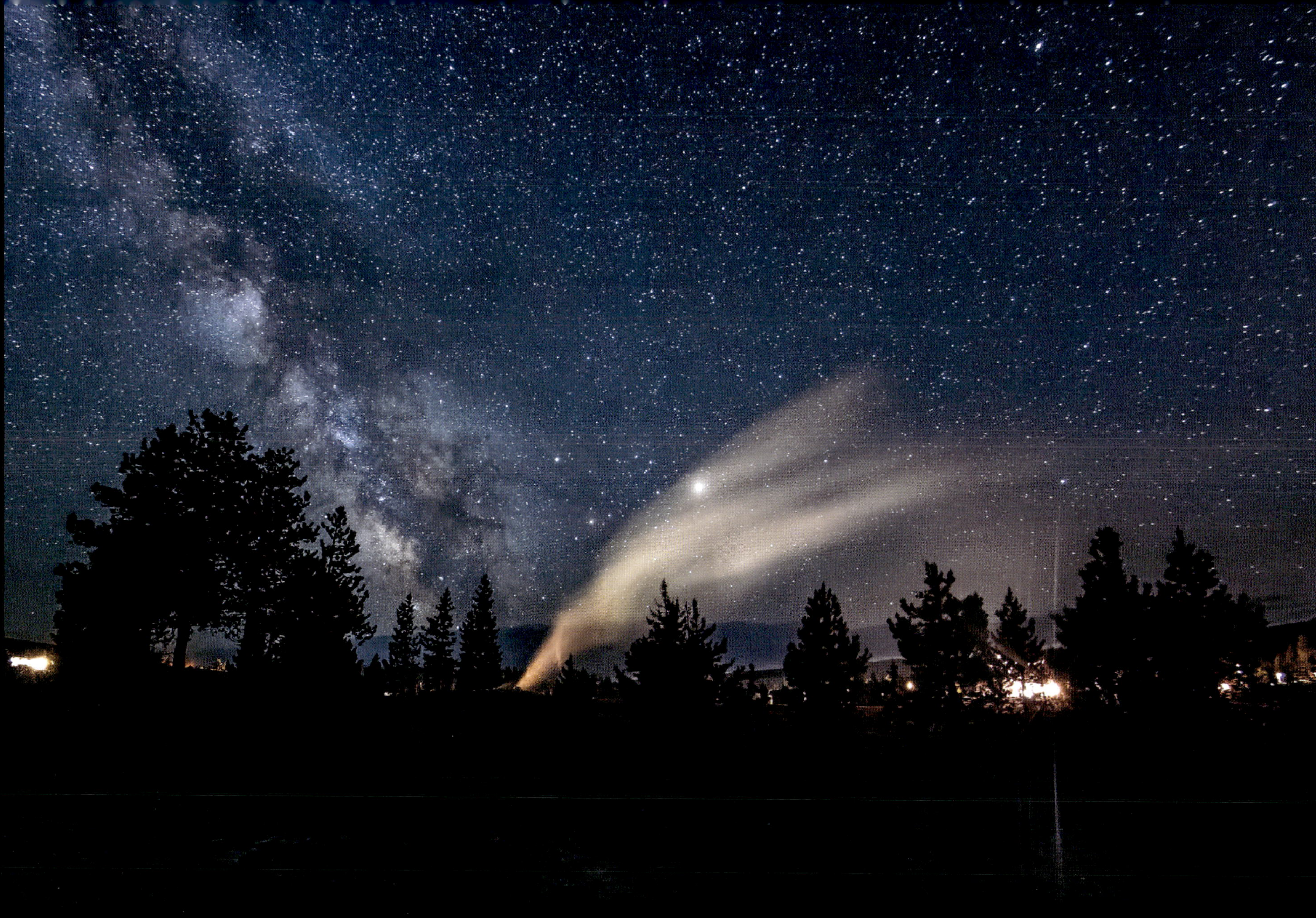

In Yellowstone National Park, as in other sanctuaries, the prevalence of light around building structures makes it impossible to achieve perfect dark skies in all areas of the park.

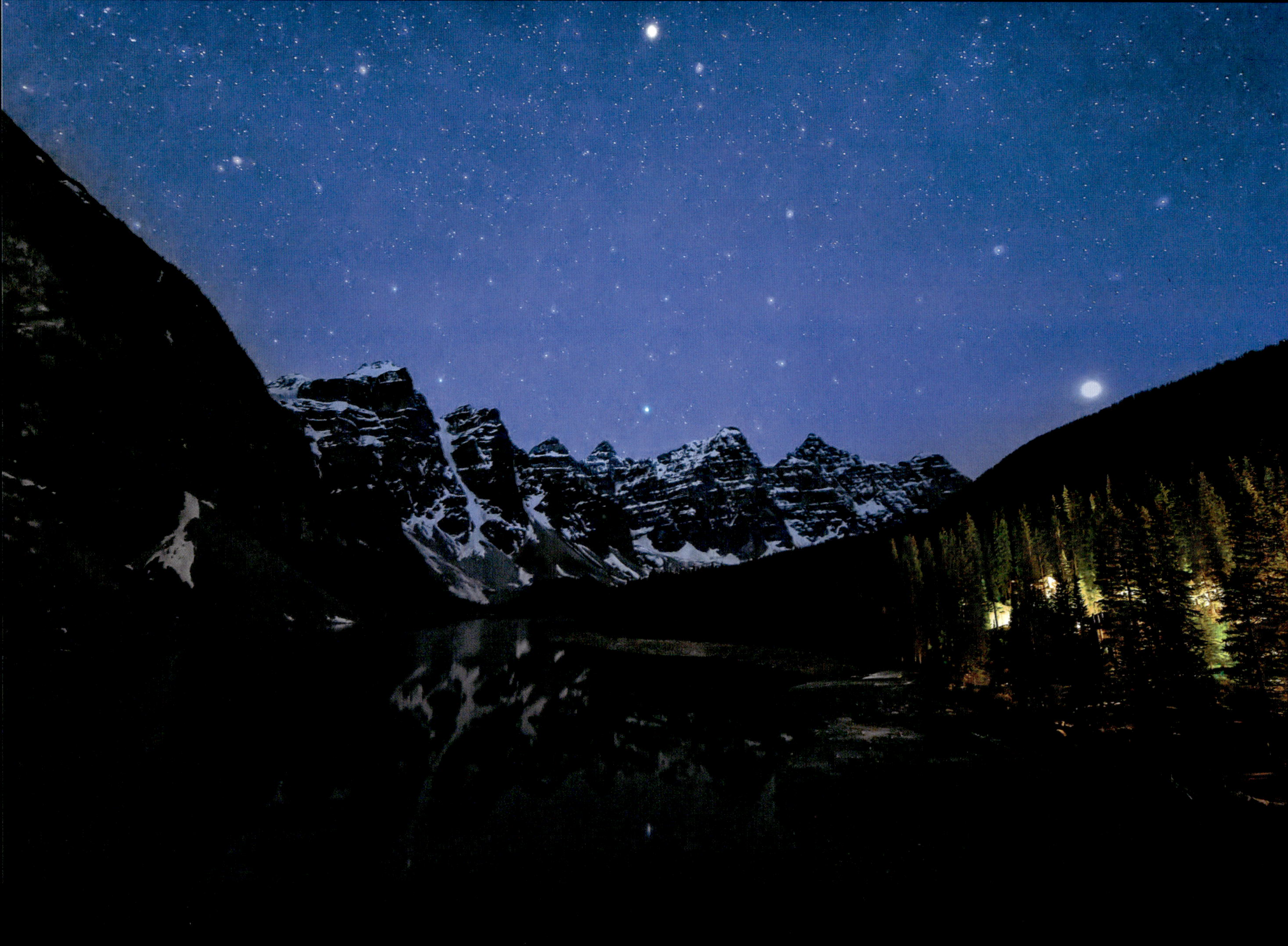

Moraine Lake, Banff National Park, Alberta, Canada

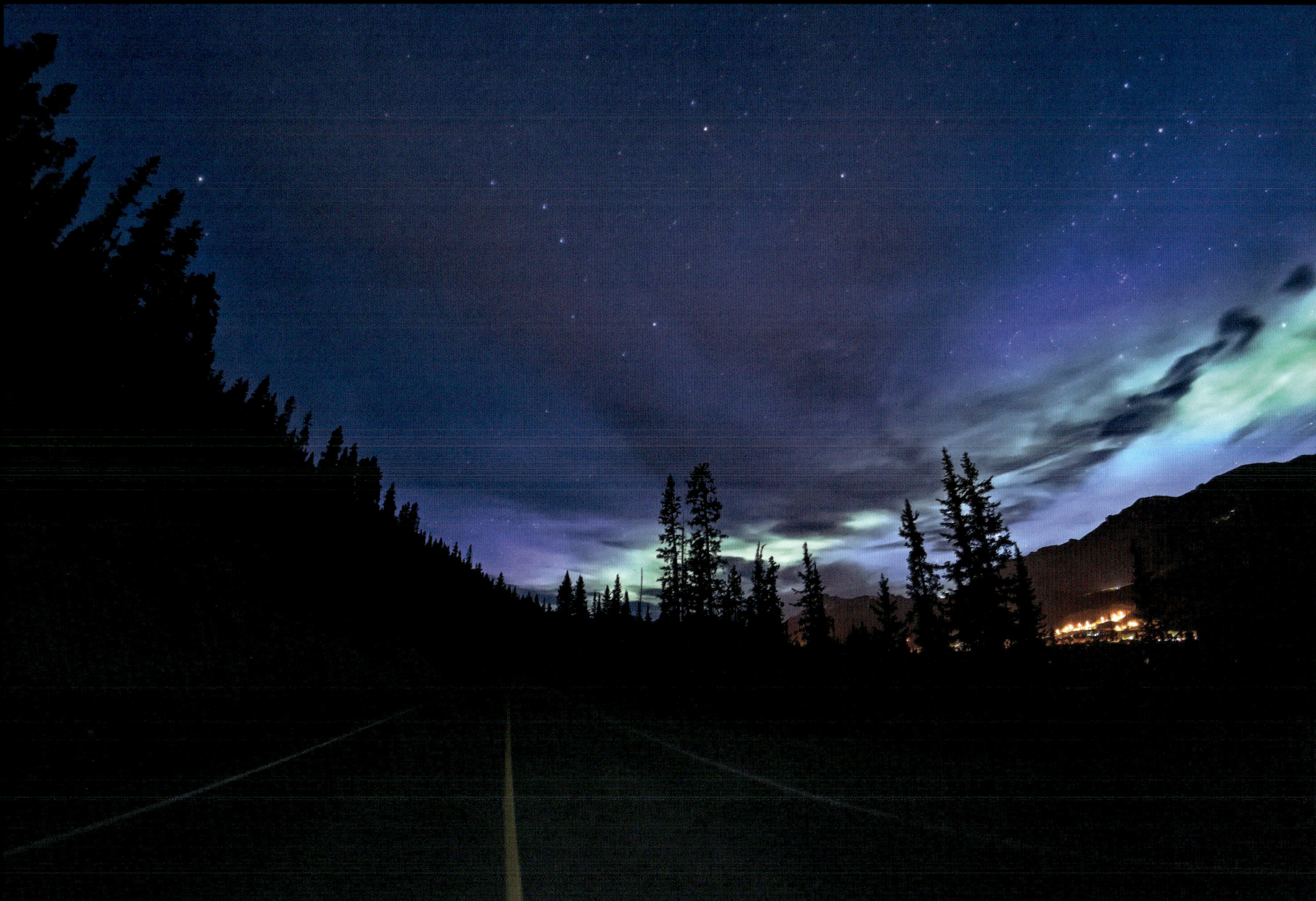

In National Parks such as Canada's Banff, light pollution impacts visitors' enjoyment of the Aurora Borealis (Northern Lights), seen here above the hamlet Lake Louise.

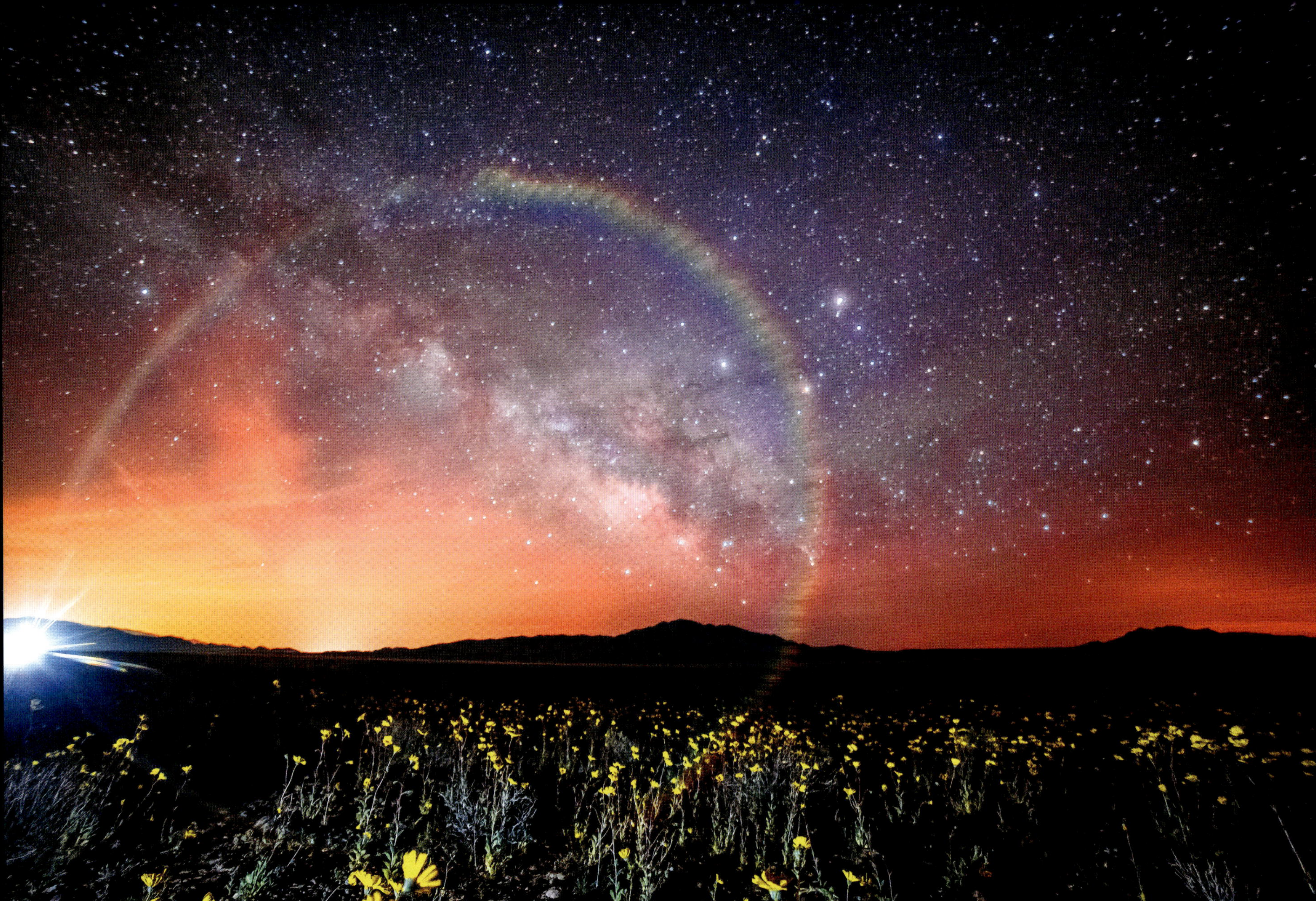

Wildflower Superbloom, Death Valley National Park, California

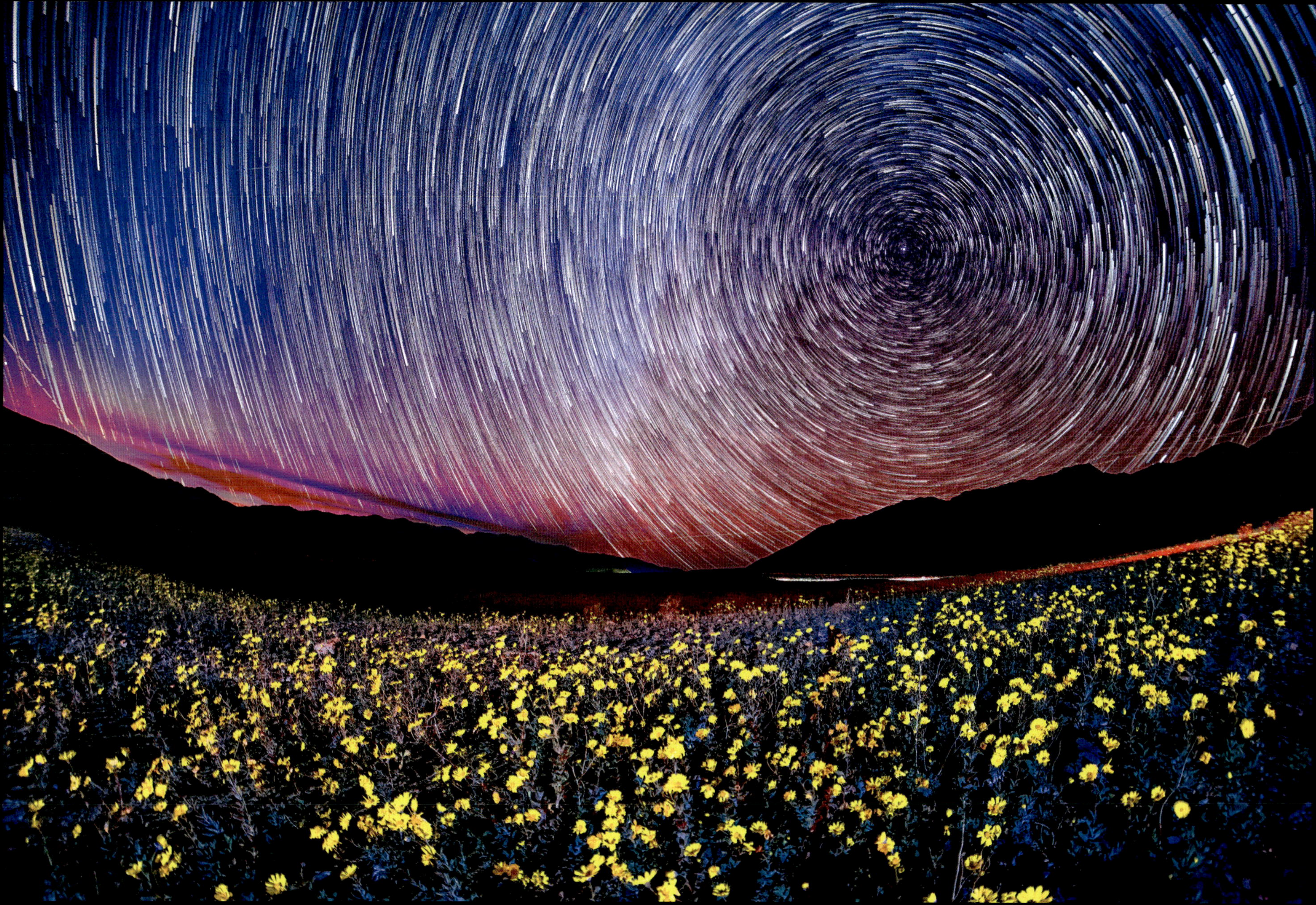

Even in the more detached parks, such as Death Valley, car lights add to a few other forms of artificial light intruding on the quality of darkness.

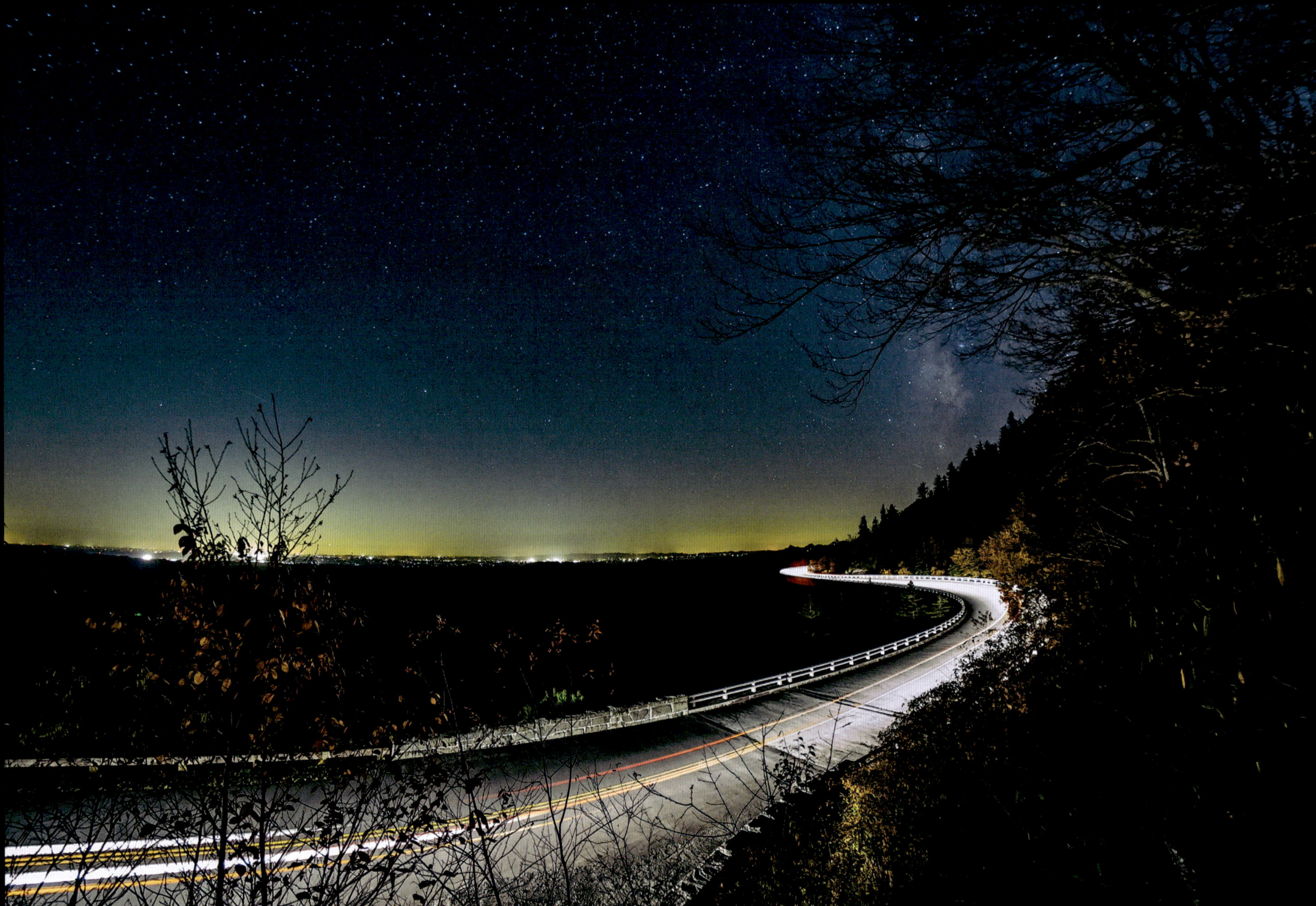

A combination of car lights and overwhelming light pollution from surrounding towns significantly impacts night sky quality over every single stretch of the 469-mile Blue Ridge Parkway.

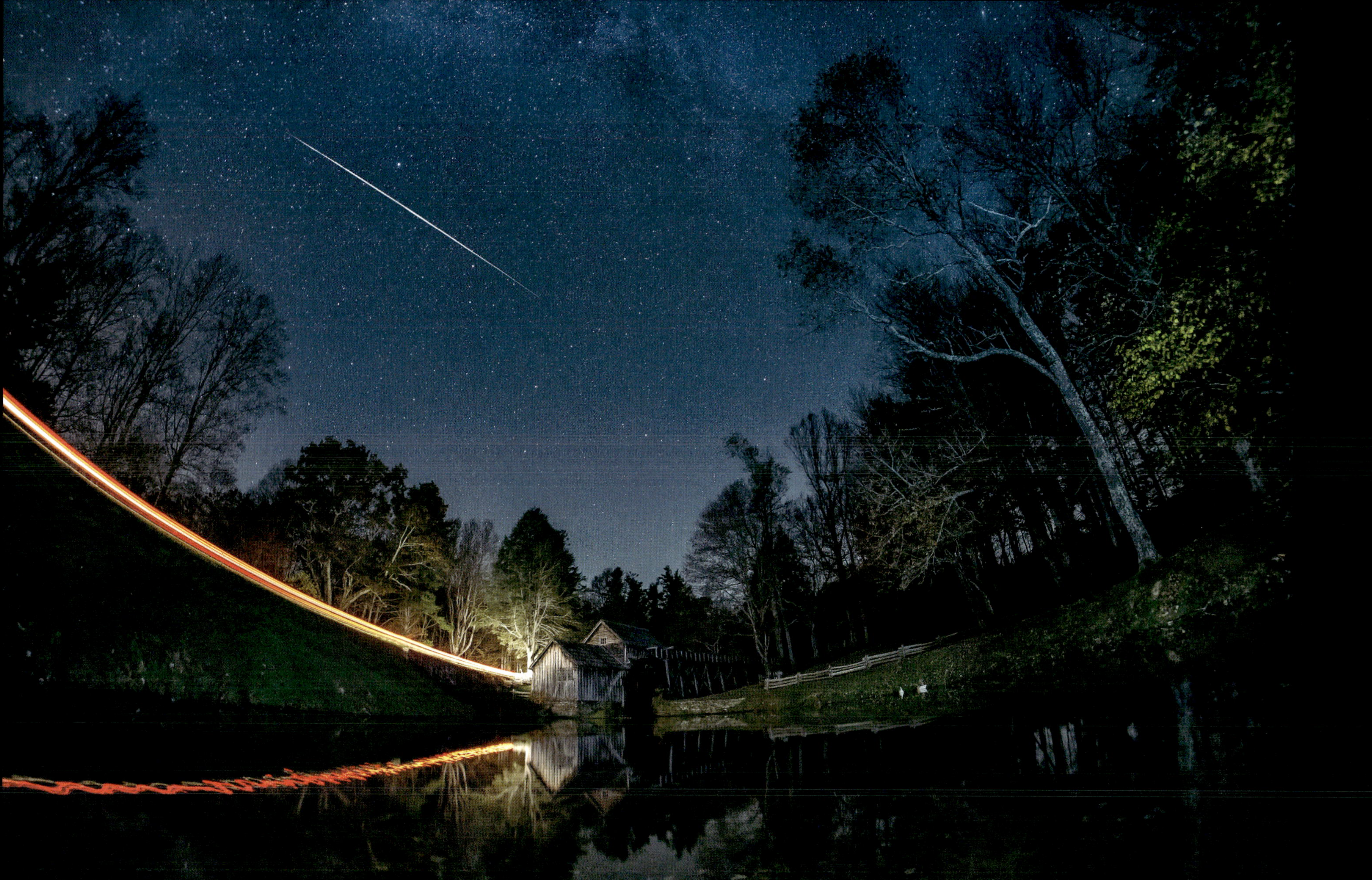

In the not-too-distant future, Blue Ridge Parkway visitors will no longer be able to enjoy phenomena such as this Orionid Meteor.

Mabry Mill, Blue Ridge Parkway, Virginia

4

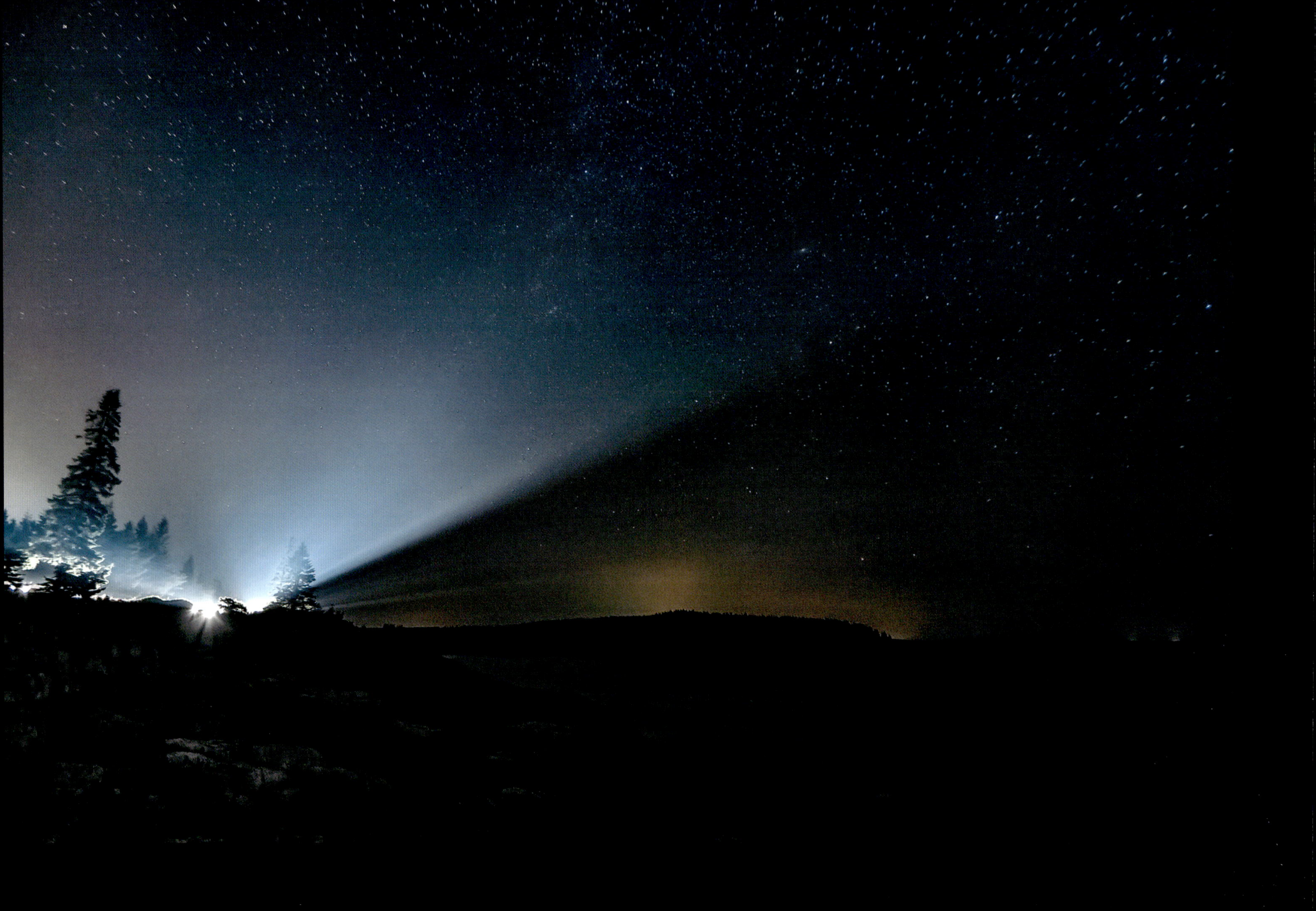

Acadia National Park, Maine

3

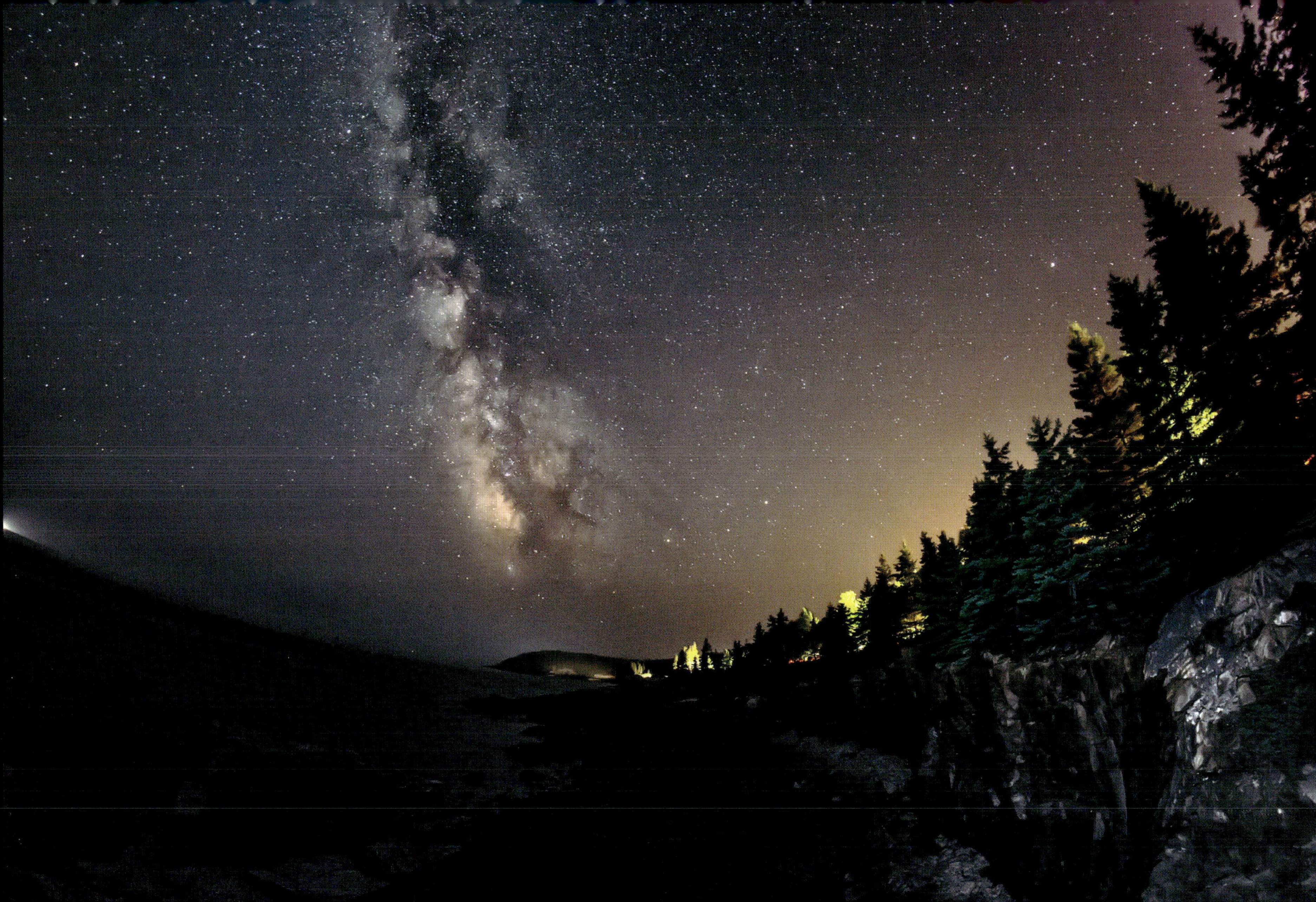

In Acadia National Park, which is surrounded by dozens of small towns, only the darkness of the vast Atlantic Ocean keeps some semblance of night skies alive

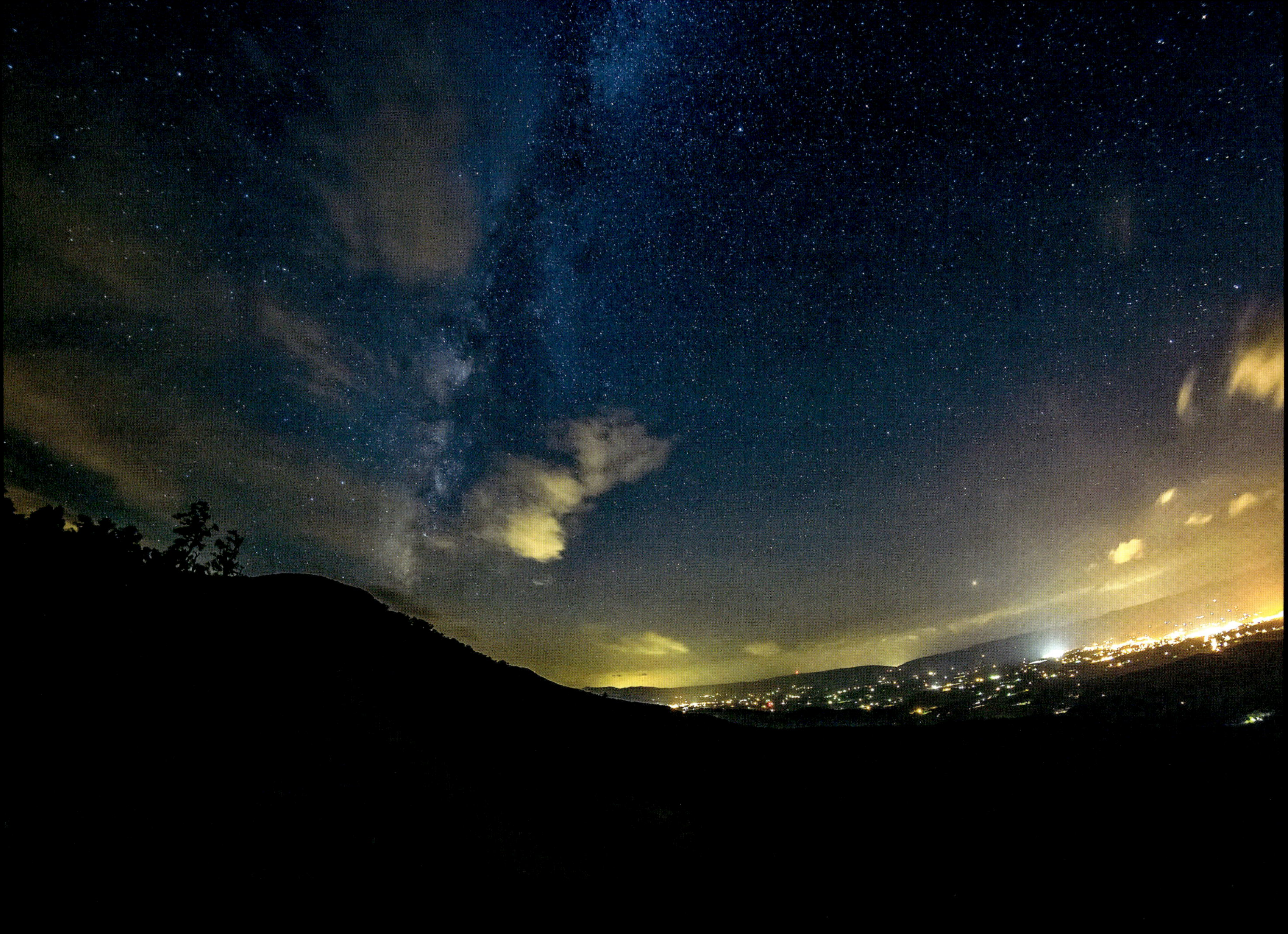

Skyline Drive overlooking Harrisonburg, Shenandoah National Park, Virginia

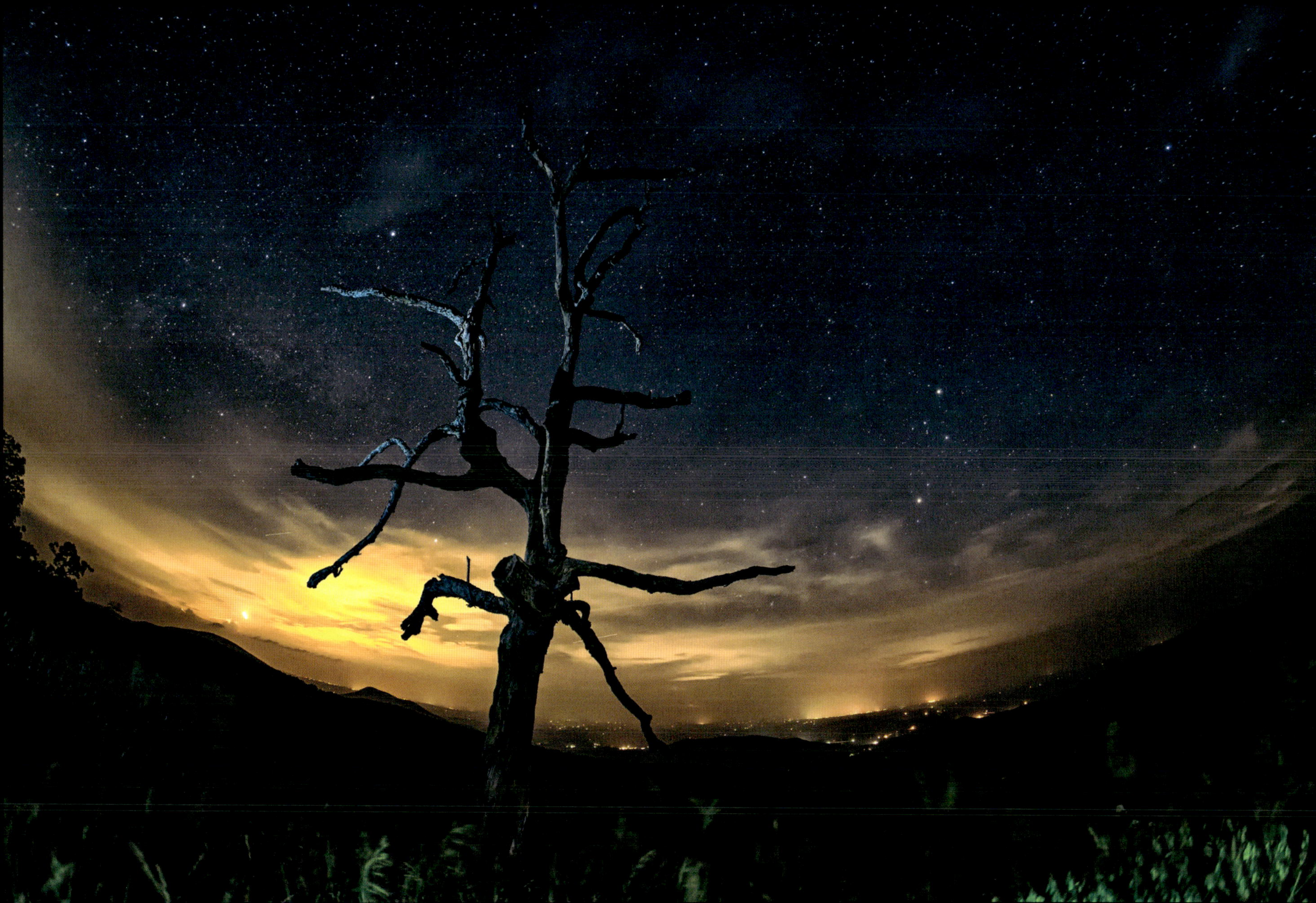

And in Shenandoah National Park, only the occasional passing clouds block enough light from the surrounding cities to offer visitors a decent view of the heavens. With an estimated light pollution growth at 6 percent a year, National Parks, along with all of the developed world, may lose their dark skies by the end of the 21st Century.

INTERNATIONAL DARK-SKY ASSOCIATION

Founded in 1988, the International Dark-Sky Association (IDA) is dedicated to protecting the nighttime environment and our heritage of dark skies through quality outdoor lighting. Light pollution is an environmental threat that harms our planet and robs us of the opportunity to experience the wonder of a natural night sky. Since its founding, IDA has taken the lead in identifying and publicizing the adverse impacts of artificial lighting on wildlife, the climate and health.

Outdoor lighting in the U.S. alone uses some 120 terawatt hours of energy per year. IDA estimates that at least 30 percent of light generated is wasted due to poor design, releasing 21 million tons of CO_2 into the air every year.

IDA advances light pollution abatement by working with manufacturers, planners, legislators and citizens to provide and implement lighting that is energy efficient, cost effective and dark-sky friendly.

IDA's conservation program, International Dark Sky Places, was created in 2001 to recognize excellent stewardship of the night sky. Designations are based on implementation of stringent lighting management plans, outreach programs to educate the public about night sky protection, and provide access to the night environment provided through stargazing parties and events. Since 2001, IDA has certified more than 70 such places in 15 countries across six continents.

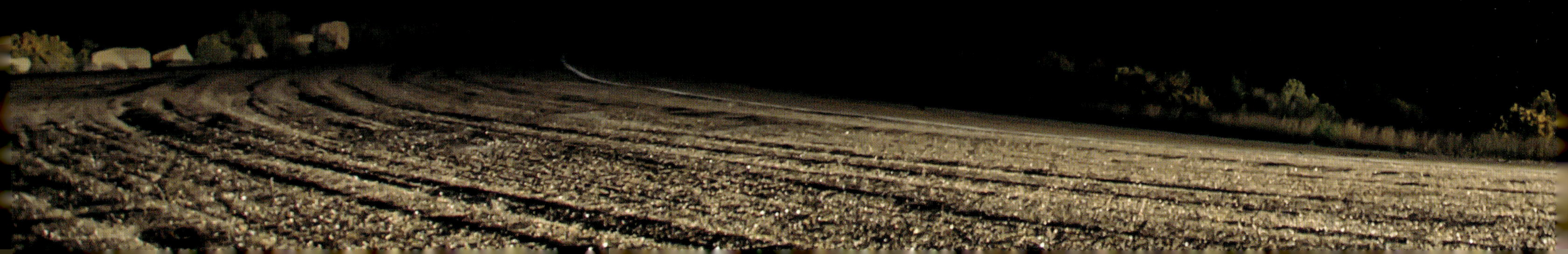

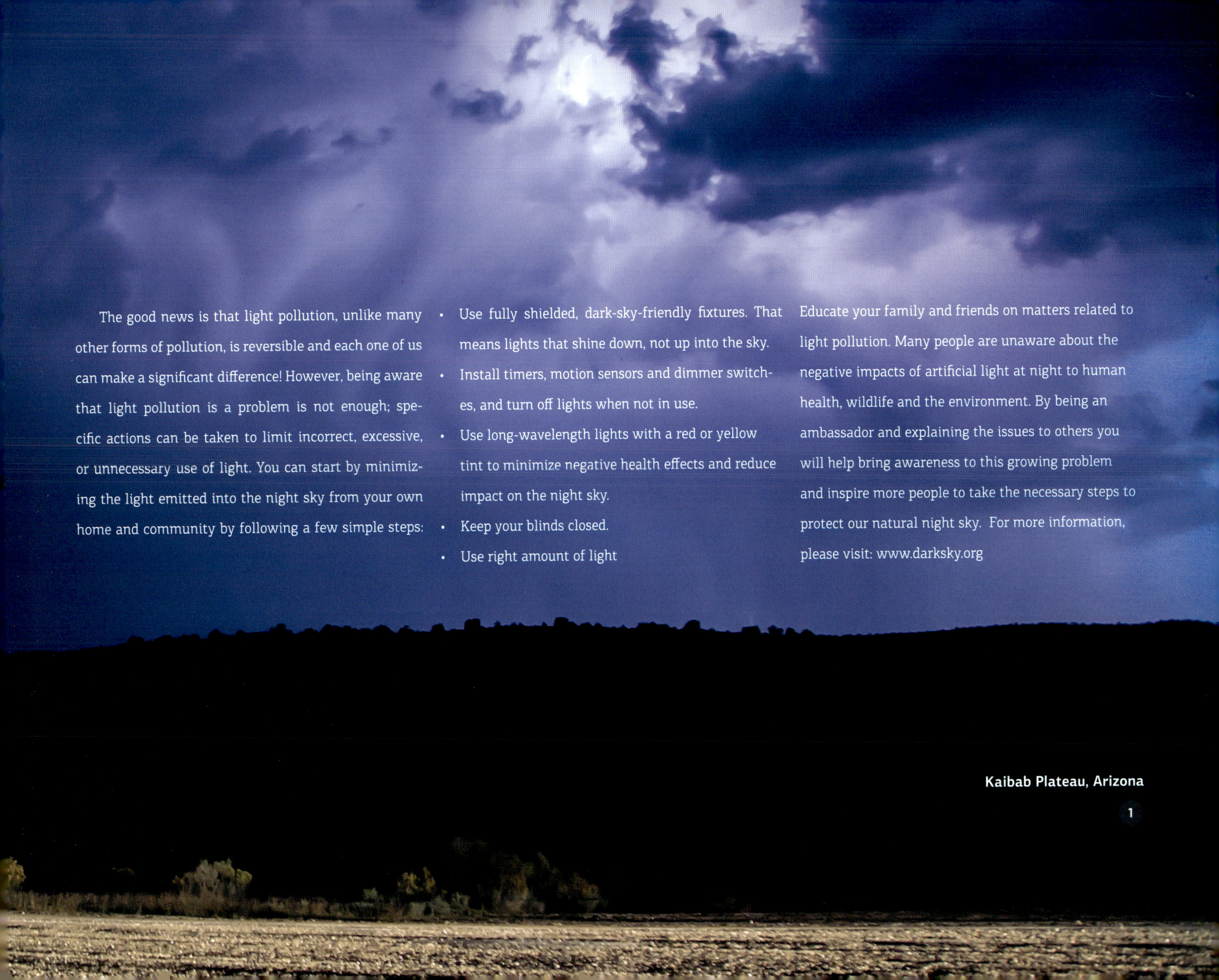

The good news is that light pollution, unlike many other forms of pollution, is reversible and each one of us can make a significant difference! However, being aware that light pollution is a problem is not enough; specific actions can be taken to limit incorrect, excessive, or unnecessary use of light. You can start by minimizing the light emitted into the night sky from your own home and community by following a few simple steps:

- Use fully shielded, dark-sky-friendly fixtures. That means lights that shine down, not up into the sky.
- Install timers, motion sensors and dimmer switches, and turn off lights when not in use.
- Use long-wavelength lights with a red or yellow tint to minimize negative health effects and reduce impact on the night sky.
- Keep your blinds closed.
- Use right amount of light

Educate your family and friends on matters related to light pollution. Many people are unaware about the negative impacts of artificial light at night to human health, wildlife and the environment. By being an ambassador and explaining the issues to others you will help bring awareness to this growing problem and inspire more people to take the necessary steps to protect our natural night sky. For more information, please visit: www.darksky.org

Kaibab Plateau, Arizona

Harun Mehmedinović

Harun took up photography as a hobby during his road trips across America. Years later, his portrait project Bloodhoney* became one of the most successful Kickstarter photography campaigns of all time.

He is a regular contributor to *BBC Earth*, and has contributed photographs to *Vogue Italia*, *National Geographic*, *Astronomy Magazine*, *BBC Travel*, *Discovery Science*, and *Blindfold*. Harun's photography work has been featured by various media outlets, including *Time*, *Wired*, *NPR*, *Los Angeles Times*, *LA Weekly*, *Vice*, and *Washington Post*. It has also been the subject of a *TEDx* Talk. His timelapse videos have been featured at various events, most notably by The Rolling Stones on their *ZIP CODE* tour and the upcoming 2016/2017 tour; Pink Floyd's Roger Waters' 2016 Tour and Desert Trip Indio Concert; Trance duo Cosmic Gate's *am2pm* music video; and National Park Service's "100 Years" centennial video; among others.

Harun's photograph of the cloud-inverted Grand Canyon was listed among the 2015 best travel photos of the year by *National Geographic*, and in 2016, he was on the *BBC Earth* Instagram team that won a Webby Award for "Best Photography and Graphics."

Prior to his venture into photography, Harun's film *In the Name of the Son* premiered at Telluride Film Festival and won over thirty international awards, including Shanghai, Savannah, and Cleveland film festivals. It was the first live action short film to receive an exclusive screening for the members of United States Congress on Capitol Hill.

Prada Marfa, Valentine, Texas

Gavin Heffernan

Gavin is a filmmaker and screenwriter whose photography and timelapse work has been featured in many venues worldwide, including The Rolling Stones *ZIP CODE* tour & 2016 Desert Trip performances; the 2016 Roger Waters tour; and trance duo Cosmic Gate's *am2pm* music video, which he co-directed with Mehmedinović, whom he met while they were both attending the American Film Institute in Los Angeles.

Gavin's timelapses can also be seen on Virgin America flights; *BBC Earth*; *Bravo's* first scripted show *Girlfriends' Guide to Divorce*; the LA 2024 Olympics Bid; the National Park Service "100 Years" centennial video; and numerous media outlets such as *Time*, *National Geographic*, *CBS News*, *USA Today*, and *Wired*.

Borrego Springs, California named the 2014 edition of their annual town festival after Heffernan's *Borrego Stardance* astrotimelapse, and October 25th was subsequently named "Gavin Heffernan Day" in San Diego County.

As a screenwriter, Gavin and writing partner Adam Robitel co-wrote Bad Hat Harry's psychological horror feature, *The Taking of Deborah Logan* (2014), and helped rewrite *Paranormal Activity: The Ghost Dimension* (2015) for Paramount Pictures. They are currently working on the upcoming Hong Kong sci-fi disaster film *Imago* (2017) for Cristal Pictures.

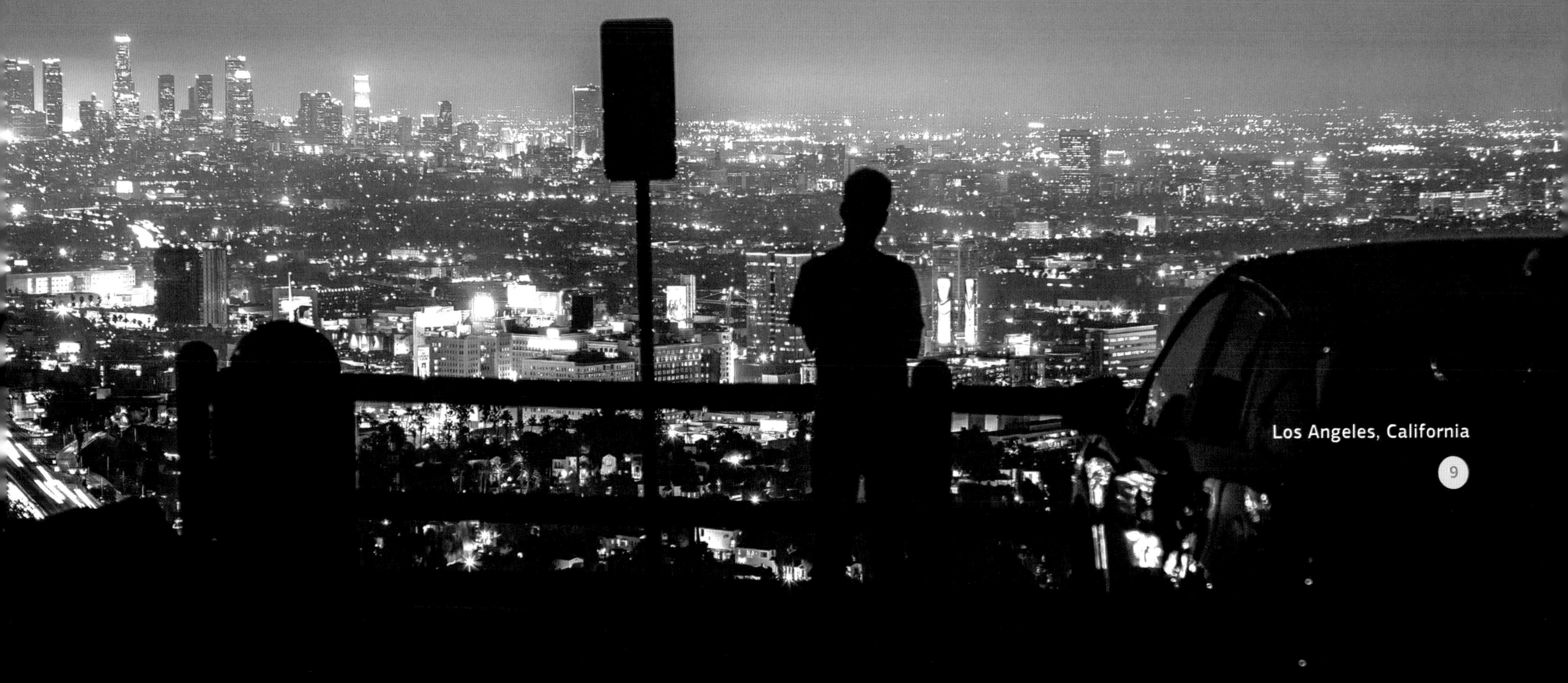

Los Angeles, California

TEAM

Writing Contributors

Eric Betz | www.ericbetz.com

Tyler Nordgren | www.tylernordgren.com

Phil Plait | www.badastronomy.com

Additional Photography

Nathaniel Smith | Pg. 4 | www.nathanielsmithphotography.com

Tom Piekunka | Pg. 53 | www.piekunkaphotographicimagery.com

NASA International Space Station Image Archive

Getty Images

Light Pollution Maps

The New World Atlas of Artificial Night Sky Brightness

Fabio Falchi, Pierantonio Cinzano, Dan Duriscoe, Christopher C. M. Kyba, Christopher D. Elvidge, Kimberly Baugh, Boris A. Portnov, Nataliya A. Rybnikova, Riccardo Furgoni

http://advances.sciencemag.org/content/2/6/e1600377

Editor: Michael Juliani

Design: Amer Mržljak | Fabrika.com

Publicity: Rebecca Feldbin | Core PR Group

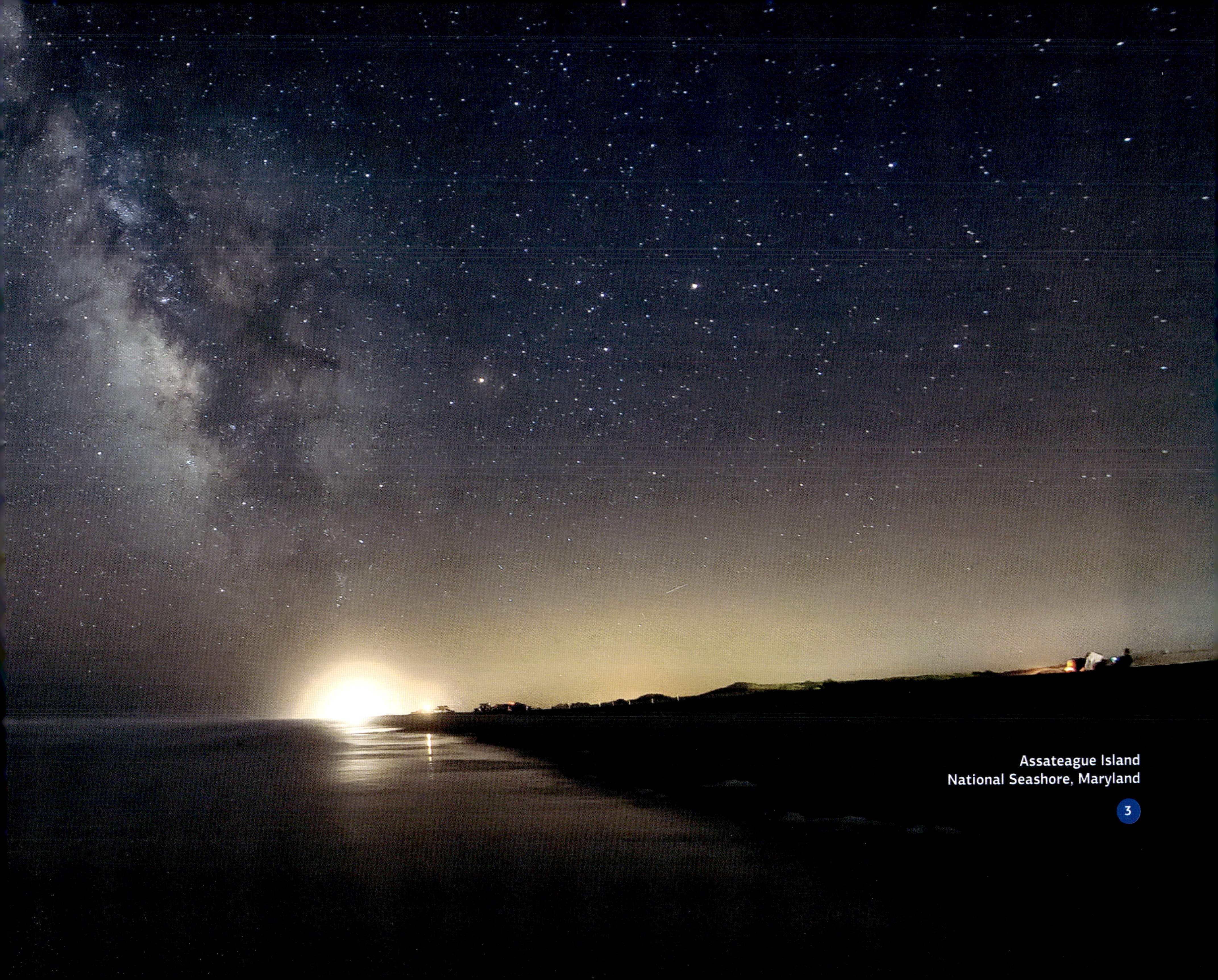

Assateague Island
National Seashore, Maryland

Dana Guerin, Asim Haracic, Tom & Vicki Williamson, Trevor Tatham, Tom Knowles, Ed Anderson, Letitia Noel, D.J. Cole, Richard Norman, Kristofer Menez, Stephen Collard, Marshall Betrand, Don Eggert, Alison Wood, Josh Jordan, Edwin Leong, Michelle Mitchell, Tatiana Lawrence, Susan Wheelin Shehata, Kendra Lee Davis, Elinor Nicholson, James Yun, Davin Lee, Audrey Fischer, Stephen Grate, Ricky Patterson, Pebbles Darwin, Erin Wolff, Mark Hessinger, Audrey Deterding, Ann Hughes, Fr. Timothy Sauppé, Steven Lee, Lorna Herf, Joey Civin, Sarah Steinberg, Sejla Holland, Tony Addario, Gerald Olszewski, Michael Weilmeier, Russell Higbee, Jason Chen, Ty McNeeley, Natalie Rose Maquidato Olaso, Andjela Bogunović, Almir Osmanović, Orna Blum, Rob Lancefield, Drew Fleshman, Chris Mathieu, David Jones, Mark Neumann, Richael K. Young, J. Ashley Odell, Rainer Schulz, Iann Wu, Suzana Ognjanović, Theresa Defino, Karen Bjorkman, Hiram Sarmiento, Scott Jacobs, Jonathan Rosenberg, Stanley McPartland, Kimberly Benjamin, Reid Simpson, Greg Bruno, Kyle Broadus, Kevin Ballinas, Patrick Nola, John Stancik, Paolo Bergamo, Harry Bui, Kirk Powell, Jon LaFollette, Joshua Lowry, Kat Robinson, Joshua Garcia, Aida Bogunić, Alia Rashid, Alejandro Sánchez de Miguel, Matthew Hodgson, Stacy Shure, Vladimir Boyko Kuznetsov, Corinna Vigier, Matt Stroh, Roberta Hoy, Faith Dincolo, Warren Tsuyuki, Talia Maynor, Brandon Barclay, Brian Elliot, Jim McKay, Wayne Giza, Emily Bernhardt, Karen Williams, Gerhard Zelenka, TheFilmArtist, Siba Kikanović, Candie Solis, Eric Ganz, Shannon Vest, Chris Miller, Kevin Cobble, Stephanie Petrie, Terran Baylor, Sri Panchalam, Gunther Jönsson, Yeojun Chun, Mark Buckler, Kimberly Nicole Smith Christine Carner Gibbons, Marie-Therese Enga, Semezdin Mehmedinović, Josh Ridderhoff, Douglas Dankel, Naida Čolakhodzić, Saw Kyaw Lwin, Christopher Goodgame, Gregory Clark, Sheila Lynne Bolda, Robert Davis, Vijay Bal, C Gabriel Traupman, Del Williams, Alicia Hoogveld, Michael Lovan, Hiroki Hagiwara, John Fitch, Wayne and Cathleen Garrison, Kurt Lancaster, Joel Charles, Kris Cheng, Len Ahgeak, Nathaniel Smith, Aaron Wood, Samir Dobrić, Dr. Valerie Sakamura, Bradley Ng, Douglas Daughaday, Ken G Baker, Jason O'Bryant, Craig French Hendry, Catherine Quayle, Trystan Trenberth, Michael Kowalski, Cynthia Lester, Dan Duriscoe, Barry Malpas, William Burden, Jonathan Sule, James Mitchell, Michelle Miller, Trace Wherry-Phillips, Herald Bardenhagen, Jules Kragen, Kurt F. Habetler

Special Thanks to International Dark-Sky Association

Cheryl Ann Bishop, J. Scott Feierabend, Scott Kardel, John Barentine

Special Thanks to Flagstaff Dark Skies Coalition

Chris & Debra Luginbuhl, Lance Diskan, William Seven

Special Thanks to National Park Service, Bureau of Land Management, and National Parks of Canada

Charles Strickfaden & Megan Urban | Fort Union National Monument, Kelly Clark & Nick Fuechsel | Dry Tortugas National Park, Geoff Goins | Bandelier National Monument, Kathy Zerkle | New River Gorge National River, Dan Duriscoe & Bob Meadows | NPS Natural Sounds & Night Skies, Sébastien Giguère | Parc national du Mont-Mégantic, Michael Haubert | Wupatki National Monument

Special Thanks to Astronomy Observatories

Sarah Scoles & Jill Tarter | SETI Institute | Allen Telescope Array, Alan Tokunaga, Lars Bergknut & Brian McOuat | NASA Infrared Telescope Facility, Mike Holstine | Green Bank Observatory, James Lamb | Caltech Owens Valley Observatory

Special Thanks – Supporters and Collaborators

Sanja & Semezdin Mehmedinović, Emina Bečirović, Briana Nadeau, Amanda Deprez, Iva Vest, Richael Young, Romina Pereira, Gala Goliani, Lejla Arapčić, Aida Bogunić, Joel Charles, Patrick Woodroffe & Sam Pattinson of The Third Company, The Rolling Stones, Sean Evans & Dead Skin Boy, Roger Waters, Carolyn Wood, Amanda Hukanovic, Ryan Christiansen, Ty McNeely, Daniel Alpert, Mary Denman, Janna Jones, Kurt Lancaster, Stephanie Petrie, Mark Neumann, Bill Carter, Josh Biggs, Alexander Anderson, John Hessinger, Brian Bradley, Paola Cutri, Vickie Campbell, Kent Wagner, Adam Robitel, NASA ISS Image Archive, Kevin Howard of Alliance Visitors Boureau, Cosmic Gate, Angelika Harden-Norman & Lodgepole Gallery and Tipi Village, Eva Osmond & The Annenberg Space for Photography, Myle Ott, Peter & Christine Heffernan, Adam Williams, Big Geek Daddy, Linda Haddock & City of Borrego Springs, Patrick Sampson & La Casa Del Zorro, Ricardo Breceda, Mia Tramz & TIME, John C. Brookins, Moby, Patty West, Michael & Rachel Darrow, Ben Dally, Will Goodman, Mike & Vickie Seybold, Shattuck Windmill Park & Museum, Thomas Langley, Joseph & Catherine Nadeau, Steve Bing, Michael Franco, Mikey Moran & Sierra Magazine, Maddie Stone & Carli Velocci of Gizmodo, Patrick Dell, David Futch & Jill Stewart of LA Weekly, Tara MacIsaac & Epoch Times, Nancy Atkinson & Universe Today, Bonnie & Evan Todoroff, Kevin & Ashley Todoroff, Joe Hanson, Bailey Stead, Heather Cheifetz Sherman, Ernie Mastroianni, Elizabeth Howell & SPACE.com, Jonathan Novack & ABC7, Ken E. Kaplan, Claudette Godfrey, Jeff Sullivan, Jeff Sullivan & Bodie State Historic Park, LACMA, Jack Kent Cooke Foundation, Coconino Astronomical Society, American Lighthouse Foundation, Grand Canyon Youth

Special Thanks to Equipment, Publicity, and Travel Sponsors

Aaron McNally | 5DSR & 5DIV Cameras & Lenses | Canon USA, Inc.

Greg Horvath | Michron, Pulse & Radian Timelapse Devices | Alpine Labs, Inc.

Kevin Noble | Vagabond Batteries | Paul C. Buff, Inc.

Neil Pugh | Digislider

Matt Walker & Pierangelo Pirak | BBC Earth

Sari Zeidler & Kara Šegedin | BBC Earth Social Media

Anne Banas | BBC Travel

Norm Medoff & Karen Pugliesi | Northern Arizona University

Canon   ALPINE LABS   Paul C. Buff   BBC earth   NORTHERN ARIZONA UNIVERSITY

PAGE NUMBER | BORTLE CLASS | CAMERA/LENS | PHOTOGRAPHER | DETAILS

4 | N/A | Canon 5DIII & Canon 24mm f1.4L II | G. Heffernan & N. Smith | 2-image composite

5 | N/A | Canon 5DIII & Canon 70-200mm f2.8L II | G. Heffernan & H. Mehmedinović | 2-image composite

6 | N/A | Canon 5DIII & Canon 16-35mm f2.8L II | G. Heffernan & H. Mehmedinović | Sculpture by Chris Burden

7 | N/A | Canon 6D & Canon 24mm f1.4L II | G. Heffernan | Multi-image composite

8 | 6 | Canon 6D & Canon 70-200mm f2.8L II | G. Heffernan |

12 | 1 | Canon 5DIII & Canon 70-200mm f2.8L II | H. Mehmedinović |

14 | 2 | Canon 5DSR & Canon 16-35mm f2.8L II | H. Mehmedinović |

16 | 2 | Canon 6D & Rokinon 12mm f2.8 | G. Heffernan | Lyrid meteor shower

17 | 2 | Canon 6D & Canon 24mm f2.8L II | G. Heffernan | Star trails composite

18 | 1 | Canon 5DIII & Rokinon 12mm f2.8 | H. Mehmedinović |

19 | 1 | Canon 5DIII & Canon 24-70mm f2.8L II | H. Mehmedinović |

20 | 1 | Canon 5DIII & Rokinon 14mm f2.8 | H. Mehmedinović | Perseid meteor shower

21 | 2 | Canon 5DIII & Rokinon 12mm f2.8 | H. Mehmedinović | Lunar eclipse

22 | 1 | Canon 6D & Rokinon 12mm f2.8 | G. Heffernan | Star trails composite, Perseid meteor shower

23 | 1 | Canon 6D & Canon 24mm f1.4L II | G. Heffernan | Perseid meteor shower

24 | 1 | Canon 5DIII & Rokinon 14mm f2.8 | H. Mehmedinović | Star trails composite

25 | 1 | Canon 6D & Canon 24mm f1.4L II | G. Heffernan | Star trails composite

26 | 1 | Canon 5DIII & Rokinon 14mm f2.8 | H. Mehmedinović |

27 | 1 | Canon 5DIII & Rokinon 14mm f2.8 | H. Mehmedinović |

28 | 3 | Canon 6D & Canon 28mm f2.8 | G. Heffernan | Sculpture by Ricardo Breceda

29 | 3 | Canon 6D & Canon 24mm f2.8L II | G. Heffernan | Sculpture by Ricardo Breceda

30 | 1 | Canon 5DIII & Canon 24-70mm f2.8L II | H. Mehmedinović |

31 | 1 | Canon 6D & Canon 24mm f1.4L II | G. Heffernan | Star trails composite

32 | 1 | Canon 5DIII & Canon 24-70mm f2.8L II | H. Mehmedinović |

33 | 1 | Canon 6D & Rokinon 12mm f2.8 | G. Heffernan | Star trails composite

34 | 3 | Canon 5DIII & Rokinon 14mm f2.8 | H. Mehmedinović |

35 | 1 | Canon 5DIII & Rokinon 14mm f2.8 | H. Mehmedinović |

36 | 1 | Canon 6D & Canon 24mm f1.4L II | G. Heffernan |

37 | 3 | Canon 5DIII & Rokinon 14mm f2.8 | H. Mehmedinović |

38 | 1 | Canon 5DSR & Canon 16-35mm f2.8L II | H. Mehmedinović |

39 | 3 | Canon 5DSR & Rokinon 14mm f2.8 | H. Mehmedinović |

40 | 1 | Canon 5DSR & Canon 16-35mm f2.8L II | H. Mehmedinović |

41 | 1 | Canon 5DSR & Rokinon 14mm f2.8 | H. Mehmedinović |

42 | 1 | Canon 5D III & Rokinon 14mm f2.8 | H. Mehmedinović |

43 | 1 | Canon 6D & Rokinon 12mm f2.8 | G. Heffernan |

44 | 1 | Canon 6D & Canon 24mm f1.4L II | G. Heffernan | Star trails composite

45 | 3 | Canon 6D & Rokinon 12mm f2.8 | G. Heffernan |

46 | 4 | Canon 5DSR & Rokinon 12mm f2.8 | H. Mehmedinović |

47 | 4 | Canon 5D Mark III & Canon 24-70mm f2.8L II | H. Mehmedinović |

48 | 1 | Canon 5D Mark III & Rokinon 14mm f2.8 | H. Mehmedinović | Star trails composite

49 | 1 | Canon 5D Mark III & Rokinon 14mm f2.8 | H. Mehmedinović |

50 | 4 | Canon 5D Mark III & Rokinon 14mm f2.8 | H. Mehmedinović |

51 | 4 | Canon 5D Mark III & Canon 24-70mm f2.8L II | G. Heffernan |

52 | 4 | Canon 6D & Canon 24mm f1.4L II | G. Heffernan |

53 | 1 | Canon 5D Mark III & Nikon 14-24mm f2.8 | T. Piekunka |

54 | 1 | Canon 6D & Canon 24mm f1.4L II | G. Heffernan | Star trails composite

55 | 1 | Canon 6D & Canon 24mm f1.4L II | G. Heffernan |

56 | 3 | Canon 5DSR & Rokinon 12mm f2.8 | H. Mehmedinović |

57 | 4 | Canon 5DIII & Rokinon 12mm f2.8 | H. Mehmedinović |

58 | 1 | Canon 5DSR & Rokinon 14mm f2.8 | H. Mehmedinović | Star trails composite

60 | 1 | Canon 5DSR & Rokinon 12mm f2.8 | H. Mehmedinović |

61 | 1 | Canon 5DSR & Canon 24-70mm f2.8L II | H. Mehmedinović | Star trails composite

62 | 1 | Canon 5DSR & Canon 16-35mm f2.8L II | H. Mehmedinović |

63 | 1 | Canon 5DSR & Canon 16-35mm f2.8L II | H. Mehmedinović |

64 | 1 | Canon 5DSR & Rokinon 14mm f2.8 | H. Mehmedinović |

65 | 1 | Canon 5DSR & 24-70mm f2.8L II | H. Mehmedinović |

66 | 1 | Canon 5D III & Rokinon 14mm f2.8 | H. Mehmedinović |

67 | 4 | Canon 6D & Canon 24mm f1.4L II | G. Heffernan |

68 | 3 | Canon 5DIII & Rokinon 14mm f2.8 | H. Mehmedinović |

69 | 1 | Canon 6D & Rokinon 12mm f2.8 | G. Heffernan |

PAGE NUMBER | BORTLE CLASS | CAMERA/LENS | PHOTOGRAPHER | DETAILS

70 | 1 | Canon 6D & Rokinon 12mm f2.8 | G. Heffernan |

70 | 1 | Canon 6D & Rokinon 12mm f2.8 | G. Heffernan |

70 | 1 | Canon 5DIII & Sigma 24-35mm f2 | H. Mehmedinović |

71 | 1 | Canon 6D & Rokinon 12mm f2.8 | G. Heffernan |

72 | 1 | Canon 6D & Rokinon 12mm f2.8 | G. Heffernan |

73 | 1 | Canon 6D & Rokinon 12mm f2.8 | G. Heffernan |

74 | 1 | Canon 6D & Canon 24mm f1.4L II | G. Heffernan |

74 | 1 | Canon 5DIII & Rokinon 14mm f2.8 | H. Mehmedinović |

74 | 1 | Canon 5DSR & Canon 16-35mm f2.8L II | H. Mehmedinović |

75 | 1 | Canon 6D & Rokinon 12mm f2.8 | G. Heffernan |

75 | 1 | Canon 5DIII & Rokinon 14mm f2.8 | H. Mehmedinović |

75 | 1 | Canon 6D & Rokinon 12mm f2.8 | G. Heffernan |

76 | 2 | Canon 5DIII & Rokinon 14mm f2.8 | H. Mehmedinović | Zodiacal Light

77 | 4.5 | Canon 5DIII & Canon 16-35mm f2.8L II | H. Mehmedinović | Sculptures by Jim Reinders

78 | 4.5 | Canon 5DIII & Rokinon 12mm f2.8 | H. Mehmedinović | Star trails composite, Sculptures by Jim Reinders

79 | 4.5 | Canon 6D & Canon 24-70mm f2.8L II | G. Heffernan | Sculptures by Jim Reinders

80 | 9 | Canon 6D & Canon 24mm f1.4L II | G. Heffernan |

82 | 8 | Canon 5DIV & Canon 16-35mm f2.8L II | H. Mehmedinović |

83 | 8 | Canon 5DIV & Canon 16-35mm f2.8L II | H. Mehmedinović |

87 | 6 | Canon 6D & Canon 24mm f1.4L II | G. Heffernan |

88 | 8 | Canon 6D & Canon 24mm f1.4L II | G. Heffernan |

90 | 8 | Canon 6D & Canon 24-70mm f2.8L II | G. Heffernan | Car light trails

91 | 8 | Canon 6D & Canon 70-200mm f2.8L II | G. Heffernan |

92 | 8 | Canon 6D & Canon 24mm f1.4L II | G. Heffernan |

93 | 8 | Canon 6D & Canon 24mm f1.4L II | G. Heffernan |

94 | 9 | Canon 6D & Canon 24mm f1.4L II | G. Heffernan | Car light trails

95 | 9 | Canon 6D & Canon 24mm f1.4L II | G. Heffernan |

96 | 9 | Canon 6D & Canon 24mm f1.4L II | G. Heffernan | Pictured: Briana Nadeau

97 | 9 | Canon 5DIV & Rokinon 12mm f2.8 | H. Mehmedinović |

98 | 6 | Canon 6D & Canon 24mm f1.4L II | G. Heffernan |

99 | 4.5 | Canon 5DIII & Rokinon 14mm f2.8 | H. Mehmedinović |

100 | 1 | Canon 5DIII & Rokinon 14mm f2.8 | H. Mehmedinović | Airplane light trails

105 | 8 | Canon 5DIV & Rokinon 12mm f2.8 | H. Mehmedinović |

106 | 8 | Canon 6D & Canon 24mm f1.4L II | G. Heffernan | Car light trails

108 | 9 | Canon 6D & Canon 24mm f1.4L II | G. Heffernan | Car light trails

109 | 9 | Canon 6D & Canon 24mm f1.4L II | G. Heffernan |

111 | 9 | Canon 6D & Rokinon 12mm f2.8 | G. Heffernan |

112 | 5 | Canon 6D & Canon 24mm f1.4L II | G. Heffernan |

113 | 9 | Canon 6D & Canon 24mm f1.4L II | G. Heffernan |

114 | 8 | Canon 5DIV & Rokinon 14mm f2.8 | H. Mehmedinović | Pictured: Emina Bečirović

114 | 8 | Canon 5DIV & Rokinon 14mm f2.8 | H. Mehmedinović |

115 | 9 | Canon 6D & Canon 24mm f1.4L II | G. Heffernan |

116 | 4.5 | Canon 5DIII & Rokinon 14mm f2.8 | H. Mehmedinović | Photinus carolinus fireflies

117 | 4.5 | Canon 5DIII & Rokinon 12mm f2.8 | H. Mehmedinović | Photinus carolinus fireflies

118 | 3 | Canon 6D & Canon 24mm f1.4L II | G. Heffernan | Star trails composite, Sculpture by Gary Greff

119 | 3 | Canon 6D & Canon 24mm f1.4L II | G. Heffernan | Sculpture by Gary Greff, car light trails

120 | 3 | Canon 6D & Canon 24mm f1.4L II | G. Heffernan | Sculpture by Gary Greff

121 | 3 | Canon 6D & Canon 24mm f1.4L II | G. Heffernan | Sculpture by Gary Greff, car light trails

122 | 4 | Canon 5DIII & Rokinon 12mm f2.8 | H. Mehmedinović | Star trails composite, Airplane Light Trails

123 | 8 | Canon 6D & Canon 24mm f1.4L II | G. Heffernan |

124 | 1 | Canon 5DSR & Canon 24-70mm f2.8L II | H. Mehmedinović |

127 | 1 | Canon 5DIII & Rokinon 12mm f2.8 | H. Mehmedinović |

128 | 1 | Canon 6D & Canon 24mm f1.4L II | G. Heffernan | Star trails composite

130 | 1 | Canon 6D & Canon 24mm f1.4L II | G. Heffernan |

131 | 1 | Canon 6D & Canon 24mm f1.4L II | H. Mehmedinović |

132 | 1 | Canon 6D & Rokinon 12mm f2.8 | G. Heffernan |

133 | 1 | Canon 6D & Rokinon 12mm f2.8 | G. Heffernan |

133 | 1 | Canon 6D & Rokinon 12mm f2.8 | G. Heffernan |

134 | 1 | Canon 5DIII & Canon 24-70mm f2.8L II | H. Mehmedinović | Aquarid meteor shower

135 | 1 | Canon 5DIII & Rokinon 12mm f2.8 | H. Mehmedinović |

PAGE NUMBER | BORTLE CLASS | CAMERA/LENS | PHOTOGRAPHER | DETAILS

135 | 1 | Canon 5DIII & Canon 24mm f1.4L II | G. Heffernan | Star trails composite

135 | 1 | Canon 5DIII & Rokinon 12mm f2.8 | H. Mehmedinović |

135 | 1 | Canon 5DIII & Canon 24-70mm f2.8L II | H. Mehmedinović |

135 | 3 | Canon 5DIII & Rokinon 12mm f2.8 | H. Mehmedinović |

136 | 3 | Canon 5DIII & Rokinon 12mm f2.8 | H. Mehmedinović | Aurora Borealis

137 | 3 | Canon 5DIII & Rokinon 12mm f2.8 | H. Mehmedinović | Aurora Borealis

138 | 1 | Canon 5DSR & Canon 16-35mm f2.8L II | H. Mehmedinović | Car Light Trails

139 | 1 | Canon 5DSR & Canon 16-35mm f2.8L II | H. Mehmedinović |

140 | 5 | Canon 5DIII & Rokinon 12mm f2.8 | H. Mehmedinović |

141 | 4 | Canon 5DIII & Rokinon 12mm f2.8 | H. Mehmedinović |

142 | 5 | Canon 6D & Canon 24mm f1.4L II | G. Heffernan & H. Mehmedinović | 2-image composite, car light trails

143 | 6 | Canon 5DIV & Canon 16-35mm f2.8L II | H. Mehmedinović | Low Pressure Sodium Vapor Lamps

144 | 3 | Canon 5DIII & Rokinon 14mm f2.8 | H. Mehmedinović | Car light trails

145 | 3 | Canon 5DIII & Rokinon 12mm f2.8 | H. Mehmedinović | Car light trails

146 | 3 | Canon 6D & Rokinon 12mm f2.8 | G. Heffernan |

147 | 3 | Canon 6D & Rokinon 12mm f2.8 | G. Heffernan |

148 | 3 | Canon 6D & Canon 24mm f1.4L II | G. Heffernan | Star trails composite, Car light trails

149 | 3 | Canon 5DIII & Rokinon 12mm f2.8 | H. Mehmedinović | Sculpture by Ricardo Breceda

149 | 3 | Canon 6D & Canon 24mm f1.4L II | G. Heffernan | Sculpture by Ricardo Breceda

149 | 3 | Canon 5DIII & Rokinon 12mm f2.8 | H. Mehmedinović | Sculpture by Ricardo Breceda

149 | 3 | Canon 5DIII & Rokinon 12mm f2.8 | H. Mehmedinović |Sculpture by Ricardo Breceda

152 | 1 | Canon 20DA & Sigma 10mm f2.8 | T. Nordgren | 360 degree Milky Way in April

154 | 3 | Canon 6D & Rokinon 12mm f2.8 | G. Heffernan| Star trails composite, Sculpture by Gary Greff

155 | 1 | Canon 6D & Canon 24mm f1.4L II | G. Heffernan | Pictured: Michael Darrow

156 | 1 | Canon 5DIII & Rokinon 14mm f2.8 | H. Mehmedinović |

157 | 1 | Canon 5DIII & Rokinon 14mm f2.8 | H. Mehmedinović |

158 | 1 | Canon 6D & Canon 24mm f1.4L II | G. Heffernan | Star trails composite

159 | 1 | Canon 6D & Canon 24mm f1.4L II | G. Heffernan | Pictured: John C. Brookins, G. Heffernan

160 | 1 | Canon 5DIII & Rokinon 12mm f2.8 | H. Mehmedinović | Quadrantid meteor shower

161 | 1 | Canon 5DIII & Rokinon 14mm f2.8 | H. Mehmedinović |

162 | 1 | Canon 5DIII & Rokinon 12mm f2.8 | H. Mehmedinović |

163 | 1 | Canon 5DIII & Rokinon 14mm f2.8 | H. Mehmedinović |

164 | 1 | Canon 5DIII & Rokinon 12mm f2.8 | H. Mehmedinović |

165 | 1 | Canon 5DIII & Rokinon 12mm f2.8 | H. Mehmedinović |

166 | 4 | Canon 6D & Canon 24mm f1.4L II | G. Heffernan |

167 | 4 | Canon 6D & Canon 24mm f1.4L II | G. Heffernan | Camelopardalid meteor shower

168 | 2 | Canon 5DIII & Rokinon 14mm f2.8 | H. Mehmedinović | Perseid meteor shower

169 | 2 | Canon 5DIII & Rokinon 12mm f2.8 | H. Mehmedinović |

170 | 3 | Canon 5DSR & Rokinon 14mm f2.8 | H. Mehmedinović |

171 | 2 | Canon 5DIII & Rokinon 14mm f2.8 | H. Mehmedinović | Boat light trails

172 | 3 | Canon 5DIII & Rokinon 14mm f2.8 | H. Mehmedinović |

173 | 3 | Canon 5DIII & Rokinon 14mm f2.8 | H. Mehmedinović |

174 | 2 | Canon 5DSR & Rokinon 14mm f2.8 | H. Mehmedinović |

175 | 2 | Canon 5DSR & Rokinon 14mm f2.8 | H. Mehmedinović |

176 | 1 | Canon 5DSR & Rokinon 14mm f2.8 | H. Mehmedinović | Lens Flare

177 | 1 | Canon 5DSR & Rokinon 12mm f2.8 | H. Mehmedinović | Star trails composite, Car light trails

178 | 4 | Canon 5DSR & Rokinon 14mm f2.8 | H. Mehmedinović | Car light trails

179 | 4 | Canon 5DSR & Rokinon 12mm f2.8 | H. Mehmedinović | Orionid meteor shower, Car light trails

180 | 3 | Canon 5DIII & Canon 24-70mm f2.8L II | H. Mehmedinović |

181 | 3 | Canon 5DIII & Rokinon 12mm f2.8 | H. Mehmedinović |

182 | 4 | Canon 5DIII & Rokinon 12mm f2.8 | H. Mehmedinović |

183 | 4 | Canon 5DIII & Rokinon 12mm f2.8 | H. Mehmedinović |

184 | 1 | Canon 6D & Canon 24mm f1.4L II | G. Heffernan |

186 | 1 | Canon 5DSR & Rokinon 14mm f2.8 | E. Bečirović | Sculpture by Michael Elmgreen & Ingar Dragset

187 | 9 | Canon 6D & Canon 24mm f1.4L II | G. Heffernan |

188 | 3 | Canon 5DSR & Rokinon 12mm f2.8 | H. Mehmedinović | Car light trails

Stock photos and graphics were omitted from the index